PRACTICAL
QUANTUM ELECTRODYNAMICS

T0203809

CRC SERIES *in* PURE *and* APPLIED PHYSICS

Dipak Basu
Editor-in-Chief

PUBLISHED TITLES

Handbook of Particle Physics
M. K. Sundaresan

High-Field Electrodynamics
Frederic V. Hartemann

Fundamentals and Applications of Ultrasonic Waves
J. David N. Cheeke

Introduction to Molecular Biophysics
Jack Tuszynski
Michal Kurzynski

Practical Quantum Electrodynamics
Douglas M. Gingrich

PRACTICAL

QUANTUM
ELECTRODYNAMICS

Douglas M. Gingrich

CRC Press
Taylor & Francis Group
Boca Raton London New York

CRC Press is an imprint of the
Taylor & Francis Group, an **informa** business

CRC Press
Taylor & Francis Group
6000 Broken Sound Parkway NW, Suite 300
Boca Raton, FL 33487-2742

First issued in paperback 2019

ISBN-13: 978-1-58488-542-9 (hbk)
ISBN-13: 978-0-367-39088-4 (pbk)

Library of Congress Cataloging-in-Publication Data

PRACTICAL

QUANTUM
ELECTRODYNAMICS

Douglas M. Gingrich

CRC Press
Taylor & Francis Group
Boca Raton London New York

CRC Press is an imprint of the
Taylor & Francis Group, an **informa** business

CRC Press
Taylor & Francis Group
6000 Broken Sound Parkway NW, Suite 300
Boca Raton, FL 33487-2742

First issued in paperback 2019

© 2006 by Taylor Francis Group, LLC
CRC Press is an imprint of Taylor & Francis Group, an Informa business

No claim to original U.S. Government works

ISBN-13: 978-1-58488-542-9 (hbk)
ISBN-13: 978-0-367-39088-4 (pbk)

Library of Congress Cataloging-in-Publication Data

Visit the Taylor & Francis Web site at
http://www.taylorandfrancis.com

and the CRC Press Web site at
http://www.crcpress.com

To Genevieve

Contents

IV Appendices 307

Preface

This book combines the principles of special relativity and quantum mechanics needed to perform calculations of the electromagnetic scattering of electrons and positrons, as well as the emission and absorption of photons. I start by introducing the wave equations for spin-0 and spin-1/2 particles. The basic principles of relativistic quantum mechanics are first introduced for spin-0 particles, where the extra spin-degree of freedom does not obscure the new concepts arising from a relativistic treatment of quantum mechanics. The formalism is then redeveloped for spin-1/2 particles in which a rich set of new concepts are revealed. Using this approach, the relativistic and spin effects may be clearly distinguished by comparing the spin-0 and spin-1/2 cases. I emphasise how the relativistic treatment of quantum mechanics and the spin-1/2 degree of freedom are necessary to describe electromagnetic interactions involving the scattering of electrons.

The field-theoretical approach to relativistic quantum mechanics is avoided in this introduction since it is conceptually quite distinct from nonrelativistic quantum mechanics covered at the undergraduate level; whereas relativistic quantum mechanics is a natural extension of the nonrelativistic concepts. The shortfalls of the wave-equation approach to relativistic quantum mechanics are pointed out, and it is mentioned how a many-particle quantized field description of the theory is necessary. The calculational formalism of field theory is also avoided by using the heuristic approach of the propagator formalism developed by Feynman and Stückelberg. The Feynman rules for quantum electrodynamics are developed by example. This is an intuitive and practical approach that gets the reader doing calculations quickly.

This book is pedagogical in nature. It is meant to serve as a first introduction to the theory of quantum electrodynamics, or interacting particles in general. The material will provide adequate preparation for further studies in relativistic quantum field theory or nonrelativistic many-body theory. For readers with no intention of pursuing modern quantum theory in any greater depth, it is hoped that the material presented in this book will give them an intuitive feel for the meaning of the theory and the power of the calculational method.

It is my belief that physics at an advanced level is learned by participation, and in the case of the theoretical topics presented here, by performing numerous calculations to develop hands-on experience. Many derivations have been worked out in detail. This will be a benefit to a reader wanting to study the subject on their own. It is hoped that by studying the material in this book and working through the problems, that the reader will gain the necessary background to pursue further study and research in theoretical and particle physics. The serious reader should be able to calculate simple diagrams, lifetimes, and cross sections correctly.

This book can serve as a textbook for a graduate course in relativistic quantum mechanics. The problems at the end of each chapter consist of filling in mathematical steps left out of the presentation, proofs of expressions in the main text, proofs of identities, or extensions to ad-hoc models or unusual cases of the theory. Solutions to selected problems in the chapters can be found on my web-site http://csr.phys.ualberta.ca/gingrich/qed/qed.html.

There is a list of books in the bibliography. In most cases, I have not explicitly cited these books in the main text. These bibliography items should therefore be regarded less as formal references than as acknowledgements of those who originated some of the ideas I use in this book. I would like to acknowledge the book by J.D. Bjorken and S.D. Drell [4], which has formed the basis for many books on the subject. I would also like to acknowledge the book by W. Greiner and J. Reinhardt [8] which came back into print in North America while this book was being written. The debt I owe to these authors will be quite obvious from the text.

I would also like to acknowledge Jos Vermaseren for writing the LaTeX style file axodaw[1] that was used to make all the diagrams in this book. He is also the author of the symbolic manipulation package FORM[2] which was used to check many of the trace calculations. Samples of these calculations can be found in appendix D.

Finally, I would like to thank Kaston Leung for his assistance in proof reading an early version of the manuscript, and James Fuite for proof reading the final manuscript.

Doug Gingrich

---

[1] J.A.M. Vermaseren, "Axodraw", Comp. Phys. Comm. 83 (1994) 45-58.

[2] J.A.M. Vermaseren, "New features of FORM", math-ph/0010025.

Part I

Introduction and Background

Chapter 1

Introduction

At the beginning of the 20th Century new branches of physics were developed to describe two extreme realms of reality: the very small and the very fast. The special theory of relativity was developed around 1905 and describes physical systems traveling at near the speed of light. Likewise, quantum mechanics was developed starting around 1900 to describe systems at the atomic level. Both theories radically transformed our understanding of science and philosophy, and now affect our everyday life.

In spite of the success of special relativity and quantum mechanics, certain properties of the electron, and the emission and absorption of radiation by atoms defied explanation until around 1926. An amalgamation of the theories of special relativity[1] and quantum mechanics, called relativistic quantum mechanics, was required before a satisfactory explanation emerged.

Since we will develop the wave-equation approach to relativistic quantum mechanics in this book, we review some of the concepts of wave functions in nonrelativistic quantum mechanics. For a given physical system there exists a state function that can be used to describe all that we can know about the system. We will usually deal directly with a coordinate realization of the state function: the wave function $\psi(q_1,\ldots,q_N;s_1,\ldots,s_N;t)$. This wave function is a complex function of all the classical degrees of freedom $q_1,\ldots,q_N$, of the time t, and of any additional degrees of freedom, such as spin $s_1,\ldots,s_N$, which are intrinsically quantum mechanical. $|\psi(q_1,\ldots,q_N;s_1,\ldots,s_N;t)|^2 \geq 0$ is interpreted as the probability of the system having values $q_1,\ldots,q_N;s_1,\ldots,s_N$ at time t. This probability interpretation requires that the sum of positive contributions $|\psi|^2$ for all values of $q_1,\ldots,q_N;s_1,\ldots,s_N$ at time t be finite for all ψ representing a physical system in the real world.

If the wave function is given at some instant, not only are all the properties of the system at that instant described, but its behavior at all subsequent instants is determined. The value of the derivative of the wave function with respect to time $\partial\psi/\partial t$ at any given instant can be determined by the value of the function itself at that instant. By the principle of superposition, the relationship between ψ and $\partial\psi/\partial t$ must be linear.

The time development of a physical system is expressed by the equation

---

[1]Notice that I explicitly say *special* relativity. As of 2005, a satisfactory theory combining general relativity and quantum mechanics does not yet exist.

$$i\hbar\frac{\partial\psi}{\partial t} = H\psi, \tag{1.1}$$

where H is a linear hermitian operator[2]. This operator is what corresponds in classical mechanics to Hamilton's function. The operator is called the Hamiltonian operator, or more briefly, the Hamiltonian, of the system. The Hamiltonian has no explicit time dependence in the Schrödinger picture, $\partial H/\partial t = 0$, for a closed physical system. This means that the eigenvalues of H are the possible stationary states of the system. If the form of the Hamiltonian is known, equation 1.1 determines the wave function of the physical system concerned. This fundamental equation of the quantum mechanics is called the wave equation.

One approach to developing nonrelativistic quantum mechanics, that leads to the Schrödinger equation from equation 1.1, is to invoke the correspondence rule. We start with a classical dynamical system represented by a Hamiltonian $H(q_1, \ldots, q_N; p_1, \ldots, p_N; t)$, which depends on the coordinates q_i of the system in configuration space, on their momenta p_i, and on the time t. The total energy E of the system is

$$E = H(q_1, \ldots, q_N; p_1, \ldots, p_N; t). \tag{1.2}$$

To this classical system corresponds a quantum system whose dynamical state is represented by a wave function $\psi(q_1, \ldots, q_N; t)$ defined in configuration space and whose wave equation can be obtained by performing on both sides of equation 1.2 the substitutions

$$E \to i\hbar\frac{\partial}{\partial t} \quad \text{and} \quad p_i \to \frac{\hbar}{i}\frac{\partial}{\partial q_i} \quad (i = 1, \ldots, N), \tag{1.3}$$

and by writing down that E and H, considered as operators, give identical results when acting on ψ.

The equation thus obtained is a wave equation of the corresponding quantum mechanical system. For a nonrelativistic Hamiltonian, the Schrödinger equation is obtained. The wave equation will have all the invariance principles of the Hamiltonian from which it was derived. For an isolated system, the Schrödinger equation will be invariant under spatial rotations and translations. It will also be invariant under a Galilean transformation between two reference systems moving relative to each other with a constant velocity (see problem 1.1). The Schrödinger equation is not invariant under a Lorentz transformation and is thus not expected to describe physical reality as the relative velocities of the particles involved becomes large.

This book embarks on the journey of relativistic quantum mechanics using the correspondence rule and the wave-function approach. This approach will

---

[2]We will normally represent operators with the symbol ˆ over top of them. When there is little chance for confusion, we drop the ˆ symbol, such as with the case of $H \simeq \hat{H}$.

be applied to particles of matter (electrons), as well as, those of radiation (photons). This is the approach put forth by Dirac[3]. Quantum electrodynamics will be developed as a result of the interaction of matter with radiation.

Because of their wave nature, it is tempting to treat radiation, or more specifically light, and matter similarly within the wave equation approach to quantum mechanics. However there is an important difference between the two in nonrelativistic quantum mechanics. Even in the simplest situations, the number of photons may vary in the course of time due to emission and absorption through interactions with matter. By contrast, the number of electrons, or more generally the number of elementary particles of matter, remain constant. Thus a photon wave function would have to depend on a variable number of parameters, and it is desirable to avoid such a situation. It must be emphasized that the wave function described here is always interpreted as representing just one particle of matter, and not the statistical distribution of a number of particles.

In reality, the conservation law of the number of particles is not an absolute conservation law, and the disparity between matter and radiation is not as pronounced as we have just stated. Because of the equivalence of mass and energy, particles can also be created or absorbed whenever the interaction gives rise to energy transfers above the rest mass of the particles involved. It is possible, under certain circumstances, to create electron-positron pairs – emission of matter and antimatter. Conversely an electron and a positron colliding can annihilate – absorption of matter and antimatter – giving off energy in the form of radiation. In addition, an electron can be emitted (created) in beta decay of atomic nuclei. Beta decay is not a process that occurs within the theory of quantum electrodynamics and will thus not be treated in this book.

If we restrict ourselves to phenomena of atomic physics, the positrons are absent, nuclei are stable, and the energy transfers lie below the threshold for electron-positron pair creation; the particle-number conservation law stated above is then obeyed. Furthermore, we can approximately describe phenomena concerning the interaction between radiation and matter, for example the emission, absorption, or scattering of photons, using a semiclassical treatment of such processes for atoms.

One of the main difficulties in elaborating relativistic quantum mechanics comes from the fact that the law of conservation of the number of particles ceases, in general, to be true. To be a complete theory, relativistic quantum mechanics must encompass in a single scheme dynamical states differing not only by the quantum state, but also by the number and the nature of the elementary particles of which they are composed. To describe these dynamical states and number, would require us to journey into the concepts of second

---

[3]Fermi teated the photon as a field and the electron as a particle. Heisenberg and Pauli treated both as fields.

quantization or quantized fields, otherwise known as quantum field theory. That journey is beyond the scope of this book.

We will instead develop the propagator approach to quantum electrodynamics first exploited by Feynman[4]. We obtain the solutions of the one-particle wave equation for free electrons and then study the scattering of one particle by another by treating the interaction as a perturbation. To incorporate the creation and annihilation of antiparticles into the theory, the negative-energy solutions of the relativistic wave equations will be used. The final formalism will be a covariant version of nonrelativistic perturbation theory using only solutions to single-particle wave equations. We will thus circumvent the enormous task of developing the formalism of quantized field theory by using this practical approach to performing calculations in quantum electrodynamics.

1.1 Problems

1. (a) Show that the Schrödinger equation is invariant under spatial rotations and translations.

    (b) Show that it is also invariant under a Galilean transformation between two reference systems moving relative to each other with a constant velocity.

    (c) Show that the Schrödinger equation is not Lorentz invariant.

[4]R.P. Feynman, "The Theory of Positrons", Phys. Rev., **76** (1949) 749-759; R.P. Feynman, "Space-Time Approach to Quantum Electrodynamics", Phys. Rev., **76** (1949) 769-789.

Chapter 2

Notation and Conventions

In this chapter, we explain the notation and conventions adopted throughout the book. Some useful definitions and formulae are also given. Only a brief review of the necessary results will be presented – mainly without proof. It is assumed the reader is familiar with the background material that leads to the results presented in this chapter. If this is not the case, we suggest readers familiarize themselves with the necessary background by referring to the mathematical literature on the subject.

2.1 Units

Most often, we will work in a natural system of units in which Planck's constant $\hbar$ and the speed of light in vacuum c are set to unity: $\hbar = c = 1$. This implies that time has the unit of length, while energy, momentum, and mass all have units of inverse length.

The final result of any calculation of a measurable quantity in electrodynamics can always be expressed in terms of the dimensionless fine-structure constant $\alpha \approx 1/137.03599911$. However, this constant is related in different ways to the elementary electric charge e depending on the units used:

$$\alpha = \begin{cases} e^2/(\hbar c) & \text{in the MKSA system,} \\ e^2/(4\pi\hbar c) & \text{in the Gaussian system, or} \\ e^2/(4\pi\varepsilon_0\hbar c) & \text{in the Heaviside-Lorentz system,} \end{cases} \tag{2.1}$$

where ε_0 is the permittivity of the vacuum (units Fm^{-1}). In the MKSA and Gaussian systems, the electric charge is dimensionless when using natural units. The Heaviside-Lorentz system of units is sometimes referred to as the rationalized Gaussian system of units.

When we state definitions or are considering the relative magnitudes of quantities, we will often reintroduce the units. We will most often, but not always, work in the Gaussian system of units.

2.2   Maxwell's Equations in Vacuum

We assume a working knowledge of Maxwell's equations. Their form in vacuum will be sufficient for our purposes:

$$\vec{\nabla} \cdot \vec{E} = 4\pi k_1 \rho, \tag{2.2}$$

where

$$k_1 = \begin{cases} 1/(4\pi\varepsilon_0) & \text{in the MKSA system,} \\ 1 & \text{in the Gaussian system, and} \\ 1/(4\pi) & \text{in the Heaviside-Lorentz system;} \end{cases} \tag{2.3}$$

$$\vec{\nabla} \cdot \vec{B} = 0; \tag{2.4}$$

$$\vec{\nabla} \times \vec{E} = -\frac{k_2}{c} \frac{\partial \vec{B}}{\partial t}, \tag{2.5}$$

where

$$k_2 = \begin{cases} c & \text{in the MKSA system,} \\ 1 & \text{in the Gaussian system, and} \\ c & \text{in the Heaviside-Lorentz system;} \end{cases} \tag{2.6}$$

$$k_2 \vec{\nabla} \times \vec{B} = \frac{4\pi k_1}{c} \vec{j} + \frac{1}{c} \frac{\partial \vec{E}}{\partial t}. \tag{2.7}$$

For Maxwell's equations, the Heaviside-Lorentz system of units is the simplest to work with, since all factors of 4π vanish.

2.3   Coordinates

A point in space-time is specified by the coordinates x^0, x^1, x^2, x^3, in which $x^0 \equiv ct, x^1 \equiv x, x^2 \equiv y, x^3 \equiv z$. A four-vector in this notation can be written as

$$x^\mu \equiv (x^0, x^k) \equiv (x^0, x^1, x^2, x^3), \tag{2.8}$$

where $\mu = 0, 1, 2, 3$ and $k = 1, 2, 3$. In general, Greek indices will run from 0 to 3, while Latin indices will run from 1 to 3.

2.4 Metric Tensor

We will always work with the flat space-time metric (pseudo-Euclidean) defined by the second-rank tensor $g_{\mu\nu}$ or $g_{00} = 1$, $g_{kk} = -1$, and $g_{\mu\nu} = 0$ if $\mu \neq \nu$. In the matrix representation,

$$g_{\mu\nu} = \begin{pmatrix} 1 & 0 & 0 & 0 \\ 0 & -1 & 0 & 0 \\ 0 & 0 & -1 & 0 \\ 0 & 0 & 0 & -1 \end{pmatrix}. \tag{2.9}$$

2.5 Covariant and Contravariant Indices

We will need to distinguish between covariant and contravariant vectors. A contravariant vector transforms like the coordinate vector, while the covariant vector transforms like the gradient. In terms of indices, we define for the vector a,

$$\text{covariant indices} \rightarrow a_{\mu} \text{ (subscript)},$$
$$\text{contravariant indices} \rightarrow a^{\mu} \text{ (superscript)}.$$

The metric tensor is used to convert from one type of vector to the other:

$$a_{\mu} = \sum_{\nu=0}^{4} g_{\mu\nu} a^{\nu} \quad \Rightarrow \quad a_0 = a^0, \ a_k = -a^k. \tag{2.10}$$

Normally the sum over identical indices is implied – Einstein summation convention – and we simply write $a_{\mu} = g_{\mu\nu} a^{\nu}$. We can also raise indices: $a^{\mu} = g^{\mu\nu} a_{\nu}$, where $g^{\mu\nu} = g_{\mu\nu}$ for a Lorentz metric. Also $g_{\mu}{}^{\nu} = g_{\mu\rho} g^{\rho\nu} = g^{\mu}{}_{\nu} = \delta_{\mu}{}^{\nu}$, where $\delta_{\mu}{}^{\nu}$ is the four-dimensional Kronecker-delta symbol:

$$\delta_{\mu}{}^{\nu} = \begin{cases} 1 & \text{if } \mu = \nu, \text{ or} \\ 0 & \text{if } \mu \neq \nu. \end{cases} \tag{2.11}$$

Also, notice that $g_{\mu\nu} g^{\mu\nu} = (g_{\mu\mu})^2 = 4$.

2.6   Three-Vector, Four-Vector, and Scalar Product

The three space components of a contravariant four-vector a^μ form a three-vector.

$$a^\mu \equiv (a^0, a^1, a^2, a^3) \equiv (a^0, \vec{a}), \qquad (2.12)$$

where $\vec{a} \equiv (a_x, a_y, a_z)$ and $a^1 = a_x, a^2 = a_y, a^3 = a_z$.

The scalar product of two three-vectors $\vec{a}$ and $\vec{b}$ is defined as

$$\vec{a} \cdot \vec{b} \equiv a_x b_x + a_y b_y + a_z b_z. \qquad (2.13)$$

Most often we omit the index and denote a four-vector a^μ by just a. We can thus write the scalar product of two four-vectors a and b as

$$a \cdot b = a_\mu b^\mu = a^\mu b_\mu = a^\mu g_{\mu\nu} b^\nu = a^0 b^0 - \vec{a} \cdot \vec{b}. \qquad (2.14)$$

This is often taken as the definition of the metric tensor $g_{\mu\nu}$. We notice that for finite four-vectors a and b, $a \cdot b$ can be positive, negative, or zero.

2.7   Classification of Four-Vectors

Three different types of four-vectors exist as shown in figure 2.1:

$$\text{If} \quad a_\mu a^\mu < 0, \quad a^\mu \text{ is a space-like vector.} \qquad (2.15)$$
$$\text{If} \quad a_\mu a^\mu = 0, \quad a^\mu \text{ is a null or light-like vector.} \qquad (2.16)$$
$$\text{If} \quad a_\mu a^\mu > 0, \quad a^\mu \text{ is a time-like vector.} \qquad (2.17)$$

For a time-like vector in configuration space, $a^0 > 0$ means the vector points towards the future; $a^0 < 0$ means the vector points towards the past.

2.8   Gradient and Differential Operators

The gradient vector operator is defined as

$$\vec{\nabla} \equiv \left(\frac{\partial}{\partial x}, \frac{\partial}{\partial y}, \frac{\partial}{\partial z} \right). \qquad (2.18)$$

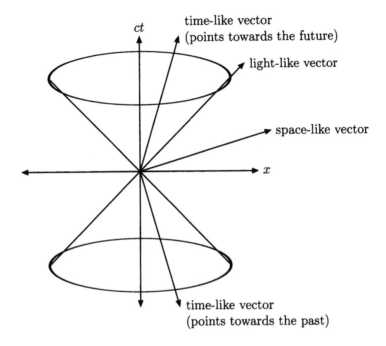

FIGURE 2.1: Classification of four-vectors relative to the light cone.

The Laplacian scalar operator is defined as $\nabla^2 \equiv \vec{\nabla} \cdot \vec{\nabla}$.

The four partial-differential operators $\partial/\partial x^\mu$ form a covariant vector, called the covariant gradient operator

$$\partial_\mu \equiv \frac{\partial}{\partial x^\mu} \equiv \left(\frac{\partial}{\partial x^0}, \frac{\partial}{\partial x^1}, \frac{\partial}{\partial x^2}, \frac{\partial}{\partial x^3} \right) \equiv \left(\frac{1}{c} \frac{\partial}{\partial t}, \vec{\nabla} \right). \tag{2.19}$$

Also, the contravariant gradient is defined as – notice the minus sign –

$$\partial^\mu \equiv g^{\mu\nu} \partial_\nu \equiv \left(\frac{1}{c} \frac{\partial}{\partial t}, -\vec{\nabla} \right). \tag{2.20}$$

Finally, the d'Alembert operator is defined as

$$\square \equiv \partial_\mu \partial^\mu \equiv \partial \cdot \partial \equiv \partial^2 \equiv \frac{1}{c^2} \frac{\partial^2}{\partial t^2} - \nabla^2. \tag{2.21}$$

Sometimes the symbol $\triangle$ or $\square^2$ is used for the d'Alembert operator.

2.9 Pauli Matrices

The Pauli matrices are

$$\sigma_1 = \begin{pmatrix} 0 & 1 \\ 1 & 0 \end{pmatrix}, \quad \sigma_2 = \begin{pmatrix} 0 & -i \\ i & 0 \end{pmatrix}, \quad \sigma_3 = \begin{pmatrix} 1 & 0 \\ 0 & -1 \end{pmatrix}. \tag{2.22}$$

The identity matrix I along with the three Pauli matrices form a basis for the set of 2×2 matrices (see problem 2.1).

The Pauli matrices are hermitian, and have the property (see problem 2.2)

$$\sigma_i \sigma_j = \delta_{ij} + i\epsilon_{ijk}\sigma_k, \tag{2.23}$$

where ϵ_{ijk} is the total antisymmetric tensor of third rank,

$$\epsilon_{ijk} = \begin{cases} +1 \text{ for an even permutation of 1, 2, 3,} \\ -1 \text{ for an odd permutation of 1, 2, 3, or} \\ 0 \text{ otherwise.} \end{cases} \tag{2.24}$$

For two arbitrary three-vectors $\vec{a}$ and $\vec{b}$, the following identity is satisfied (see problem 2.3).

$$(\vec{a} \cdot \vec{\sigma})(\vec{b} \cdot \vec{\sigma}) = \vec{a} \cdot \vec{b} I + i\vec{\sigma} \cdot (\vec{a} \times \vec{b}), \tag{2.25}$$

where $\vec{\sigma} \equiv (\sigma_1, \sigma_2, \sigma_3)$ and I is the 2×2 identity matrix.

2.10 Groups

A group is a set of distinct elements, $\mathcal{G} \equiv \{E, A, B, C, D, \ldots\}$, endowed with a law of composition – for example, addition, multiplication, matrix multiplication, etc. – such that the following properties are satisfied:

1. The composition of any two elements A and B of $\mathcal{G}$ under the given law results in an element which also belongs to $\mathcal{G}$. Thus,

$$A \circ B \in \mathcal{G}, \quad B \circ A \in \mathcal{G}, \tag{2.26}$$

where we have denoted the composition of two elements of $\mathcal{G}$ by the symbol $\circ$. This property is known as the closure property of the group and the set is said to be closed under the given law of composition.

2. There exists an identity element $E \in \mathcal{G}$ such that for all $A \in \mathcal{G}$,

$$E \circ A = A \circ E = A. \tag{2.27}$$

E is known as the identity element of $\mathcal{G}$.

3. For any element $A \in \mathcal{G}$, there exists a unique element $B \in \mathcal{G}$ such that

$$A \circ B = B \circ A = E. \tag{2.28}$$

B is called the inverse of A, and vice versa.

4. The law of composition of the group is associative, i.e. for any $A, B, C \in \mathcal{G}$,

$$A \circ (B \circ C) = (A \circ B) \circ C. \tag{2.29}$$

Groups in which the commutative law of composition also holds, i.e. in which $A \circ B = B \circ A$, are called Abelian groups.

Two groups $\mathcal{G}$ and $\mathcal{G}'$ are isomorphic if their elements can be put into a one-to-one correspondence which is preserved under composition. If we indicate the correspondence of A in $\mathcal{G}$ by A' in $\mathcal{G}'$, so that the prime on an element signifies its corresponding element, then $\mathcal{G}$ and $\mathcal{G}'$ are isomorphic if $A' \circ B' = (A \circ B)'$, where $A' \circ B'$ means the product of A' and B' according to the law of composition of $\mathcal{G}'$, while $A \circ B$ means the product of A and B according to the law of composition of the group $\mathcal{G}$. Isomorphic groups are essentially identical; the individual elements are merely labeled differently.

Homomorphism resembles isomorphism except that the correspondence is not required to be one-to-one. A group $\mathcal{G}$ is homomorphic onto another group $\mathcal{G}'$ if one and only one element of $\mathcal{G}'$ corresponds to every element of $\mathcal{G}$ and if at least one element of $\mathcal{G}$ corresponds to every element of $\mathcal{G}'$, and if the correspondence is such that the product of A and B of $\mathcal{G}$ corresponds to the product $A' \circ B' = (A \circ B)'$ of the corresponding elements A' and B' of $\mathcal{G}'$. Accordingly, homomorphism is not a reciprocal property. If $\mathcal{G}$ is homomorphic to $\mathcal{G}'$, then $\mathcal{G}'$ is not necessarily homomorphic to $\mathcal{G}$. The number of elements of $\mathcal{G}$ must be equal to or greater than the number of elements of $\mathcal{G}'$; if the number is equal, the homomorphism becomes an isomorphism, which is then reciprocal.

If we map an arbitrary group $\mathcal{G}$ homomorphically onto a group of operators $D(\mathcal{G})$ in a linear space R_N, we say that the operator group $D(\mathcal{G})$ is a representation of the group $\mathcal{G}$ in the representation space R_N. If the dimension of R_N is N, we say that the representation has degree N, or is an N-dimensional representation of the group $\mathcal{G}$. The group $\mathcal{G}$ can have in general many representations of different dimensionality.

Any linear transformation can be looked upon as a linear operator of R_N and can be represented by an $N \times N$ square matrix. Therefore the representation of an abstract group $\mathcal{G}$ means in fact a homomorphic mapping of the elements $\mathcal{G}_i$ onto a set of $N \times N$ matrices $D(\mathcal{G}_i)$ acting on the vectors of R_N and forming a group.

We shall primarily study representations in a linear space with a finite number of dimensions for the Lorentz group. It is also possible to construct representations in an infinite dimensional Hilbert space. In fact, we shall see the important role these infinite dimensional representations play in relativistic quantum mechanics.

2.11   Useful Definitions

The following notation will be handy when dealing with scalar wave functions. For two scalar functions $\phi_1(x)$ and $\phi_2(x)$,

$$\phi_1 \overleftrightarrow{\partial}_\mu \phi_2 \equiv \phi_1 \left(\frac{\partial \phi_2}{\partial x^\mu}\right) - \left(\frac{\partial \phi_1}{\partial x^\mu}\right) \phi_2. \tag{2.30}$$

The Dirac delta function δ can be defined using

$$\int_{-\infty}^{+\infty} dx \; e^{i(p_f - p_i)x} = 2\pi \delta(p_f - p_i). \tag{2.31}$$

A useful property of the Dirac delta function is

$$\delta[f(x)] = \sum_i \frac{1}{|f'(x)|_{x=x_i}} \delta(x - x_i), \tag{2.32}$$

where x_i is the ith root of the function $f(x)$: $f(x_i) = 0$. One particularly useful example of the above general relationship is

$$\delta(x^2 - a^2) = \frac{1}{2|a|}[\delta(x - a) + \delta(x + a)], \tag{2.33}$$

where a is a constant.

The delta function in four dimensions is written as

$$\begin{aligned}
\delta^4(x - x') &\equiv \delta(x_0 - x_0')\delta^3(\vec{x} - \vec{x}\,') \\
&\equiv \delta(x_0 - x_0')\delta(x_1 - x_1')\delta(x_2 - x_2')\delta(x_3 - x_3').
\end{aligned} \tag{2.34}$$

We define $f(z)$ to be a complex function of a complex variable z. If $f(z)$ is analytic on a closed contour C and within the interior region bounded by C, Cauchy's integral formula is

$$\oint_C dz \frac{f(z)}{z - z_0} = 2\pi i f(z_0), \tag{2.35}$$

where z_0 is some point in the interior region bounded by C. The direction of the contour of integration is clockwise. A counter-clockwise direction of integration results in an overall minus sign.

For two functions u and v, integration by parts gives

$$\int_{-\infty}^{+\infty} dx u \frac{dv}{dx} = uv|_{-\infty}^{+\infty} - \int_{-\infty}^{+\infty} dx \frac{du}{dx} v. \tag{2.36}$$

The "surface term" $uv|_{-\infty}^{+\infty}$ can usually be neglected by physical arguments.

2.12 Problems

1. Show that any matrix with two rows and two columns can be expressed as a linear combination of σ_1, σ_2, σ_3, and I. Use this result to show that there is no matrix that anticommutes with each of the first three of these.

2. Using the explicit forms for the 2×2 Pauli matrices, verify the commutation (square brackets) and anticommutation (curly brackets) relationships

$$[\sigma_i, \sigma_j] = 2i\epsilon_{ijk}\sigma_k \quad \text{and} \quad \{\sigma_i, \sigma_j\} = 2\delta_{ij}I,$$

where I is the 2×2 unit matrix. Hence show that

$$\sigma_i\sigma_j = \delta_{ij}I + i\epsilon_{ijk}\sigma_k.$$

3. Use the identity in the previous question to prove the result

$$(\vec{\sigma} \cdot \vec{a})(\vec{\sigma} \cdot \vec{b}) = \vec{a} \cdot \vec{b}I + i\vec{\sigma} \cdot \vec{a} \times \vec{b},$$

were $\vec{a}$ and $\vec{b}$ are arbitrary three-vectors.

Using the explicit 2×2 form for $\vec{\sigma} \cdot \vec{a}$, show that

$$(\vec{\sigma} \cdot \vec{a})^2 = \vec{a}\,^2 I.$$

Chapter 3

Lorentz Covariance

In this book, we will develop a theory which has as its very foundation the principles of the theory of special relativity. The mathematical formalism of special relativity is embodied in the Lorentz transformation. A theory describing physical reality must be invariant under a Lorentz transformation. In future chapters, we will require Lorentz covariance at each step of the development. Historically it was only after a fully Lorentz covariant realization of quantum electrodynamics was formulated in the late 1940s, that physical effects could be untangled from the meaningless divergencies, allowing the development of the theory to progress further. We begin by summarizing a few properties of the Lorentz transformation.

3.1 Lorentz Group

According to the relativity principle, the form of a theory describing nature has to be invariant under a transformation from one inertial reference frame to another. Consider two observers, S and S', in different inertial reference frames. They describe the same physical event with their particular, different, space-time coordinates. Let the coordinates of the event be x^μ for observer S and x'^μ for observer S'. The Lorentz transformation is a real linear transformation of the coordinates that conserves the norm of the intervals between all points in space-time.

The Lorentz transformation $\Lambda^\nu{}_\mu$ of S's coordinates to S''s coordinates is given by

$$x'^\nu = \Lambda^\nu{}_\mu x^\mu + a^\nu, \tag{3.1}$$

or in matrix form $x' = \Lambda x + a$. The real four-vector a^ν represents a translation of the space-time axes. In what follows, we shall treat the translations separately and give the name Lorentz transformation to the homogeneous transformations: $a^\nu = 0$. The group formed by all Lorentz transformations including translations is called the inhomogeneous Lorentz group, or Poincaré group.

If the underlying space is real, x^μ real, a mapping of the real space onto itself requires all transformation coefficients to be real. All the components of $\Lambda^\nu{}_\mu$ are thus real, $\Lambda^\nu{}_\mu = (\Lambda^\nu{}_\mu)^*$, and depend only on the relative velocities and spatial orientations of reference frames.

Since the components of $\Lambda^\nu{}_\mu$ are constant for inertial reference frames, the distance between two space-time points in S'''s reference frame is

$$ds^2 = dx'^\mu dx'_\mu = \Lambda^\mu{}_\alpha dx^\alpha \Lambda_\mu{}^\beta dx_\beta. \tag{3.2}$$

Since the norm must be conserved under a Lorentz transformation, $ds^2 = dx'^\mu dx'_\mu = dx^\mu dx_\mu$, we require

$$\Lambda^\mu{}_\alpha \Lambda_\mu{}^\beta = \delta_\alpha{}^\beta. \tag{3.3}$$

This is an orthogonality relationship for Lorentz transformations. Since we are dealing with four-dimensional space-time, there are 16 elements of the transformation. The orthogonality conditions give 10 relationships between the elements. Thus there are six independent parameters of the Lorentz transformation.

Using the orthogonality relationship, we can write (see problem 3.1)

$$x^\mu = \Lambda_\nu{}^\mu x'^\nu, \tag{3.4}$$

where $\Lambda_\nu{}^\mu$ is the inverse Lorentz transformation. Since the inverse Lorentz transformation exists, and the product of two Lorentz transformations is again a Lorentz transformation, the set of all homogenous Lorentz transformations form a group $\mathcal{L}$ called the homogeneous Lorentz group (see problem 3.2).

3.1.1   Classification of the Lorentz Subgroups

There are several types of transformations we can think of that preserve the Minkowski interval $s^2 = (ct)^2 - \vec{x}^2$, for example, pure space rotations, spatial reflections, and time reversal. To clarify the distinction between these different transformations, we examine subgroups of the Lorentz group.

Equation 3.3 can be written as (see problems 3.3)

$$\Lambda^\mu{}_\alpha g_{\mu\nu} \Lambda^\nu{}_\beta = g_{\alpha\beta}. \tag{3.5}$$

In matrix form, we can write

$$\Lambda^T g \Lambda = g, \tag{3.6}$$

where T denotes the transpose of a matrix: $\Lambda^\mu{}_\alpha = (\Lambda^T)^\alpha{}_\mu$. Equation 3.6 gives the multiplication rule for elements of the Lorentz group.

The determinant of the Lorentz transformation can be determined by using the orthogonality relationship equation 3.6:

$$\det(\Lambda^T g \Lambda) = \det \Lambda^T \det g \det \Lambda = \det g. \tag{3.7}$$

Since $\det \Lambda^T = \det \Lambda$, equation 3.7 becomes

$$(\det \Lambda)^2 = 1. \tag{3.8}$$

Since Λ is real,

$$\det \Lambda = \pm 1. \tag{3.9}$$

Transformations with determinants ± 1 are called unimodular.

The Lorentz group thus has two subsets, the transformations which are characterized by $\det \Lambda = +1$ and those transformations characterized by $\det \Lambda = -1$. There is no continuous path from one subset of the group to the other. Since the subset of transformations satisfying $\det \Lambda = +1$ contains the identity transformation, it forms a group: the proper Lorentz group $\mathcal{L}_p$. The subset of transformations satisfying $\det \Lambda = -1$ does not form a group. A typical member of the subset with $\det \Lambda = -1$ is coordinate inversion, e.g. time or space inversion.

We can further classify the set of Lorentz transformations by looking at, for example, the $\alpha = \beta = 0$ component of equation 3.5:

$$(\Lambda^0{}_0)^2 - \sum_{i=1}^{3}(\Lambda^i{}_0)^2 = 1 \quad \Rightarrow \quad (\Lambda^0{}_0)^2 \geq 1, \tag{3.10}$$

so that $\Lambda^0{}_0 \geq 1$ or $\Lambda^0{}_0 \leq -1$. These two subsets are also disjoint, and are not continuously connected.

A Lorentz transformation for which $\Lambda^0{}_0 \geq 1$ is called an orthochronous Lorentz transformation. A Lorentz transformation is orthochronous if and only if it transforms every positive time-like vector into a positive time-like vector (see problem 3.4). The set of all orthochronous Lorentz transformations forms a group: the orthochronous Lorentz group $\mathcal{L}^{\uparrow}$.

The sign of the determinant of Λ and the sign of $\Lambda^0{}_0$ can be used to classify the elements of the Lorentz group. We denote these classifications by $L^{\uparrow}_{\pm}$, where the subscript represents the sign of the determinant, and the superscript represents the domain of $\Lambda^0{}_0$. In the following, we represent all the elements of transformation $L^{\uparrow}_{\pm}$ by the set $\mathcal{L}^{\uparrow}_{\pm}$. The disjoint subsets are

$$\mathcal{L}^{\uparrow}_{+} : \quad \det \Lambda = +1 \quad \text{and} \quad \Lambda^0{}_0 \geq +1,$$
$$\mathcal{L}^{\uparrow}_{-} : \quad \det \Lambda = -1 \quad \text{and} \quad \Lambda^0{}_0 \geq +1,$$
$$\mathcal{L}^{\downarrow}_{-} : \quad \det \Lambda = -1 \quad \text{and} \quad \Lambda^0{}_0 \leq -1,$$
$$\mathcal{L}^{\downarrow}_{+} : \quad \det \Lambda = +1 \quad \text{and} \quad \Lambda^0{}_0 \leq -1.$$

We will now describe each disjoint subset of the Lorentz group.

The transformations $L^{\uparrow}_{+}$ are called the proper orthochronous Lorentz transformations, sometimes referred to as the restricted Lorentz transformations.

The elements consist of the identity transformation, the infinitesimal Lorentz transformations, and their iterations to build up finite transformations, i.e. all spatial rotations and special[1] Lorentz transformations. These form an invariant subgroup: the proper orthochronous Lorentz group $\mathcal{L}_+^\uparrow$.

The proper orthochronous group of Lorentz transformations contains a subgroup which is isomorphic to the three-dimensional rotation group. This subgroup consists of all the Λ^μ_ν of the form

$$\Lambda(R) = \begin{pmatrix} 1 & 0 \\ 0 & R \end{pmatrix}, \tag{3.11}$$

where R is the 3×3 matrix with $RR^T = R^T R = 1$. Such a Λ is called a spatial rotation. Similarly, the proper orthochronous group of Lorentz transformations contains a subgroup which is isomorphic to the special Lorentz group. This subgroup consists of Λ^μ_ν of the form, for example,

$$\Lambda(L_1) = \begin{pmatrix} L_1 & 0 & 0 \\ 0 & 1 & 0 \\ 0 & 0 & 1 \end{pmatrix}, \tag{3.12}$$

where L_1 is a 2×2 matrix representing a special Lorentz transformation in the x^1-direction. Every proper orthochronous Lorentz transformation can be decomposed as follows (see problem 3.5):

$$\Lambda = \Lambda(R_2)\Lambda(L_1)\Lambda(R_1), \tag{3.13}$$

where $\Lambda(R_1)$ and $\Lambda(R_2)$ are two spatial rotations.

A typical element of $\mathcal{L}_-^\uparrow$ is spatial reflection: $x_0 \to x_0$ and $\vec{x} \to -\vec{x}$. Space reflection reverses the handedness of space. In the matrix representation

$$\Lambda(P) = \begin{pmatrix} 1 & 0 & 0 & 0 \\ 0 & -1 & 0 & 0 \\ 0 & 0 & -1 & 0 \\ 0 & 0 & 0 & -1 \end{pmatrix}. \tag{3.14}$$

All elements of $\mathcal{L}_-^\uparrow$ can be reached continuously from the spatial reflection element P when multiplied by elements of $\mathcal{L}_+^\uparrow$. However, since $L_-^\uparrow = PL_+^\uparrow$ will not contain the identity element, it is not a group.

The basic element of $\mathcal{L}_-^\downarrow$ is time reflection: $x_0 \to -x_0$ and $\vec{x} \to \vec{x}$. Time reflection interchanges the forward and backward light cones. In the matrix representation

---
[1]The special Lorentz transformations are the transformations in four-space that mix the time and space coordinates.

$$\Lambda(T) = \begin{pmatrix} -1\,0\,0\,0 \\ 0\,1\,0\,0 \\ 0\,0\,1\,0 \\ 0\,0\,0\,1 \end{pmatrix}. \tag{3.15}$$

Starting with time reflection T and applying it successively to any element of $\mathcal{L}_+^\uparrow$, we obtain in a continuous way all elements of this subset $L_+^\downarrow = TL_+^\uparrow$, which again, do not form a group.

The typical element of $\mathcal{L}_+^\downarrow$ is total reflection of four-space: $x \to -x$. In the matrix representation

$$\Lambda(PT) = \Lambda(P)\Lambda(T) = \begin{pmatrix} -1 & 0 & 0 & 0 \\ 0 & -1 & 0 & 0 \\ 0 & 0 & -1 & 0 \\ 0 & 0 & 0 & -1 \end{pmatrix}. \tag{3.16}$$

Multiplying total reflection PT with all the elements of $\mathcal{L}_+^\uparrow$, we obtain all the elements of this subset and exhaust it. Again, $L_-^\downarrow = PTL_+^\uparrow$ is not a group.

Since the set of proper orthochronous Lorentz transformations form a group, we can form Lorentz subgroups from it. The combination of the proper orthochronous Lorentz group $\mathcal{L}_+^\uparrow$ with the elements of $\mathcal{L}_+^\downarrow$ form the group called the proper Lorentz group: $\mathcal{L}_p \equiv \mathcal{L}_+$. It contains the identity transformation, the spatial rotations, the special Lorentz transformations, time inversion, and all combinations thereof.

In addition, joining all elements of the proper orthochronous group $\mathcal{L}_-^\uparrow$ to those of $\mathcal{L}_+^\uparrow$ we obtain another subgroup, $\mathcal{L}^\uparrow$, which we call the orthochronous group, or the full Lorentz group. It contains the identity transformation, the spatial rotations, the special Lorentz transformations, space inversion, and all combinations thereof.

If time reflection is included with the orthochronous Lorentz group, or space reflection is included with the proper Lorentz group, we obtain the homogenous Lorentz group $\mathcal{L}$, often called the complete Lorentz group, or the extended Lorentz group. A summary of the Lorentz subgroups is shown in table 3.1.

Any Lorentz transformation is either proper and orthochronous, or may be written as the product of an element of the proper orthochronous Lorentz group, with one of the discrete transformations P, T, or PT; this is shown in figure 3.1. Thus the study of the complete Lorentz group reduces to the study of its proper orthochronous subgroup, plus space inversion and time reversal.

3.1.2 Infinitesimal Lorentz Transformations

A simple approach to working with continuous Lorentz transformations is to build them up from infinitesimal transformations. This means we are consider-

TABLE 3.1:   The Lorentz Subgroups.

Lorentz Group	Symbol	Composition
proper orthochronous (restricted)	$\mathcal{L}_+^\uparrow$	$\mathcal{L}_+^\uparrow$
proper	$\mathcal{L}_p$	$\mathcal{L}_+^\uparrow \cup \mathcal{L}_+^\downarrow$
orthochronous (full)	$\mathcal{L}^\uparrow$	$\mathcal{L}_+^\uparrow \cup \mathcal{L}_-^\uparrow$
homogeneous (complete or extended)	$\mathcal{L}$	$\mathcal{L}_+^\uparrow \cup \mathcal{L}_-^\uparrow \cup \mathcal{L}_-^\downarrow \cup \mathcal{L}_+^\downarrow$

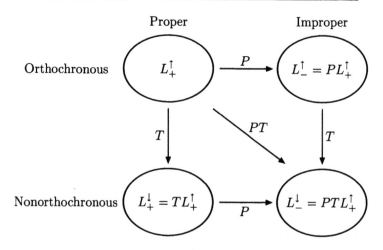

FIGURE 3.1:   Generation of the disconnected Lorentz subsets by the space and time inversion operators.

ing only the proper orthochronous Lorentz transformations since the discrete transformations can not be generated from infinitesimal transformations.

We can write

$$\Lambda_{\mu\nu} = g_{\mu\nu} + \Delta\omega_{\mu\nu}, \tag{3.17}$$

where the components of $\Delta\omega_{\mu\nu}$ are infinitesimally small.

Since $\Lambda_{\mu\nu}$ and $g_{\mu\nu}$ are real,

$$\Delta\omega_{\mu\nu} = \Delta\omega_{\mu\nu}^*. \tag{3.18}$$

Using the orthogonality relationship equation 3.3, and only keeping terms to first order in $\Delta\omega$, gives

$$\Lambda^\mu{}_\nu \Lambda_\mu{}^\lambda = \Lambda_{\mu\nu}\Lambda^{\mu\lambda} = (g_{\mu\nu} + \Delta\omega_{\mu\nu})(g^{\mu\lambda} + \Delta\omega^{\mu\lambda}),$$

$$\delta_\nu{}^\lambda = g_{\mu\nu}g^{\mu\lambda} + g_{\mu\nu}\Delta\omega^{\mu\lambda} + g^{\mu\lambda}\Delta\omega_{\mu\nu}$$
$$= g_{\nu\mu}g^{\mu\lambda} + g_{\nu\mu}\Delta\omega^{\mu\lambda} + g^{\lambda\mu}\Delta\omega_{\mu\nu}$$
$$= \delta_\nu{}^\lambda + \Delta\omega_\nu{}^\lambda + \Delta\omega^\lambda{}_\nu. \qquad (3.19)$$

Therefore

$$\Delta\omega_\nu{}^\lambda + \Delta\omega^\lambda{}_\nu = 0$$
$$\Delta\omega_{\nu\mu} + \Delta\omega_{\mu\nu} = 0. \qquad (3.20)$$

We see that $\Delta\omega_{\mu\nu}$ is required to be an antisymmetric tensor. Since $\Delta\omega_{\mu\nu}$ has dimensions 4×4, no diagonal components, and is antisymmetric, it will have only six independent components. A general Lorentz transformation thus has six parameters: three relative velocity parameters and three rotation parameters. This is consistent with the similar statements we made after equation 3.3.

For now, we will leave our discussion on general Lorentz transformations. We could go on and develop representations of the Lorentz group, generators of Lie groups, Poincaré algebra, and spinor calculus. However, to avoid formalism and keep the treatment practical, we will develop these formal, but interesting, concepts of the Lorentz group as needed.

3.2 Lorentz Boost

We first define a Lorentz boost; we have previously referred to these transformations as the special Lorentz transformations. Let S and S' be two inertial reference frames, each defined with an orthogonal set of space-time coordinates (ct, x, y, z) and (ct', x', y', z'), respectively (see figure 3.2). Let the constant velocity of S' relative to S be $\vec{v}$. Let the coordinate systems of S and S' be parallel to each other, and let their x- and x'-axes be collinear with each other and with the relative velocity. Then the transformation from S to S' is a Lorentz boost along the common x-, x'-axis. Under this Lorentz boost, four-dimensional space-time transforms as three irreducible subspaces: one two-dimensional irreducible subspace (ct, x), and two one-dimensional irreducible subspaces, y and z.

A Lorentz boost not only applies to the transformation of coordinates but to any four-vector, such as the energy-momentum four-vector. The energy and momentum $(E', \vec{p}')$ viewed from a reference frame moving with velocity $\vec{\beta} \equiv \vec{v}/c$ relative to $(E, \vec{p})$ are given by

$$\begin{pmatrix} E' \\ p'_{||} \end{pmatrix} = \begin{pmatrix} \gamma & -\gamma\beta \\ -\gamma\beta & \gamma \end{pmatrix} \begin{pmatrix} E \\ p_{||} \end{pmatrix} \quad \text{and} \quad p'_T = p_T, \qquad (3.21)$$

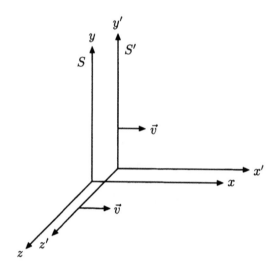

FIGURE 3.2: Two frames of reference in uniform translation. The x- and x'-axes are supposed to be collinear. The ct-axis can not be drawn.

where $\gamma = (1-\beta^2)^{-1/2}$, and p_T and $p_{||}$ are the components of $\vec{p}$ perpendicular and parallel to $\vec{\beta}$.

When the relative velocity is taken as an independent parameter, the transformation matrix represented by the Lorentz boost $L(\beta)$ is

$$L(\beta) = \gamma(\beta) \begin{pmatrix} 1 & \beta \\ \beta & 1 \end{pmatrix}. \tag{3.22}$$

The determinant, inverse, adjoint, and transpose of the Lorentz boost are given by (see problem 3.7)

$$\det L(\beta) = 1, \quad L^{-1}(\beta) = L(-\beta), \quad L^{\dagger}(\beta) = L^{T}(\beta) = L(\beta). \tag{3.23}$$

Since β and γ are related by $\gamma^2 - \beta^2\gamma^2 = 1$, we can define a single "rapidity" parameter ω using

$$\gamma = \cosh\omega \quad \text{and} \quad \gamma\beta = \sinh\omega, \tag{3.24}$$

and thus

$$\beta = \tanh\omega, \tag{3.25}$$

where $-\infty < \omega < +\infty$ as $-1 < \beta < +1$. Using the properties of the hyperbolic functions, it is sometimes useful to use

$$e^\omega = \sqrt{\frac{1+\beta}{1-\beta}} \quad \text{and} \quad \omega = \frac{1}{2}\ln\left(\frac{1+\beta}{1-\beta}\right). \tag{3.26}$$

In terms of the single rapidity parameter, we have

$$\begin{pmatrix} E' \\ p'_{\|} \end{pmatrix} = \begin{pmatrix} \cosh\omega & -\sinh\omega \\ -\sinh\omega & \cosh\omega \end{pmatrix} \begin{pmatrix} E \\ p_{\|} \end{pmatrix}$$

$$= \cosh\omega \begin{pmatrix} 1 & -\tanh\omega \\ -\tanh\omega & 1 \end{pmatrix} \begin{pmatrix} E \\ p_{\|} \end{pmatrix}. \tag{3.27}$$

Other four-vectors, such as the space-time coordinates of events transform in the same way. If $p_{\|} = p_z$, we can show that the Lorentz boost may be regarded as a rotation through an imaginary angle $i\omega$ in the ict-z plane.

3.3 Lorentz Covariance and Conservation Laws

According to the theory of special relativity, all physical laws of nature must have the same form in any two coordinate systems which arise from each other by a proper orthochronous Lorentz transformation. A relativistic invariant theory need not be invariant under P and T. In addition to P and T, we will encounter a third non-spacetime discrete operation: particle-antiparticle conjugation, denoted by C. Under this operation, particles and antiparticles are interchanged. All observations indicate that the combination CPT is a perfect symmetry of nature.

A theory is Lorentz covariant if its equations are invariant in form under a Lorentz transformation. It is not Lorentz covariant if it changes its structure under a transformation from one inertial system to another. For an equation to be Lorentz covariant, we must ensure that all unrepeated indices, upper and lower separately, balance on each side of an equation, and that all repeated indices appear once as upper and once as lower indices. An equation such as $\partial_\mu F^{\mu\nu} = j^\nu$ is Lorentz covariant.

The invariance of a closed system under subsets of the Lorentz transformation lead to certain conservation laws. We will state them without proof here (see Gross [10] or Schwabl [28]). If we assume space is homogenous, spatial translational invariance implies the conservation of three-momentum. If we assume time is homogenous, temporal translational invariance implies the conservation of energy. If we assume space is isotropic, three-space rotational invariance implies the conservation of angular momentum. Space-reflection invariance implies that parity is conserved. When we say that nature is invariant under time reflection we mean that there is nothing intrinsically inherent in directing the time axis along the increasing positive numbers – into the future – rather than directing it in the opposite way.

3.4 Problems

1. Given the orthogonality condition for the Lorentz transformation, derive the inverse Lorentz transformation.

2. Show that the set of homogenous Lorentz transformations form a group.

3. Prove the orthogonality relationship

$$\Lambda^{\mu}{}_{\alpha} g_{\mu\nu} \Lambda^{\nu}{}_{\beta} = g_{\alpha\beta}.$$

4. Show that a Lorentz transformation is orthochronous if and only if it transforms every positive time-like vector into a positive time-like vector.

5. Prove equation 3.13.

6. Show that all three types of inversions satisfy the orthogonality relationship.

7. Prove the four identities given in equation 3.23.

8. We have seen that the Lorentz boost can be represented in terms of the relativistic scale factor $\gamma(\beta)$ or the rapidity using $\beta = \tanh \omega$. An alternative expression for the Lorentz boost uses the relativistic Doppler factor $\lambda(\beta)$ defined by

$$\lambda(\beta) = \sqrt{\frac{1+\beta}{1-\beta}}.$$

   (a) Derive the relationship between λ and γ.

   (b) Derive an expression for β in terms of λ.

   (c) Derive an expression for λ in terms of the hyperbolic functions of the rapidity parameter, and

   (d) in terms of the exponential as a function of the rapidity parameter.

   (e) Show that the following identities are satisfied:

$$\lambda(\beta)\lambda(-\beta) = 1, \quad \lambda(\beta) + \lambda(-\beta) = 2\gamma(\beta), \quad \lambda(\beta) - \lambda(-\beta) = 2\beta\gamma(\beta).$$

Part II

Relativistic Quantum Mechanics

Chapter 4

Klein-Gordon Equation

In this chapter, and chapters 5 and 7, we develop relativistic wave equations for particles of spin 0, spin 1/2, and spin 1, respectively. These wave equations take the form of partial differential equations. We view the wave as the carrier of dynamical observables such as energy, momentum, and angular momentum. These waves carry the dynamical information by propagating according to the wave equations. In principle, the only restriction we know on the form of a wave equation is that it be Lorentz covariant; but Lorentz covariance is not sufficient to completely determine the form of a possible wave equation. We will see that a good description of nature can be achieved by requiring the wave equations for free particles to be linear in the wave functions and their derivatives, be at most second order in the differentials, and be local. This latter requirement means that the state of a particle at a given space-time point, is completely determined by the wave function and its derivatives evaluated at that particular point, and not on neighbouring points.

We first develop the wave equation for a relativistic spin-0 particle. Starting with a spinless particle allows us to elucidate some of the new physical properties resulting from requiring relativistic covariance, while at the same time delaying the development of the more involved mathematics needed to handle particles with spin.

4.1 Wave Equation for a Spin-0 Particle

The simplest physical system is an isolated free particle. The nonrelativistic classical Hamiltonian for such a system is

$$H = \frac{\vec{p}^2}{2m},\tag{4.1}$$

where m is the mass of the particle and $\vec{p}$ is its three-momentum. Using the operator formalism developed in chapter 1 (equations 1.2 and 1.3) we write

$$H \rightarrow \hat{E} \equiv i\hbar \frac{\partial}{\partial t}\tag{4.2}$$

and

$$\vec{p} \to \hat{p} \equiv -i\hbar\vec{\nabla}. \tag{4.3}$$

When operating on a wave function $\psi(\vec{x}, t)$, we have

$$\hat{E}\psi(\vec{x}, t) = \frac{\hat{p}^2}{2m}\psi(\vec{x}, t), \tag{4.4}$$

$$i\hbar\frac{\partial\psi}{\partial t} = -\frac{\hbar^2}{2m}\nabla^2\psi, \tag{4.5}$$

which is the familiar Schrödinger equation.

The left-hand and right-hand sides of the Schrödinger equation 4.5 transform differently under a Lorentz transformation (see problem 1.1). To obtain a Lorentz covariant wave equation, it is natural to repeat the above derivation, but this time starting with the classical Hamiltonian for a relativistic free particle:

$$H = \sqrt{\vec{p}^2 c^2 + m^2 c^4}. \tag{4.6}$$

In terms of operators, we write

$$i\hbar\frac{\partial\psi}{\partial t} = \sqrt{-\hbar^2 c^2 \nabla^2 + m^2 c^4}\,\psi \equiv \hat{H}\psi. \tag{4.7}$$

You might wonder how the square-root operator in equation 4.7 should be interpreted. This operator can be defined by expanding ψ in terms of the eigenfunctions of ∇, the momentum eigenfunctions, after which the operation $\hat{H}$ can be carried out (see problem 4.2). Another approach is to define $\hat{H}$ by its power series expansion

$$\hat{H} \approx mc^2 \left[1 - \frac{(\hbar c\nabla)^2}{2(mc^2)^2} - \frac{(\hbar c\nabla)^4}{8(mc^2)^4} \cdots\right]. \tag{4.8}$$

By expanding equation 4.7 in a power series, we obtain all powers of the gradient operator, and this will lead to a non-local theory which violates causality (see problem 4.2). The requirement of locality is the requirement that physical processes cannot influence each other if they are outside each other's light cone, i.e. if speeds larger than that of light are needed to connect the events. In addition, relativistic invariance is not clearly exhibited in equation 4.7 since there is a lack of symmetry between the space and time coordinates.

We can avoid these difficulties by starting with the square of the classical Hamiltonian, and thus removing the square-root:

$$H^2 = \vec{p}^2 c^2 + m^2 c^4. \tag{4.9}$$

This Hamiltonian is equivalent to $H = \pm\sqrt{\vec{p}^2 c^2 + m^2 c^4}$. By squaring the Hamiltonian we have introduced an extraneous solution. We will see later that the negative sign will give rise to solutions of negative energy, which at first sight have no physical significance. After some hindsight, we will see in section 4.7 that the negative-energy solutions can be interpreted physically in terms of antiparticles.

Applying the operator form of equation 4.9 to a wave function ψ, we obtain

$$-\hbar^2 \frac{\partial^2 \psi}{\partial t^2} = \left(-\hbar^2 c^2 \nabla^2 + m^2 c^4\right)\psi, \qquad (4.10)$$

$$\frac{1}{c^2}\frac{\partial^2 \psi}{\partial t^2} - \nabla^2 \psi + \frac{m^2 c^2}{\hbar^2}\psi = 0$$

$$\frac{\partial^2 \psi}{\partial (x^0)^2} - \frac{\partial^2 \psi}{\partial (x^k)^2} + \frac{m^2 c^2}{\hbar^2}\psi = 0$$

$$\left(\Box + \frac{m^2 c^2}{\hbar^2}\right)\psi = 0. \qquad (4.11)$$

Using natural units, $c = \hbar = 1$, we obtain

$$\boxed{(\Box + m^2)\psi = 0}, \qquad (4.12)$$

which is the Klein-Gordon equation. The Klein-Gordon equation describes relativistic particles with a unique rest mass m. This form is not surprising since the only two Lorentz invariants with the dimensions of inverse length squared available to us are $\partial_\mu \partial^\mu$ and m^2. If the particle has no rest mass, $m = 0$, we obtain the classical wave equation, which is also relativistically invariant.

4.2   Lorentz Covariance of the Klein-Gordon Equation

We now examine the Lorentz covariance of the Klein-Gordon equation. The d'Alembert operator $\Box \equiv \partial^\mu \partial_\mu$ is invariant under a Lorentz transformation because it is a scalar product of four-vectors ∂_μ. Also, the mass m is a scalar so the Klein-Gordon operator $(\Box + m^2)$ is Lorentz invariant. It remains to show that $\psi(x)$ is invariant under a Lorentz transformation. Transforming from the primed to the unprimed system[1], we write

$$\left(\frac{\partial}{\partial x'^\mu}\frac{\partial}{\partial x'_\mu} + m'^2\right)\psi'(x'_\mu) = 0$$

---

[1] When speaking about a system we mean a system described by an inertial reference frame.

$$\left(\frac{\partial}{\partial x^\mu}\frac{\partial}{\partial x_\mu} + m^2\right)\psi'(x'_\mu) = 0$$

$$(\Box + m^2)\psi'(x'_\mu) = 0. \tag{4.13}$$

If $\psi(x_\mu) = \psi'(x'_\mu)$, the Klein-Gordon equation is Lorentz covariant. This is true if ψ is a scalar function – $\psi(x_\mu)$ and $\psi'(x'_\mu)$ refer to the same space-time point. On the other hand, $\psi'(x'_\mu)$ and $\psi'(x_\mu)$ are different, and are related by $x'_\mu = \Lambda_\mu{}^\nu x_\nu$. Also $\psi(x_\mu)$ and $\psi'(x_\mu)$ refer to two different points with coordinates x_μ in the old and new system, respectively. Since $\psi(x)$ is a Lorentz scalar, from now on we will denote solutions to the Klein-Gordon equations as $\phi(x)$, since $\psi(x)$ will be used for the Dirac equation in chapter 5.

Because of the form of the Klein-Gordon equation, we reason that the wave function may be multiplied by a phase factor, with absolute value of unity, under a Lorentz transformation. In other words, under a transformation from the unprimed to the primed system,

$$\phi(x) \rightarrow \phi'(x') = \lambda\phi(x), \tag{4.14}$$

with $|\lambda| = 1$. Since the Lorentz transformation operator is real, λ must be real[2] and hence $\lambda = \pm 1$.

If the Lorentz transformation is continuously connected to the identity transformation, i.e. a pure rotation in four-space, it will depend continuously on the relative velocity and rotation angles, say α_i. For all $\alpha_i \rightarrow 0$, we must approach the identity transformation and thus $\lambda = 1$ holds in this case. A wave function which does not change under spatial rotations describes a scalar particle with spin 0.

For space inversion, which is a discrete Lorentz transformation, $x'^k = -x^k$ and $x'^0 = x^0$. Applying the space inversion operator twice leads to the identity transformation, regardless if ϕ is a real or a complex wave function. Therefore $\lambda^2 = 1$ or $\lambda = \pm 1$. We define two states under space inversion:

for the case $\lambda = +1$,

$$\phi(\vec{x}, t) \rightarrow \phi'(\vec{x}', t') = \phi'(-\vec{x}, t) = \phi(\vec{x}, t) \quad \text{is a scalar}, \tag{4.15}$$

for the case $\lambda = -1$,

$$\phi(\vec{x}, t) \rightarrow \phi'(\vec{x}', t') = \phi'(-\vec{x}, t) = -\phi(\vec{x}, t) \quad \text{is a pseudoscalar.} \tag{4.16}$$

Therefore, solutions of the Klein-Gordon equation are scalar or pseudoscalar, i.e. invariant under proper orthochronous Lorentz transformations, and invariant (scalar) or change sign (pseudoscalar) under space inversion. The

---

[2]Strictly speaking this is only true for real wave functions. If the wave function is complex, λ is an arbitrary phase factor. A consistent convention can be picked so that the phase factor is real.

value of λ is called the intrinsic parity of the particle. The difference between the two parity states can only be revealed by studying their interactions. The π-meson (pion) is an example of a pseudoscalar meson that under certain circumstances can be described using the Klein-Gordon equation.

4.3   Plane-Wave Solutions of the Klein-Gordon Equation

Having developed a relativistic invariant wave equation for a spin-0 particle, we now develop the plane-wave solutions of the Klein-Gordon equation. Like other plane-wave solutions, they will form a complete set. We try (with $\hbar = c = 1$)[3]

$$\phi(x) = N e^{-i(p_0 t - \vec{p} \cdot \vec{x})}, \tag{4.17}$$

where, for now, N is an arbitrary normalization. The choice of a relative minus sign between the two terms in the exponential is also arbitrary since the Klein-Gordon equation is second order in the derivatives. Our convention is in accordance with the nonrelativistic quantum theory. Because of the exponential, wave functions 4.17 are eigenfunctions of the operators $\hat{E}$ and $\hat{p}$ with eigenvalues p_0 and $\vec{p}$.

We still need to show that the wave function 4.17 is a solution of the Klein-Gordon equation. Operating with the Klein-Gordon operator on this wave function gives

$$(\Box + m^2)\phi(x) = (-p_0^2 + \vec{p}^2 + m^2)\phi(x) = 0. \tag{4.18}$$

$\phi(x)$ must vanish for all x unless $p_0^2 = \vec{p}^2 + m^2$ or

$$p_0 = \pm\sqrt{\vec{p}^2 + m^2} = \pm E. \tag{4.19}$$

The constraint equation 4.19 – relativistic energy statement – shows there are two classes of plane-wave solutions of the Klein-Gordon equation: positive-energy and negative-energy.

Since $\vec{p}$ and the sign of E are arbitrary (also N), we use them to label the plane-wave solutions $\phi(x)$. For $p_0 = E$, we have positive-energy solutions

$$\phi_p^{(+)}(x) = N e^{-i(Et - \vec{p} \cdot \vec{x})}. \tag{4.20}$$

For $p_0 = -E$, it is standard to reverse the arbitrary relative minus-sign between the two terms in the exponential of equation 4.17 and write the negative-energy solution as

---

[3]If natural units were not being used, there would be a factor of $1/\hbar$ in the argument of the exponential in equation 4.17.

$$\phi_p^{(-)}(x) = N e^{i(Et - \vec{p} \cdot \vec{x})}, \qquad (4.21)$$

where $E = \sqrt{\vec{p}^2 + m^2}$. This enables us to write the plane-wave solutions in the compact Lorentz covariant form

$$\boxed{\phi_p^{(\pm)}(x) = N e^{\mp i p \cdot x}}, \qquad (4.22)$$

where p is the energy-momentum four-vector $(E, \vec{p})$ and x is the space-time four-vector $(t, \vec{x})$. According to this convention, the positive-energy solutions have momentum eigenvalues $\vec{p}$, while the negative-energy solutions have momentum eigenvalues $-\vec{p}$. The positive- and negative-energy solutions will not mix amongst each other. In the absence of interactions, once a wave is an eigenstate of one type of energy it will not become an eigenstate of the other type of energy.

We have not yet specified the normalization of the plane waves. Many different normalizations of the wave function are in use. Some common amplitudes N for plane waves (equation 4.22) are

$$\frac{1}{(2\pi)^{3/2}}, \quad \frac{1}{(2\pi)^{3/2}\sqrt{2}}, \quad \frac{1}{(2\pi)^{3/2}\sqrt{2E}}, \quad \frac{1}{(2\pi)^{3/2}}\sqrt{\frac{m}{E}}. \qquad (4.23)$$

The $(2\pi)^{-3/2}$ factor[4] is common in the so called "continuum language" or the "delta-function normalization". This factor is often replaced in the "box normalization" convention by $V^{-1/2}$, where V is the volume of a box in which the physical interaction is confined. For the remainder of this chapter we will use the third normalization $1/[(2\pi)^{3/2}\sqrt{2E}]$.

Since plane waves are solutions of the Klein-Gordon equation, equation 4.22 describes waves associated with the unique spin of zero. According to de Broglie, these waves can be associated with particles: plane waves with free particles. Any particle that can be described by a plane wave has a definite moment and sign of the energy (and thus energy). The occurrence of negative-energy solutions does not present any difficulty for a free particle. A particle originally in a positive energy state will always remain in a positive energy state in the absence of any interactions. Relativistic spin-0 particles in the absence of interactions are physical solutions of the Klein-Gordon equation.

---

[4]If natural units were not being used, the corresponding normalization would be $(2\pi\hbar)^{-3/2}$.

4.4 General Solution of the Klein-Gordon Equation

The plane-wave solutions are eigenfunctions of a definite momentum and energy. There is also a class of solutions which are the Fourier transform of momentum functions, say $\tilde{\phi}^{(\pm)}(p)$.

These solutions have either positive or negative energy, and can be written as (using $\hbar = c = 1$)

$$\phi_a^{(\pm)}(x) = \frac{1}{\sqrt{(2\pi)^3}} \int \frac{d^3p}{\sqrt{2E}} e^{\mp ip \cdot x} \tilde{\phi}_a^{(\pm)}(p) \ , \qquad (4.24)$$

where the index a is used to label different solutions of identical mass. Because of d^3p, the relativistic invariance of solutions of this form is not immediately obvious. To reveal the relativistic invariance of equation 4.24, we write the general solution as

$$\phi_a^{(\pm)} = \frac{\sqrt{2}}{\sqrt{(2\pi)^3}} \int d^4p \sqrt{E} e^{-ip \cdot x} \delta(p^2 - m^2) \theta(\pm p) \tilde{\phi}_a^{(\pm)}(p). \qquad (4.25)$$

This form is manifestly Lorentz invariant if the Lorentz invariant step function is defined as

$$\theta(p) \equiv \theta(p_0) \equiv \begin{cases} 1 \text{ if } p^0 > 0 \quad \text{or} \\ 0 \text{ if } p^0 < 0. \end{cases} \qquad (4.26)$$

Lorentz invariance of $\delta(p^2 - m^2)$ restricts p to be a time-like vector, and $\theta(p)$ distinguishes between the past and future (see problem 4.4). Thus the expression $\int dp_0 \delta(p^2 - m^2)\theta(\pm p)$ ensures the condition $p^0 = \pm\sqrt{\vec{p}^2 + m^2}$.

We can show that the general solution (equation 4.25) gives rise to the usual form of equation 4.24 by rewriting the delta function in the general solution as

$$\delta(p^2 - m^2) = \delta(p_0^2 - \vec{p}^2 - m^2) = \delta(p_0^2 - E^2) \qquad (4.27)$$

and then applying the identity given by equation 2.33.

For positive energy,

$$\phi_a^{(+)}(x) = \frac{2^{1/2}}{(2\pi)^{3/2}} \int d^3p \, dp_0 \sqrt{E} \left[\frac{\delta(p_0 - E)}{2E} \right.$$

$$\left. + \frac{\delta(p_0 + E)}{2E} \right] \theta(p_0) e^{-i(p_0 t - \vec{p} \cdot \vec{x})} \tilde{\phi}_a^{(+)}(p)$$

$$= \frac{1}{(2\pi)^{3/2}} \int \frac{d^3p}{\sqrt{2E}} e^{-i(Et - \vec{p} \cdot \vec{x})} \tilde{\phi}_a^{(+)}(E, \vec{p})$$

$$= \frac{1}{\sqrt{(2\pi)^3}} \int \frac{d^3p}{\sqrt{2E}} e^{-ip \cdot x} \tilde{\phi}_a^{(+)}(p). \tag{4.28}$$

$\tilde{\phi}_a^{(+)}(E, \vec{p})$ is written as $\tilde{\phi}_a^{(+)}(p)$ for convenience. It is a function of the three-momentum only with the condition that $p_0 = E$.

For negative energy,

$$\phi_a^{(-)}(x) = \frac{2^{1/2}}{(2\pi)^{3/2}} \int d^3p \, dp_0 \sqrt{E} \left[\frac{\delta(p_0 - E)}{2E} \right.$$
$$\left. + \frac{\delta(p_0 + E)}{2E} \right] \theta(-p_0) e^{i(p_0 t - \vec{p} \cdot \vec{x})} \tilde{\phi}_a^{(-)}(p)$$
$$= \frac{1}{(2\pi)^{3/2}} \int \frac{d^3p}{\sqrt{2E}} e^{-i(Et + \vec{p} \cdot \vec{x})} \tilde{\phi}_a^{(-)}(-E, \vec{p})$$
$$= \frac{1}{\sqrt{(2\pi)^3}} \int \frac{d^3p}{\sqrt{2E}} e^{-ip \cdot x} \tilde{\phi}_a^{(-)}(-p). \tag{4.29}$$

In the last step, we have applied the transformation $\vec{p} \to -\vec{p}$ to the dummy variable, and redefined $\tilde{\phi}_a^{(-)}(-E, -\vec{p})$ as $\tilde{\phi}_a^{(-)}(-p)$ for convenience. It is a function of $-\vec{p}$ only, with the condition that $p_0 = -E$.

Equations 4.28 and 4.29 are identical to the initial solutions given by equation 4.24. Thus solutions 4.24 are Lorentz covariant. These solutions can be considered as wave packets, i.e. a superposition of plane waves with wave vector $\vec{p}/\hbar$, which satisfy the restriction given by equation 4.19. Relativistic spin-0 wave packets are thus solutions of the Klein-Gordon equation.

4.5   Conserved Charge and Current

Now that we have found the general solution to the Klein-Gordon equation, we look for a physical interpretation of the wave function. Following the approach used in nonrelativistic quantum mechanics, we define the position probability density ρ and current probability density $\vec{j}$, which are required to satisfying the continuity equation

$$\frac{\partial \rho}{\partial t} + \vec{\nabla} \cdot \vec{j} = 0. \tag{4.30}$$

Using the four-vector notation, $j^\mu \equiv (j^0, \vec{j}) \equiv (c\rho, \vec{j})$, we write the continuity equation in the more compact form:

$$\partial_\mu j^\mu = \partial \cdot j = 0. \tag{4.31}$$

Integrating equation 4.30 over a volume in space V large enough to contain all the probability, and using the divergence theorem, we have

$$\int_V d^3x \frac{\partial \rho}{\partial t} = -\int_V d^3x \vec{\nabla} \cdot \vec{j},$$
$$\frac{d}{dt}\int_V d^3x \rho = -\int_S \vec{j} \cdot d\vec{S}, \tag{4.32}$$

where S is the surface which bounds the volume V. Any probability which flows out of the volume V must pass through the surface S. The continuity equation 4.30 implies that probability cannot be created or destroyed at any point; it can only flow from one point to another. Thus probability that satisfies the continuity equation will be locally conserved.

We must now show that we can construct a position probability density and current probability density which satisfy the continuity equation. In general, solutions to the Klein-Gordon equation will be complex, and we want our probability densities to be real. We start by multiplying the Klein-Gordon equation for ϕ on the left by ϕ^*, and write

$$\phi^*(\partial_\mu \partial^\mu + m^2)\phi = 0$$
$$\partial_\mu(\phi^* \partial^\mu \phi) - (\partial_\mu \phi^*)(\partial^\mu \phi) + m^2 \phi^* \phi = 0. \tag{4.33}$$

Similarly, multiplying the Klein-Gordon equation for ϕ^* on the left by ϕ, we write

$$\phi(\partial_\mu \partial^\mu + m^2)\phi^* = 0$$
$$\partial_\mu(\phi \partial^\mu \phi^*) - (\partial_\mu \phi)(\partial^\mu \phi^*) + m^2 \phi \phi^* = 0. \tag{4.34}$$

Subtracting equation 4.33 from 4.34 gives

$$\partial_\mu(\phi^* \partial^\mu \phi) - \partial_\mu(\phi \partial^\mu \phi^*) = 0$$
$$\partial_\mu(\phi^* \partial^\mu \phi - \phi \partial^\mu \phi^*) = 0$$
$$\partial_\mu(\phi^* \overleftrightarrow{\partial^\mu} \phi) = 0, \tag{4.35}$$

where in the last step we have used the notation introduced by equation 2.30. Equation 4.35 is a continuity equation. Since $(\phi^* \overleftrightarrow{\partial^\mu} \phi)^* = -(\phi^* \overleftrightarrow{\partial^\mu} \phi)$, the resulting conserved four-current is purely imaginary.

We want to interpret the zero component of the conserved four-current as the probability density, so we multiply it by $i\hbar/2m$. This normalization makes the current real and have dimensions of inverse volume, if the wave functions are normalized to have amplitude $\sqrt{m/(EV)}$. The current normalization also causes the three-current probability density to have the same form as in nonrelativistic quantum mechanics. Applying this normalization, we write

$$\partial_\mu \left[\frac{i\hbar}{2m} (\phi^* \partial^\mu \phi - \phi \partial^\mu \phi^*) \right] = 0 \tag{4.36}$$

$$\frac{\partial}{\partial(ct)} \frac{i\hbar}{2m} \left(\phi^* \frac{\partial \phi}{\partial(ct)} - \phi \frac{\partial \phi^*}{\partial(ct)} \right) - \vec{\nabla} \frac{i\hbar}{2m} (\phi^* \vec{\nabla} \phi - \phi \vec{\nabla} \phi^*) = 0. \tag{4.37}$$

We define

$$\frac{j^0}{c} = \rho \equiv \frac{i\hbar}{2mc^2} \left(\phi^* \frac{\partial \phi}{\partial t} - \phi \frac{\partial \phi^*}{\partial t} \right) \tag{4.38}$$

and

$$j^k = \vec{j} \equiv \frac{\hbar}{2im} \left(\phi^* \vec{\nabla} \phi - \phi \vec{\nabla} \phi^* \right), \tag{4.39}$$

such that ρ and $\vec{j}$ satisfy the continuity equation 4.30. The expression for $\vec{j}$ is identical with the nonrelativistic form and the expression for ρ can be shown to reduce to the nonrelativistic form in the nonrelativistic limit (see problem 4.5).

Since the Klein-Gordon equation is second order in the time derivative, the initial values of ϕ and $\partial \phi / \partial t$ can be chosen independently. So at any given time both ϕ and $\partial \phi / \partial t$ may have arbitrary values, and since ρ contains two terms it can be either positive or negative. Although we wanted to interpret ρ as a position probability density, it is not positive definite.

To circumvent the problem of a non-positive definite probability density, we need a wave equation first order in the time derivative. The Dirac equation, developed in chapter 5, is first order in the time derivative but we will find that it still proves impossible to retain a positive definite probability density for a single particle, while at the same time providing a physical interpretation of the negative-energy solutions. This means that the Klein-Gordon equation is no worse than the Dirac equation with respect to physically interpreting the probability density for a single particle.

As an example, let us calculate the current four-vector using unnormalized plane waves (equation 4.17) and using natural units ($\hbar = c = 1$). We obtain

$$j^\mu = \left(\frac{p_0}{m}, \frac{\vec{p}}{m} \right) = \frac{p^\mu}{m}. \tag{4.40}$$

Since p^μ is a four-vector, so is j^μ. We notice that $\rho = p_0/m = \pm E/m$ can be either positive or negative, depending on the sign of the energy.

In obtaining equation 4.40, we have used unnormalized plane waves. Often the exponential of the plane-wave solutions is multiplied by the amplitude $\sqrt{m/E}$ so that the time component of the current is normalized to unity.

Equation 4.40 suggests that we could interpret $|q|j^\mu$ as the current density four-vector[5]. The continuity equation thus becomes a statement of the conservation of charge, assuming that j^μ has been normalized to be real. As we will see, for a theory representing reality, the number of particles many not be conserved and thus the interpretation of $|q|j^\mu$ as a current density only applies to a single-particle, single-charge, theory. This single-particle interpretation is also true in nonrelativistic quantum mechanics. A general treatment of the conserved current requires a formulation in terms of quantum field theory.

Let us return to the continuity equation and the divergence theorem. Integration of the current over the configuration space gives

$$\frac{\partial}{\partial t} \int_V d^3x \rho = -\int_S \vec{j} \cdot d\vec{S}. \qquad (4.41)$$

If we now take our volume of integration arbitrarily large, and if $\vec{j} \to 0$ as $|\vec{x}| \to \infty$ sufficiently rapidly, then $\int \vec{j} \cdot d\vec{S} \to 0$ on the surface. This is required of a physical current. Thus $\int d^3x \rho$ is constant in time. We can interpret this constant as the charge q of the particle satisfying the Klein-Gordon equation and define it using

$$\boxed{q = i \int d^3x \phi^* \overleftrightarrow{\partial_0} \phi}, \qquad (4.42)$$

where the i makes q purely real. At this point q is not required to have the same magnitude as the fundamental charge e. If it does not, we should normalize the wave function ϕ in equation 4.42 such that $|q| = e$.

4.5.1 Charge of a Klein-Gordon Particle

In deriving the current density, we had assumed the solutions to the Klein-Gordon equation were complex. Using the definition of charge in equation 4.42, we see that ϕ and ϕ^* have opposite charge. Real solutions are also possible. However, in this case the current density does not exist, since equation 4.35 is zero when $\phi = \phi^*$; a real wave function has zero charge. This in turn means that there is no conservation law for neutral spin-0 particles. The absence of a conserved current is a general property of neutral particles, and does not require the spin to be zero. In general, we have

complex scalar solutions are charged,
real scalar solutions are uncharged.

---

[5]We define q as the charge of a particle and e as the fundamental electric charge, which is always positive. For an electron $q = -e$.

If the Klein-Gordon equation is applied to the description of a charged particle, the norm can be interpreted as a charge density, with positive norm states describing + charge, and negative norm states − charge. The conservation of charge then appears as a consequence of the invariance of the norm. If the particle has no electric charge but some other quantum number − a generalized charge − which satisfies an additive conservation law, the norm can be interpreted as the density of this generalized charge. In either case, the existence of two states, one carrying positive charge and one carrying negative charge, is assumed.

We now examine the charge for a superposition of positive-energy and a superposition of negative-energy solutions to the Klein-Gordon equation. If $\phi_p^{(+)}(x)$ is a positive-energy plane-wave solution to the Klein-Gordon equation with momentum $\vec{p}$, the superposition of all such positive-energy solutions is

$$\phi^{(+)}(x) = \int d^3p \, a_+(p) \phi_p^{(+)}(x), \tag{4.43}$$

where $a_+(p)$ is a weighting function of three-momentum only and $\phi_p^{(+)}(x)$ is given by equation 4.22. Using equation 4.42, the charge for this general positive-energy solution is

$$q = i \int d^3x \, \phi^{(+)*}(x) \overset{\leftrightarrow}{\partial_0} \phi^{(+)}(x)$$

$$= i \int d^3x \, d^3p \, d^3p' \, a_+^*(p) a_+(p') \phi_p^{(+)*}(x) \overset{\leftrightarrow}{\partial_0} \phi_{p'}^{(+)}(x)$$

$$= \int \frac{d^3x \, d^3p \, d^3p'}{(2\pi)^3} \frac{E+E'}{2\sqrt{EE'}} a_+^*(p) a_+(p') e^{it(E-E')} e^{-i\vec{x}\cdot(\vec{p}-\vec{p}')}$$

$$= \int d^3p \, d^3p' \frac{E+E'}{2\sqrt{EE'}} a_+^*(p) a_+(p') \delta^3(\vec{p}-\vec{p}') e^{it(E-E')}$$

$$= \int d^3p \, |a_+(p)|^2 > 0, \tag{4.44}$$

where the last step follows because $\vec{p} = \vec{p}'$ implies $E = E'$.

For a superposition of negative-energy solutions, $q = -\int d^3p \, |a_-(p)|^2 < 0$ (see problem 4.6). Thus $\phi^{(+)}(x)$ specifies a particle with positive charge and $\phi^{(-)}(x)$ a particle with the same mass but negative charge.

For zero charge spin-0 particles, the wave function $\phi(x)$ must be real: $\phi^*(x) = \phi(x)$. For a single momentum, we write

$$\phi^{(0)}(x) = \frac{1}{\sqrt{2}} [\phi_p^{(+)}(x) + \phi_{-p}^{(-)}(x)]$$

$$= \frac{1}{\sqrt{2}} \left[\frac{1}{(2\pi)^{3/2}} \frac{1}{\sqrt{2E}} e^{-i(Et-\vec{p}\cdot\vec{x})} + \frac{1}{(2\pi)^{3/2}} \frac{1}{\sqrt{2E}} e^{i(Et-\vec{p}\cdot\vec{x})} \right]$$

$$= \frac{1}{(2\pi)^{3/2}} \frac{1}{\sqrt{2E}} \frac{2}{\sqrt{2}} \frac{[e^{-i(Et-\vec{p}\cdot\vec{x})} + e^{i(Et-\vec{p}\cdot\vec{x})}]}{2}$$

$$= \frac{\sqrt{2}}{(2\pi)^{3/2}} \frac{1}{\sqrt{2E}} \cos(Et - \vec{p}\cdot\vec{x}). \tag{4.45}$$

From the relativistic wave equation for spin-0 particles and the interpretation of its wave functions, we are led to three solutions which correspond to the charges $+$, $-$, and 0 for every momentum $\vec{p}$. The relativistic quantum theory thus reveals the charge degree of freedom of particles.

According to our current observations, there are no fundamental spin-0 particles in nature. However, if we do not probe the internal structure of mesons, they can be considered as Klein-Gordon particles. By convention it is the positive-charged meson which is the particle, and the negative-charged meson which is the antiparticle.

4.6 Normalization and Orthogonality

The quantity $\int d^3x \phi^* \overset{\leftrightarrow}{\partial}_0 \phi$ is time independent. If ϕ is a Lorentz scalar than so is $\int d^3x \phi^* \overset{\leftrightarrow}{\partial}_0 \phi$, and it can be used for normalization. Since $\int d^3x \phi^* \overset{\leftrightarrow}{\partial}_0 \phi$ is purely imaginary, we can define the normalization and orthogonality relationships for plane waves as

$$\int d^3x \phi_{p'}^{(\pm)*}(x) i \overset{\leftrightarrow}{\partial}_0 \phi_p^{(\pm)}(x) = \pm\delta^3(\vec{p} - \vec{p}'), \tag{4.46}$$

$$\int d^3x \phi_{p'}^{(\pm)*}(x) i \overset{\leftrightarrow}{\partial}_0 \phi_p^{(\mp)}(x) = 0. \tag{4.47}$$

If we have a set of wave functions with the same mass $\phi_a(x)$, $a = 1, 2, 3, ...,$ they may be orthonormalized via

$$\int d^3x \phi_a^{(\pm)*}(x) i \overset{\leftrightarrow}{\partial}_0 \phi_b^{(\pm)}(x) = \pm\delta_{ab}, \tag{4.48}$$

$$\int d^3x \phi_a^{(\pm)*}(x) i \overset{\leftrightarrow}{\partial}_0 \phi_b^{(\mp)}(x) = 0. \tag{4.49}$$

The real solutions are normalized using $\phi^{(0)} = (\phi^{(+)} + \phi^{(-)})/\sqrt{2}$, and ensuring that the complex solutions are normalized as above.

4.7 Interaction with the Electromagnetic Field

Normally the theories of scalar particles and electromagnetic fields are developed separately. These two systems are considered free and independent of each other. Interactions between them allow the transfer from one system to the other of observable quantities such as energy and momentum. This transfer of dynamic observables gives rise to all kinds of radiative processes.

We will treat interactions between the two systems by postulating the existence of coupling terms in the wave equation. The coupling terms will depend on the wave functions of both systems, but leave the energy and momentum operators unchanged. An obvious restriction which must be satisfied by any interaction theory is the requirement of relativistic invariance. An arbitrary requirement, which would simplify things, is that the wave equations shall be modified only by terms which are lowest order in the wave functions and contain no derivatives. In addition, the interaction terms should be local. This means that the wave functions in the interaction terms are evaluated at the same point in space-time. Interactions for which this is not the case are said to be non-local, and if used, care must be taken to insure that the non-locality is constructed in such a way that Lorentz invariance is not violated. The final test of the correctness of the interaction terms will be that their observable effects agree with experiments.

4.7.1 Gauge Invariance

Gauge invariance in quantum mechanics is directly related to an invariance of the wave equation under a local phase transformation. The gauge invariance principle allows us to describe the interaction between the electromagnetic field and the particle represented by the wave function. A generalized form of this phase invariance also underlies theories of the weak and strong interactions. It is thus natural to look for coupling terms in our theory of scalars and electromagnetic fields that satisfy the same invariance principles. Such theories are known as gauge theories.

The origin of gauge invariance in classical electromagnetism lies in the fact that the vector potential $\vec{A}$ and scalar potential V are not unique for given physical fields $\vec{E}$ and $\vec{B}$. The transformations that $\vec{A}$ and V may undergo, while preserving $\vec{E}$ and $\vec{B}$, and hence Maxwell's equations, are called gauge transformations. Details on gauge invariance in electromagnetism will be presented in chapter 7.

In the following, we will distinguish between global and local transformations. Global invariance means that the same transformation is carried out at all space-time points; the transformation is carried out simultaneously everywhere. Local invariance means that different transformations are carried out at different space-time points. For example, the conservation of electric charge must be satisfied locally. A process in which charge is created at one point and destroyed at a distant point is not allowed, despite the fact that it conserves charge overall, or globally. The reason for not allowing a global form of charge conservation is that it would necessitate the instantaneous propagation of signals, and this conflicts with special relativity.

The free-particle Klein-Gordon equation is invariant under a phase transformation of the wave function

$$\phi(x) \rightarrow \phi'(x) = \exp\left(\frac{iq}{\hbar c}\chi\right)\phi(x) \qquad (4.50)$$

provided that χ is independent of the space-time coordinate x. The reason for factorizing the constants $q/\hbar c$ from χ will become apparent shortly[6].

If χ is a function of x, the phase factor in equation 4.50 is the same for all states ϕ, but not necessarily the same for all space-time points. A derivative term transforms as

$$\partial^\mu \phi(x) \rightarrow \partial^\mu \phi'(x) = \exp\left(\frac{iq}{\hbar c}\chi(x)\right)\left[\partial^\mu + \frac{iq}{\hbar c}\partial^\mu \chi(x)\right]\phi(x) \qquad (4.51)$$

and thus the Klein-Gordon equation is no longer invariant under the phase transformation. Invariance may be restored by introducing the classical electromagnetic field as a four-vector $A^\mu = (V, \vec{A})$ into the Klein-Gordon equation and requiring that it transforms as

$$A^\mu(x) \rightarrow A'^\mu(x) = A^\mu(x) - \partial^\mu \chi(x) \qquad (4.52)$$

under a gauge transformation. This transformation is allowed in the standard Maxwell equations (see chapter 7). A vector field such as A^μ, introduced to guarantee local phase invariance, is called a gauge field.

As in classical Hamiltonian theory and nonrelativistic quantum mechanics, we introduce A^μ into the Klein-Gordon equation using the "minimal coupling" prescription

$$\hat{p}^\mu \rightarrow \hat{p}'^\mu = \hat{p}^\mu - \frac{q}{c}A^\mu \qquad (4.53)$$

or equivalently, by using the gauge-covariant derivative

$$\partial^\mu \rightarrow D^\mu = \partial^\mu + \frac{iq}{\hbar c}A^\mu \qquad (4.54)$$

---

[6]In this section, we explicity show the constants $\hbar$ and c since it can be tricky to reintroduce them in the following calculations if numerical results are needed.

Equation 4.54 is usually just called the covariant derivative. It allows us to write down the wave equation for the interaction directly from the free particle equation.

The modified Klein-Gordon equation is

$$\left[D^\mu D_\mu + \frac{m^2 c^2}{\hbar^2} \right] \phi = 0$$

$$\left[\left(\partial^\mu + \frac{iq}{\hbar c} A^\mu \right) \left(\partial_\mu + \frac{iq}{\hbar c} A_\mu \right) - \frac{m^2 c^2}{\hbar^2} \right] \phi = 0. \tag{4.55}$$

Besides A^μ there is an extra constant q in the equation, which characterizes the coupling of charged Klein-Gordon solutions ϕ, representing single particles, to the electromagnetic field A^μ. The sign of q can be positive or negative, but does not appear to be of any fundamental significance in this equation. When the Klein-Gordon equation is used to describe charged mesons $q = \pm e$.

Gauge invariance can be demonstrated by simultaneously transforming $\phi(x)$ and $A^\mu(x)$ according to equation 4.50 and equation 4.52. The first-order covariant derivative acting on the wave function gives

$$
\begin{aligned}
D'^\mu \phi' &= \left[\partial^\mu + \frac{iq}{\hbar c} A'^\mu \right] \phi' \\
&= \left[\partial^\mu + \frac{iq}{\hbar c} A^\mu - \frac{iq}{\hbar c} \partial^\mu \chi \right] \exp\left(\frac{iq}{\hbar c} \chi \right) \phi \\
&= \exp\left(\frac{iq}{\hbar c} \chi \right) \left[\partial^\mu + \frac{iq}{\hbar c} A^\mu \right] \phi \\
&= \exp\left(\frac{iq}{\hbar c} \chi \right) D^\mu \phi.
\end{aligned}
\tag{4.56}
$$

The second-order covariant derivative acting on the wave function gives

$$
\begin{aligned}
D'^\mu D'_\mu \phi' &= D'^\mu \exp\left(\frac{iq}{\hbar c} \chi \right) D_\mu \phi \\
&= \exp\left(\frac{iq}{\hbar c} \chi \right) D^\mu D_\mu \phi.
\end{aligned}
\tag{4.57}
$$

Thus the invariance of the Klein-Gordon equation under this gauge transformation is easily seen:

$$\left[D'^\mu D'_\mu + \frac{m^2 c^2}{\hbar} \right] \phi' = 0$$

$$\exp\left(\frac{iq}{\hbar c} \chi \right) \left[D^\mu D_\mu + \frac{m^2 c^2}{\hbar^2} \right] \phi = 0$$

$$\left[D^\mu D_\mu + \frac{m^2 c^2}{\hbar^2}\right]\phi = 0. \qquad (4.58)$$

The Klein-Gordon equation is gauge invariant for the minimal coupling substitution given by equation 4.53. General gauge invariance can be shown by applying the gauge transformation n times[7]

$$(D'^\mu)^n \phi' = (D'^\mu)^{n-1} \exp\left(\frac{iq}{\hbar c}\chi\right) D_\mu \phi$$

$$= \exp\left(\frac{iq}{\hbar c}\chi\right)(D^\mu)^n \phi. \qquad (4.59)$$

For an arbitrary operator function f which can be expanded into a power series of the D^μ operator, we write

$$[f(D'^\mu)]\,\phi' = \exp\left(\frac{ie}{\hbar c}\chi\right)[f(D^\mu)]\,\phi. \qquad (4.60)$$

Thus the minimal coupling prescription is gauge invariant in a very general sense. This shows that not only $|\phi'|^2 = |\phi|^2$, but also forms such as $\phi'^* f(D'^\mu)\phi' = \phi^* f(D^\mu)\phi$. Since physical observables are represented by bilinear forms with the structure $\phi^* \ldots \phi$, the common phase factor does not play any role in the physics.

4.7.2 Electromagnetic Coupling

The coupling of scalar particles to an electromagnetic field is done using the gauge-invariant approach explained in the previous section. Using minimal coupling and writing the Klein-Gordon equation so that all terms involving the electromagnetic potential appear on one side of the equation, we have

$$(\Box + m^2)\phi = -\hat{V}\phi, \qquad (4.61)$$

where the potential operator $\hat{V}$ is

$$\hat{V} = iq\,(\partial \cdot A + A \cdot \partial) - q^2 A^2$$
$$= i(\partial \cdot V + V \cdot \partial) + S, \qquad (4.62)$$

where $V^\mu = qA^\mu$ and $S = -q^2 A^\mu A_\mu$.

Since the Klein-Gordon equation is second order in the derivatives, the coupling term has a complicated structure, contains gradients, and is nonlinear in A^μ. The first derivative in equation 4.62 operates on both A^μ and ϕ. All

---

[7]The covariant derivative four-vector operator to the nth power is understood to be contracted pairwise, so that it is a scalar for even n and a four-vector for odd n.

terms are scalars. The first two terms consist of a vector potential which is contracted with the ∂_μ operator to make up an overall scalar. The last term is a scalar by itself. In the most general case, the scalar S and vector V^μ parts of the potential could be independent interactions, but in electromagnetism they are related. The potential in equation 4.62 is a far cry from our aesthetic requirements mentioned at the beginning of this section (page 42). Aesthetics aside, gauge theories have been very successful in describing nature.

Using minimal coupling, the conserved current becomes (see problem 4.9)

$$j^\mu = \phi^* \left(i\hbar\partial^\mu - \frac{q}{c}A^\mu \right)\phi - \phi\left(i\hbar\partial^\mu + \frac{q}{c}A^\mu \right)\phi^*. \tag{4.63}$$

Multiplying by $|q|/2m$ to get the usual normalization gives the time and space components of the conserved current

$$\rho = \frac{|q|i\hbar}{2mc^2}\left(\phi^*\frac{\partial\phi}{\partial t} - \phi\frac{\partial\phi^*}{\partial t} - \frac{2q}{i\hbar}A_0\phi^*\phi \right) \tag{4.64}$$

and

$$\vec{j} = \frac{|q|\hbar}{2im}\left(\phi^*\vec\nabla\phi - \phi\vec\nabla\phi^* + \frac{2q}{i\hbar c}\vec{A}\phi^*\phi \right). \tag{4.65}$$

Returning to our initial normalization in equation 4.63 and using natural units, we have

$$\boxed{\rho = \phi^*(x)[i\overleftrightarrow{\partial_0} - 2qA^0(x)]\phi(x)}. \tag{4.66}$$

For a stationary state, we can write

$$\phi(\vec{x}, t) = \phi(\vec{x})\exp\left(-\frac{iqEt}{|q|\hbar} \right) \tag{4.67}$$

and equation 4.64 gives

$$\rho = \frac{|q|i\hbar}{2mc^2}\left(-\frac{iqE}{|q|\hbar} - \frac{iqE}{|q|\hbar} - \frac{2q}{i\hbar}A_0(x) \right)|\phi(\vec{x})|^2 \tag{4.68}$$

$$= q\frac{E - |q|A_0(x)}{mc^2}|\phi(\vec{x})|^2. \tag{4.69}$$

We notice that the charge density can have either sign depending on the relative values of E and $|q|A_0$ at a particular point in space-time. If $E > |q|A_0$, the charge density has the same sign as q of the particle. But if $E < |q|A_0$, the charge density has the opposite sign as q of the particle. In this case, the field is strong and we would need to invoke second quantization to show that particles are created in this case. Strong fields will be discussed further in section 4.10.

4.7.3 Charge Conjugation

There are various symmetries that have nothing to do with Lorentz invariance and appear the same in all inertial reference frames. An example is the symmetry under interchange of neutrons and protons in nuclear physics. Of interest here is the charge-conjugation symmetry between particles and antiparticles.

Consider a positive-energy solution with charge q. The equivalent Klein-Gordon equation satisfied by the positive-energy solution is

$$i\hbar\frac{\partial\phi_q^{(+)}}{\partial t} = \sqrt{m^2c^4 + (-i\hbar c\vec{\nabla} - q\vec{A})^2}\,\phi_q^{(+)} + qA_0\phi_q^{(+)}. \qquad (4.70)$$

The negative energy solution with charge q satisfies

$$i\hbar\frac{\partial\phi_q^{(-)}}{\partial t} = -\sqrt{m^2c^4 + (-i\hbar c\vec{\nabla} - q\vec{A})^2}\,\phi_q^{(-)} + qA_0\phi_q^{(-)}. \qquad (4.71)$$

Taking the complex conjugate of the negative energy equation gives

$$-i\hbar\frac{\partial\phi_q^{(-)*}}{\partial t} = -\sqrt{m^2c^4 + (i\hbar c\vec{\nabla} - q\vec{A})^2}\,\phi_q^{(-)*} + qA_0\phi_q^{(-)*},$$

$$i\hbar\frac{\partial\phi_q^{(-)*}}{\partial t} = \sqrt{m^2c^4 + (-i\hbar c\vec{\nabla} + q\vec{A})^2}\,\phi_q^{(-)*} - qA_0\phi_q^{(-)*}. \qquad (4.72)$$

Comparing equation 4.70 with equation 4.72 gives the proportionality relationship

$$\boxed{\phi_{-q}^{(-)*} \propto \phi_q^{(+)}}\,. \qquad (4.73)$$

Thus $\phi_{-q}^{(-)*}$ is the charge-conjugate solution and represents the charge conjugate state of $\phi_q^{(+)}$. Similarly, $\phi_q^{(+)*}$ is the charge conjugate state of $\phi_{-q}^{(-)}$. If we (arbitrarily) call the particle described by $\phi_q^{(+)}$ "the particle", then we call the particle described by $\phi_{-q}^{(-)*}$ the antiparticle. For example, if we call the π^+ meson the particle, then the π^- meson is the antiparticle. The undesirable negative-energy solutions have now been interpreted as antiparticles.

Neutral particles also fit into this picture, in that the charge-conjugate state is the state itself. In other words, neutral spin-0 particles are their own antiparticles. Let $\phi \equiv \phi_q^{(+)}$ and $\phi_C \equiv \phi_{-q}^{(-)*}$ so that we can write

$$\phi_C = \alpha\phi, \qquad (4.74)$$

where α is a proportionality constant which has to be real. α can be deduced since for neutral particles both ϕ and $\alpha\phi$ are real. Realizing that

$$(\phi_C)_C = \phi \qquad (4.75)$$

it follows that

$$(\alpha\phi)_C = \alpha^2\phi = \phi \qquad (4.76)$$

so that $\alpha^2 = 1$ or $\alpha = \pm 1$. Accordingly there exist two different kinds of neutral particles, namely

1. neutral particles with positive charge parity, i.e. $\alpha = +1$,

$$\phi_C = \phi,$$

2. neutral particles with negative charge parity, i.e. $\alpha = -1$,

$$\phi_C = -\phi.$$

Neutral particles are thus eigenfunctions of the charge conjugation operator, while charged particles are not.

Neutral particles by definition do not interact with the electromagnetic field. However, the relativistic doubling of states still occurs, and we should rename these two degrees of freedom so as to apply equally to both charged and neutral particles. The process of transforming from the particle to the antiparticle state, or vice versa, should be called particle-antiparticle conjugation rather than charge conjugation.

The convenience of such a terminology depends upon whether or not interactions with other fields distinguish between particles and antiparticles as, for example, the electromagnetic field distinguishes between positive and negative charge. If the interaction does distinguish between them, then the particle and antiparticle have an analog of the charge associated with each. In nature there occurs spin-0 particle, for example, the K^0 and $\overline{K}^0$, which are electrically neutral and are each other's antiparticle, differing by the sign of their "strangeness charge". On the other hand, if the interactions with other fields is identical for both particle and antiparticle it becomes possible to make an abbreviation of the theory so that only one degree of freedom enters. An example of a neutral spin-0 particle that is its own antiparticle is the π^0.

4.8   Hamiltonian Form of the Klein-Gordon Equation

It can be advantageous to transform the Klein-Gordon equation into the form of a Schrödinger equation, like equation 1.1. This would allow an easy comparison with the nonrelativistic results. Both equations are second order in the space derivative. The Klein-Gordon equation is second order in the time derivative as well. We thus need to absorb one order of the time derivative into the wave function of the Klein-Gordon equation. The transformation will come at a price. We will see that a single second-order differential equation in time will become two first-order differential equations in time. We will perform this transformation with the A^μ fields present.

We will now develop the Schrödinger form of the Klein-Gordon equation, examine the charge density ρ, use ρ to state the normalization, develop free-particle solutions, and take the nonrelativistic limit.

Using intuition, we form the linear combinations of wave functions

$$\chi_1 = \frac{1}{2}\left[\phi + \frac{i}{m}\left(\partial^0 + iqA^0\right)\phi\right] \tag{4.77}$$

and

$$\chi_2 = \frac{1}{2}\left[\phi - \frac{i}{m}\left(\partial^0 + iqA^0\right)\phi\right], \tag{4.78}$$

where ϕ is a solution of the Klein-Gordon equation. We see that $\phi = \chi_1 + \chi_2$ and $\chi_1 - \chi_2 = \frac{i}{m}\left(\partial^0 + iqA^0\right)\phi$.

χ_1 and χ_2 obey the coupled equations

$$
\begin{aligned}
(i\partial^0 - qA^0)\chi_1 &= \frac{1}{2}\left(i\partial^0 - qA^0\right)\phi + \frac{i}{2m}\left(i\partial^0 - qA^0\right)\left(\partial^0 + iqA^0\right)\phi \\
&= \frac{m}{2}(\chi_1 - \chi_2) + \frac{1}{2m}\left(i\partial^0 - qA^0\right)^2\phi \\
&= \frac{m}{2}(\chi_1 - \chi_2) + \frac{1}{2m}\left[\left(-i\vec{\nabla} - q\vec{A}\right)^2 + m^2\right](\chi_1 + \chi_2) \\
&= \frac{1}{2m}\left(-i\vec{\nabla} - q\vec{A}\right)^2(\chi_1 + \chi_2) + m\chi_1
\end{aligned}
\tag{4.79}
$$

and

$$
\begin{aligned}
(i\partial^0 - qA^0)\chi_2 &= \frac{m}{2}(\chi_1 - \chi_2) - \frac{1}{2m}\left[\left(-i\vec{\nabla} - q\vec{A}\right)^2 + m^2\right](\chi_1 + \chi_2) \\
&= -\frac{1}{2m}\left(-i\vec{\nabla} - q\vec{A}\right)^2(\chi_1 + \chi_2) - m\chi_2,
\end{aligned}
\tag{4.80}
$$

where the Klein-Gordon equation has been used in the third step of equation 4.79 and the first step of equation 4.80. The differential equations now mix χ_1 and χ_2; we have two coupled differential equations. If we define the two-component spinor

$$\chi = \begin{pmatrix} \chi_1 \\ \chi_2 \end{pmatrix}, \tag{4.81}$$

we can combine equations 4.79 and 4.80 to obtain

$$(i\partial^0 - qA^0) \begin{pmatrix} \chi_1 \\ \chi_2 \end{pmatrix} = \frac{1}{2m} \left(-i\vec{\nabla} - q\vec{A}\right)^2 \begin{pmatrix} \chi_1 + \chi_2 \\ -\chi_1 - \chi_2 \end{pmatrix} + m \begin{pmatrix} \chi_1 \\ -\chi_2 \end{pmatrix}, \quad (4.82)$$

$$(i\partial^0 - qA^0) \begin{pmatrix} 1 & 0 \\ 0 & 1 \end{pmatrix} \begin{pmatrix} \chi_1, \\ \chi_2 \end{pmatrix} = \frac{1}{2m} \left(-i\vec{\nabla} - q\vec{A}\right)^2 \begin{pmatrix} 1 & 1 \\ -1 & -1 \end{pmatrix} \begin{pmatrix} \chi_1 \\ \chi_2 \end{pmatrix}$$
$$+ m \begin{pmatrix} 1 & 0 \\ 0 & -1 \end{pmatrix} \begin{pmatrix} \chi_1 \\ \chi_2 \end{pmatrix}. \quad (4.83)$$

Using the Pauli matrices σ_i previously defined by equation 2.22, we write

$$\sigma_3 + i\sigma_2 = \begin{pmatrix} 1 & 1 \\ -1 & -1 \end{pmatrix} \quad (4.84)$$

and

$$(i\partial^0 - qA^0)\chi = \left[\frac{1}{2m}(-i\vec{\nabla} - q\vec{A})^2(\sigma_3 + i\sigma_2) + m\sigma_3\right]\chi, \quad (4.85)$$

where a 2×2 identity matrix is assumed on the left-hand side of the equation. This is a first-order Schrödinger equation (cf. $i\partial^0\chi = H\chi$). The quantity in square brackets is $H - qA^0$.

The charge density for these states is (setting $\hbar = c = 1$)

$$\rho = \frac{iq}{2m} \left[\phi^*(\partial^0 + iqA^0)\phi - \phi(\partial^0 - iqA^0)\phi^*\right]$$
$$= \frac{iq}{2m}\frac{m}{i}[(\chi_1 + \chi_2)^*(\chi_1 - \chi_2) + (\chi_1 + \chi_2)(\chi_1 - \chi_2)^*]$$
$$= \frac{q}{2}(\chi_1^*\chi_1 - \chi_2^*\chi_2 + \chi_1^*\chi_1 - \chi_2^*\chi_2)$$
$$= q(|\chi_1|^2 - |\chi_2|^2), \quad (4.86)$$

which is rather simple and somewhat similar to the nonrelativistic case. Since we are describing particles of both signs of charge, it is not surprising that the density ρ appears as the difference of two positive definite densities.

Also in this notation, the charge density can be written as

$$\rho = q(|\chi_1|^2 - |\chi_2|^2) = q\chi^\dagger\sigma_3\chi, \quad (4.87)$$

where $\chi^\dagger$ is the hermitian conjugate of χ. Since $\chi^\dagger\sigma_3\chi$ is time independent, we can use it for normalization:

$$\langle\chi|\chi\rangle \equiv \int d^3x\,\chi^\dagger(x)\sigma_3\chi(x) = \pm 1. \quad (4.88)$$

It will be shown later that the sign is determined by whether we start with particles $(+)$ or antiparticles $(-)$: $\phi(t = 0) = \phi^{(\pm)}$.

Returning to equation 4.85, the Klein-Gordon Hamiltonian operator is

$$H = \frac{1}{2m}(-i\vec{\nabla} - q\vec{A})^2(\sigma_3 + i\sigma_2) + m\sigma_3 + qA^0. \tag{4.89}$$

The Hamiltonian appears to be non-hermitian, since

$$(\sigma_3 + i\sigma_2)^\dagger = \sigma_3 - i\sigma_2 \neq \sigma_3 + i\sigma_2. \tag{4.90}$$

However

$$\langle\chi'|H|\chi\rangle = \int d^3x\chi'^\dagger(x)\sigma_3 H\chi(x) \tag{4.91}$$

and

$$\langle\chi'|H|\chi\rangle^* = \left(\int d^3x\chi'^\dagger(x)\sigma_3 H\chi(x)\right)^\dagger$$

$$= \int d^3x\chi^\dagger(x)H^\dagger\sigma_3\chi'(x)$$

$$= \int d^3x\chi^\dagger(x)\sigma_3(\sigma_3 H^\dagger\sigma_3)\chi'(x)$$

$$= \langle\chi|\sigma_3 H^\dagger\sigma_3|\chi'\rangle. \tag{4.92}$$

We notice that

$$\sigma_3(\sigma_3 + i\sigma_2)^\dagger\sigma_3 = \sigma_3(\sigma_3 - i\sigma_2)\sigma_3 = \sigma_3 - i\sigma_3\sigma_2\sigma_3 \tag{4.93}$$

and

$$\sigma_3\sigma_2\sigma_3 = \sigma_3(i\sigma_1) = i(i\sigma_2) = -\sigma_2 \tag{4.94}$$

gives

$$\sigma_3(\sigma_3 + i\sigma_2)^\dagger\sigma_3 = \sigma_3 + i\sigma_2. \tag{4.95}$$

Therefore

$$\sigma_3 H^\dagger\sigma_3 = H. \tag{4.96}$$

Because of the normalization condition, the Hamiltonian is effectively hermitian when calculating its expectation value.

Consider the free particle solutions which follow from equations 4.77 and 4.78:

$$\chi_1 = \frac{1}{2}\left[\phi + \frac{i}{m}\partial^0\phi\right] \quad \text{and} \quad \chi_2 = \frac{1}{2}\left[\phi - \frac{i}{m}\partial^0\phi\right]. \tag{4.97}$$

The positive-energy plane-wave solution normalized to unit density (equation 4.40) is

$$\phi(x) = \sqrt{\frac{m}{E}}\, e^{-i(Et - \vec{p}\cdot\vec{x})} = \sqrt{\frac{m}{E}}\, e^{-ip\cdot x}. \tag{4.98}$$

Since $\partial^0 \phi = -iE\phi$,

$$\chi_1 = \frac{1}{2}\left(1 + \frac{E}{m}\right)\phi \quad \text{and} \quad \chi_2 = \frac{1}{2}\left(1 - \frac{E}{m}\right)\phi. \tag{4.99}$$

We write

$$\chi \equiv \begin{pmatrix} \chi_1 \\ \chi_2 \end{pmatrix} = \frac{1}{2}\sqrt{\frac{m}{E}}\frac{1}{m}\begin{pmatrix} m + E \\ m - E \end{pmatrix} e^{-ip\cdot x} \equiv \chi^{(+)}(\vec{p})e^{-ip\cdot x}, \tag{4.100}$$

where

$$\chi^{(+)}(\vec{p}) = \frac{1}{2\sqrt{mE}}\begin{pmatrix} m + E \\ m - E \end{pmatrix}. \tag{4.101}$$

The corresponding negative-energy plane-wave solution is

$$\phi(x) = \sqrt{\frac{m}{E}}\, e^{-i(-Et - \vec{p}\cdot\vec{x})} \tag{4.102}$$

giving

$$\chi^{(-)}(\vec{p}) = \frac{1}{2\sqrt{mE}}\begin{pmatrix} m - E \\ m + E \end{pmatrix}. \tag{4.103}$$

Orthogonality follows:

$$\begin{aligned}
\langle \chi^{(+)}(\vec{p})|\chi^{(+)}(\vec{p})\rangle &\equiv \chi^{(+)\dagger}(\vec{p})\sigma_3\chi^{(+)}(\vec{p}) \\
&= \frac{1}{4mE}\begin{pmatrix} m + E \\ m - E \end{pmatrix}^{\dagger}\begin{pmatrix} m + E \\ -m + E \end{pmatrix} \\
&= \frac{1}{4mE}[(m + E)(m + E) + (m - E)(-m + E)] \\
&= \frac{1}{4mE}(m^2 + 2mE + E^2 - m^2 + 2mE - E^2) \\
&= 1,
\end{aligned} \tag{4.104}$$

and it can also be shown (see problem 4.10) that

$$\langle \chi^{(-)}(\vec{p})|\chi^{(-)}(\vec{p})\rangle = -1, \tag{4.105}$$

$$\langle \chi^{(+)}(\vec{p})|\chi^{(-)}(\vec{p})\rangle = 0, \tag{4.106}$$

$$\langle \chi^{(-)}(\vec{p}) | \chi^{(+)}(\vec{p}) \rangle = 0. \tag{4.107}$$

In the nonrelativistic limit, we have

$$E = (p^2 + m^2)^{1/2} = m \left(1 + \frac{p^2}{m^2}\right)^{1/2} \approx m \left(1 + \frac{1}{2}\frac{p^2}{m^2}\right) = m + \frac{p^2}{2m}. \tag{4.108}$$

The components of equation 4.101 become

$$\frac{m+E}{2\sqrt{mE}} \approx \frac{m+m+p^2/2m}{2m^{1/2}\left(m+\frac{p^2}{2m}\right)^{1/2}} = \frac{2m+p^2/2m}{2m\left(1+\frac{1}{2}\frac{p^2}{2m^2}\right)} = 1, \tag{4.109}$$

$$\frac{m-E}{2\sqrt{mE}} \approx \frac{m-m-p^2/2m}{2m^{1/2}\left(m+\frac{p^2}{2m}\right)^{1/2}} = \frac{-p^2/2m}{2m\left(1+\frac{1}{2}\frac{p^2}{2m^2}\right)} = \frac{-p^2}{4m^2+p^2} \approx -\frac{p^2}{4m^2}. \tag{4.110}$$

Equation 4.101 in the nonrelativistic limit is

$$\chi^{(+)}(\vec{p}) \approx \begin{pmatrix} 1 \\ -p^2/4m^2 \end{pmatrix} = \begin{pmatrix} 1 \\ -1/4(v/c)^2 \end{pmatrix} \approx \begin{pmatrix} 1 \\ 0 \end{pmatrix}, \tag{4.111}$$

which holds to second order in the velocity. Similarly (see problem 4.11)

$$\chi^{(-)}(\vec{p}) \approx \begin{pmatrix} 0 \\ 1 \end{pmatrix}. \tag{4.112}$$

Since the exponential functions form a complete set, any wave packet can be expanded in terms of a linear combination of positive- and negative-energy solutions

$$\phi(\vec{x}, t) = \int \frac{d^3 p}{(2\pi)^3} e^{i\vec{p}\cdot\vec{x}} \left[a_{\vec{p}}^{(+)}(t)\chi^{(+)}(\vec{p}) + a_{-\vec{p}}^{(-)}(t)\chi^{(-)}(-\vec{p}) \right]$$

$$= \int \frac{d^3 p}{(2\pi)^3} \left[a_{\vec{p}}^{(+)}(t)\chi^{(+)}(\vec{p})e^{i\vec{p}\cdot\vec{x}} + a_{\vec{p}}^{(-)}(t)\chi^{(-)}(\vec{p})e^{-i\vec{p}\cdot\vec{x}} \right], \tag{4.113}$$

where $\chi(\vec{p})$ depends only on the magnitude of $\vec{p}$, and $a_{\vec{p}}^{(\pm)}(t)$ is a function of time and the magnitude of $\vec{p}$. The entire time dependence of $\phi(\vec{x}, t)$ is contained in the functions $a_{\vec{p}}^{(\pm)}(t)$.

If the wave packet is normalized to ± 1, we have

$$\pm 1 \equiv \langle \phi | \phi \rangle = \int d^3 x \phi^\dagger(\vec{x}, t) \sigma_3 \phi(\vec{x}, t)$$

$$= \int d^3 x \frac{d^3 p'}{(2\pi)^3} \frac{d^3 p}{(2\pi)^3} \left[[a_{\vec{p}}^{(+)*}(t) \chi^{(+)\dagger}(\vec{p}') e^{-i\vec{p}' \cdot \vec{x}} + a_{\vec{p}}^{(-)*}(t) \chi^{(-)\dagger}(\vec{p}') e^{i\vec{p}' \cdot \vec{x}}] \right.$$

$$\left. \cdot \sigma_3 \left[a_{\vec{p}}^{(+)}(t) \chi^{(+)}(\vec{p}) e^{i\vec{p} \cdot \vec{x}} + a_{\vec{p}}^{(-)}(t) \chi^{(-)}(\vec{p}) e^{-i\vec{p} \cdot \vec{x}} \right] \right]$$

$$= \int d^3 x \frac{d^3 p'}{(2\pi)^3} \frac{d^3 p}{(2\pi)^3} [a_{\vec{p}}^{(+)*}(t) a_{\vec{p}}^{(+)}(t) \chi^{(+)\dagger}(\vec{p}') \sigma_3 \chi^{(+)}(\vec{p}) e^{i\vec{x} \cdot (\vec{p} - \vec{p}')}$$

$$+ a_{\vec{p}}^{(-)*}(t) a_{\vec{p}}^{(-)}(t) \chi^{(-)\dagger}(\vec{p}') \sigma_3 \chi^{(-)}(\vec{p}) e^{-i\vec{x} \cdot (\vec{p} - \vec{p}')}$$

$$+ a_{\vec{p}}^{(+)*}(t) a_{\vec{p}}^{(-)}(t) \chi^{(+)\dagger}(\vec{p}') \sigma_3 \chi^{(-)}(\vec{p}) e^{-i\vec{x} \cdot (\vec{p} + \vec{p}')}$$

$$+ a_{\vec{p}}^{(-)*}(t) a_{\vec{p}}^{(+)}(t) \chi^{(-)\dagger}(\vec{p}') \sigma_3 \chi^{(+)}(\vec{p}) e^{i\vec{x} \cdot (\vec{p} + \vec{p}')}]$$

$$= \int \frac{d^3 p}{(2\pi)^3} \left[|a_{\vec{p}}^{(+)}(t)|^2 - |a_{\vec{p}}^{(-)}(t)|^2 \right]. \tag{4.114}$$

The integral of $|a_{\vec{p}}^{(+)}(t)|^2$ gives the relative amount of positive energy, or positive charge, which is spatially distributed according to eigenstate $\chi^{(+)}(\vec{p})$, while $|a_{\vec{p}}^{(-)}(t)|^2$ gives the relative amount of negative energy, or negative charge, distrubuted according to $\chi^{(-)}(\vec{p})$. The total amount of charge is ± 1. For the normalization to be time independent, $a_{\vec{p}}^{(\pm)}(t) = e^{\mp iEt} f^{(\pm)}(|\vec{p}|)$, where $f^{(\pm)}(|\vec{p}|)$ are scalar functions of the magnitude of $\vec{p}$ only.

4.9 Free-Particle Solutions and Wave Packets

The de Broglie relationship between momentum and wavelength of a particle suggests that it might be possible to use concentrated bunches of waves, called wave packets, to describe localized particles of matter. Based on experimental observation, the wave packet should have three properties: 1) it can interfere with itself, so that it can account for the results of diffraction experiments, 2) it is large in magnitude where the particle is likely to be and small elsewhere, and 3) the wave function will be regarded as describing the behavior of a single particle, not the statistical distribution of a number of such quanta.

It is possible to imagine configurations of waves that are very localized. Such localized wave packets can be achieved by superposing continuous waves with different frequencies in a particular way, so that they destructively interfere with each other almost completely outside of a specified spatial region. Since the Klein-Gordon equation is a linear wave equation, the wave packets are also solutions of the Klein-Gordon equation. The mathematical tools for dealing with wave packets involve Fourier integrals.

The wave-packet description of the motion of a particle agrees with the classical description when the circumstances are such that we can ignore the size and internal structure of the packet. The uncertainty principle for position and momentum of a quantum of matter follows directly from the wave-packet description, as we shall see.

When dealing with covariant wave packets some new features, which were absent in the nonrelativistic theory, will arise from the fact that the wave packet includes negative-energy solutions. It is not possible to exclude the negative-energy states simply by arguing that they are not realized in nature. The positive-energy states alone do not represent a complete set of functions in which to expand the wave-packet solutions.

Consider a wave packet containing only positive-energy components

$$\phi(\vec{x}, t) = \int \frac{d^3p}{(2\pi)^3} e^{i\vec{p}\cdot\vec{x}} a_{\vec{p}}^{(+)}(t)\chi^{(+)}(\vec{p}), \tag{4.115}$$

where

$$a_{\vec{p}}^{(+)}(t) = e^{-iE_pt} f([\vec{p} - \vec{p}_0]^2) \tag{4.116}$$

is some function which peaks at the origin in momentum space. For example, a narrow Gaussian distribution.

Let us consider the position operator $\hat{x}$ operating on a positive-energy only wave packet (equation 4.101)

$$\hat{x}\phi(\vec{x}, t) = \int \frac{d^3p}{(2\pi)^3} \vec{x} e^{i\vec{p}\cdot\vec{x}} a_{\vec{p}}^{(+)}(t)\chi^{(+)}(\vec{p}). \tag{4.117}$$

Noticing that $\vec{x}e^{i\vec{p}\cdot\vec{x}} = -i\vec{\nabla}_{\vec{p}} \, e^{i\vec{p}\cdot\vec{x}}$ [8] and integrating by parts, we write

$$\hat{x}\phi(\vec{x}, t) = -i \int \frac{d^3p}{(2\pi)^3} \left(\vec{\nabla}_{\vec{p}} \, e^{i\vec{p}\cdot\vec{x}}\right) a_{\vec{p}}^{(+)}(t)\chi^{(+)}(\vec{p})]$$

$$= i \int \frac{d^3p}{(2\pi)^3} e^{i\vec{p}\cdot\vec{x}} \left[\chi^{(+)}(\vec{p})\vec{\nabla}_{\vec{p}} a_{\vec{p}}^{(+)}(t) + a_{\vec{p}}^{(+)}(t)\vec{\nabla}_{\vec{p}}\chi^{(+)}(\vec{p})\right]$$

$$= i \int \frac{d^3p}{(2\pi)^3} e^{i\vec{p}\cdot\vec{x}} \left[\chi^{(+)}(\vec{p})\vec{\nabla}_{\vec{p}} a_{\vec{p}}^{(+)}(t) - \frac{\vec{p}}{2E^2} a_{\vec{p}}^{(+)}(t)\chi^{(-)}(\vec{p})\right], \tag{4.118}$$

where we have made use of the following identities:

---

[8] We are defining $\vec{\nabla}_{\vec{p}} \equiv \left(\frac{\partial}{\partial p_x}, \frac{\partial}{\partial p_y}, \frac{\partial}{\partial p_z}\right)$.

$$\vec{\nabla}_{\vec{p}} E = \vec{\nabla}_{\vec{p}} (m^2 + \vec{p}^2)^{1/2} = 1/2(m^2 + \vec{p}^2)^{-1/2} 2\vec{p} = \frac{\vec{p}}{E} \qquad (4.119)$$

and

$$
\begin{aligned}
\vec{\nabla}_{\vec{p}} \chi^{(+)}(\vec{p}) &= \vec{\nabla}_{\vec{p}} \frac{1}{2\sqrt{mE}} \begin{pmatrix} m+E \\ m-E \end{pmatrix} \\
&= -\frac{1}{2} \frac{E^{-3/2}}{2\sqrt{m}} \frac{\vec{p}}{E} \begin{pmatrix} m+E \\ m-E \end{pmatrix} + \frac{E^{-1/2}}{2\sqrt{m}} \frac{\vec{p}}{E} \begin{pmatrix} 1 \\ -1 \end{pmatrix} \\
&= \frac{\vec{p}}{4\sqrt{mE}E^2} \begin{pmatrix} 2E - m - E \\ -2E - m + E \end{pmatrix} = \frac{\vec{p}}{4\sqrt{mE}E^2} \begin{pmatrix} -m+E \\ -m-E \end{pmatrix} \\
&= -\frac{\vec{p}}{2E^2} \chi^{(-)}(\vec{p}).
\end{aligned} \qquad (4.120)
$$

Also notice

$$
\begin{aligned}
\vec{\nabla}_{\vec{p}} \chi^{(-)}(\vec{p}) &= \vec{\nabla}_{\vec{p}} \frac{1}{2\sqrt{mE}} \begin{pmatrix} m-E \\ m+E \end{pmatrix} \\
&= -\frac{1}{2} \frac{E^{-3/2}}{2\sqrt{m}} \frac{\vec{p}}{E} \begin{pmatrix} m-E \\ m+E \end{pmatrix} + \frac{E^{-1/2}}{2\sqrt{m}} \frac{\vec{p}}{E} \begin{pmatrix} -1 \\ 1 \end{pmatrix} \\
&= \frac{\vec{p}}{4\sqrt{mE}E^2} \begin{pmatrix} -2E - m + E \\ 2E - m - E \end{pmatrix} = \frac{\vec{p}}{4\sqrt{mE}E^2} \begin{pmatrix} -m-E \\ -m+E \end{pmatrix} \\
&= -\frac{\vec{p}}{2E^2} \chi^{(+)}(\vec{p}).
\end{aligned} \qquad (4.121)
$$

In equation 4.118, we see that the position operator has introduced a negative-energy piece into the wave function even though there was none present originally. Equivalently, multiplication by the potential energy $qA^0(\vec{x})$, which is a function of $\vec{x}$ only, introduces such negative-energy states. Thus whenever a wave packet interacts with a potential, we should not be surprised to find negative-energy states appearing.

There are other consequences of the behavior of wave packets that do not appear in nonrelativistic quantum mechanics. A wave packet beginning at the origin at $t = 0$ will move with a uniform velocity $\vec{p}c/E$. If we pick a functional form for f, we will find that there is a minimum width of the wave packet, and that it cannot be made smaller so long as only positive-energy components are present (see problem 4.15).

We can also attempt to construct a wave packet localized at the origin. If one wishes to localize a wave packet within a distance smaller than $\hbar/mc$, negative-energy components are required (see problem 4.15). The characteristic length $\lambda_c = \hbar/mc$ is called the Compton wavelength. For an electron $m = m_e$ and $\lambda_c = 3.9 \times 10^{-11}$ cm, which is a typical scale in atomic processes.

4.9.1 Zitterbewegung

As a final example, let us evaluate the expectation value of the position operator for a wave packet that is a mixture of positive- and negative-energy components (equation 4.113).

$$
\begin{aligned}
\langle\phi(t)|\hat{x}|\phi(t)\rangle &= \int d^3x\,\phi^\dagger(\vec{x},t)\sigma_3\hat{x}\phi(\vec{x},t)\\[4pt]
&= \int d^3x \int \frac{d^3p'}{(2\pi)^3} \int \frac{d^3p}{(2\pi)^3}\\
&\quad\cdot \Big[a_{\vec{p}'}^{(+)*}(t)\chi^{(+)\dagger}(\vec{p}')e^{-i\vec{p}'\cdot\vec{x}} + a_{\vec{p}'}^{(-)*}(t)\chi^{(-)\dagger}(\vec{p}')e^{+i\vec{p}'\cdot\vec{x}} \Big]\\
&\quad\cdot \sigma_3\hat{x} \Big[a_{\vec{p}}^{(+)}(t)\chi^{(+)}(\vec{p})e^{+i\vec{p}\cdot\vec{x}} + a_{\vec{p}}^{(-)}(t)\chi^{(-)}(\vec{p})e^{-i\vec{p}\cdot\vec{x}} \Big]\\[4pt]
&= i\int d^3x \int \frac{d^3p'}{(2\pi)^3} \int \frac{d^3p}{(2\pi)^3}\\
&\quad\cdot \Big[a_{\vec{p}'}^{(+)*}(t)\chi^{(+)\dagger}(\vec{p}')e^{-i\vec{p}'\cdot\vec{x}} + a_{\vec{p}'}^{(-)*}(t)\chi^{(-)\dagger}(\vec{p}')e^{+i\vec{p}'\cdot\vec{x}} \Big]\\
&\quad\cdot \sigma_3 \Big[e^{+i\vec{p}\cdot\vec{x}} \big(\chi^{(+)}(\vec{p})\vec{\nabla}_{\vec{p}}a_{\vec{p}}^{(+)}(t) + a_{\vec{p}}^{(+)}(t)\vec{\nabla}_{\vec{p}}\chi^{(+)}(\vec{p}) \big)\\
&\quad+ e^{-i\vec{p}\cdot\vec{x}} \big(\chi^{(-)}(\vec{p})\vec{\nabla}_{\vec{p}}a_{\vec{p}}^{(-)}(t) - a_{\vec{p}}^{(-)}(t)\vec{\nabla}_{\vec{p}}\chi^{(-)}(\vec{p}) \big) \Big],
\end{aligned}
\tag{4.122}
$$

where we have again used $\vec{x}e^{i\vec{p}\cdot\vec{x}} = -i\vec{\nabla}_{\vec{p}}\,e^{i\vec{p}\cdot\vec{x}}$ and integrated by parts. Continuing, we have

$$
\begin{aligned}
\langle\phi(t)|\hat{x}|\phi(t)\rangle &= i\int \frac{d^3p}{(2\pi)^3} \Big[a_{\vec{p}}^{(+)*}(t)\chi^{(+)\dagger}(\vec{p})\sigma_3\chi^{(+)}(\vec{p})\vec{\nabla}_{\vec{p}}a_{\vec{p}}^{(+)}(t)\\
&\quad+ a_{\vec{p}}^{(-)*}(t)\chi^{(-)\dagger}(\vec{p})\sigma_3\chi^{(-)}(\vec{p})\vec{\nabla}_{\vec{p}}a_{\vec{p}}^{(-)}(t)\\
&\quad+ a_{-\vec{p}}^{(-)*}(t)\chi^{(-)\dagger}(-\vec{p})\sigma_3 a_{\vec{p}}^{(+)}(t)\vec{\nabla}_{\vec{p}}\chi^{(+)}(\vec{p})\\
&\quad+ a_{-\vec{p}}^{(+)*}(t)\chi^{(+)\dagger}(-\vec{p})\sigma_3 a_{\vec{p}}^{(-)}(t)\vec{\nabla}_{\vec{p}}\chi^{(-)}(\vec{p}) \Big]\\[4pt]
&= i\int \frac{d^3p}{(2\pi)^3} \Big[a_{\vec{p}}^{(+)*}(t)\chi^{(+)\dagger}(\vec{p})\sigma_3 \Big(-it\frac{\vec{p}}{E}\Big) a_{\vec{p}}^{(+)}(t)\chi^{(+)}(\vec{p})\\
&\quad+ a_{\vec{p}}^{(-)*}(t)\chi^{(-)\dagger}(\vec{p})\sigma_3 \Big(it\frac{\vec{p}}{E}\Big) a_{\vec{p}}^{(-)}(t)\chi^{(-)}(\vec{p})\\
&\quad+ a_{\vec{p}}^{(-)*}(t)\chi^{(-)\dagger}(\vec{p})\sigma_3 \Big(-\frac{\vec{p}}{2E^2}\Big) a_{\vec{p}}^{(+)}(t)\chi^{(-)}(\vec{p})\\
&\quad+ a_{\vec{p}}^{(+)*}(t)\chi^{(+)\dagger}(\vec{p})\sigma_3 \Big(-\frac{\vec{p}}{2E^2}\Big) a_{\vec{p}}^{(-)}(t)\chi^{(+)}(\vec{p}) \Big].
\end{aligned}
\tag{4.123}
$$

The first two terms in the last expression above are obtained by noticing that

$$\int \frac{d^3p}{(2\pi)^3} a_{\vec{p}}^{(\pm)^*}(t) \vec{\nabla}_{\vec{p}} a_{\vec{p}}^{(\pm)}(t)$$

$$= \int \frac{d^3p}{(2\pi)^3} \left[\mp it \frac{\vec{p}}{E} f^2([\vec{p} - \vec{p}_0]^2) + 2(\vec{p} - \vec{p}_0) f([\vec{p} - \vec{p}_0]^2) f'([\vec{p} - \vec{p}_0]^2) \right].$$

$$(4.124)$$

Since $f([\vec{p} - \vec{p}_0]^2)$ is very narrow, it approximates a Dirac delta function and thus

$$\int \frac{d^3p}{(2\pi)^3} a_{\vec{p}}^{(+)^*}(t) \vec{\nabla}_{\vec{p}} a_{\vec{p}}^{(+)}(t) \approx \mp it \frac{\vec{p}}{E}. \tag{4.125}$$

Using the normalization of χ,

$$\langle \phi(t) | \vec{x} | \phi(t) \rangle = \int \frac{d^3p}{(2\pi)^3} \frac{\vec{p}}{E} t \left(|a_{\vec{p}}^{(+)}(t)|^2 - |a_{\vec{p}}^{(-)}(t)|^2 \right)$$

$$- \Re \left[i \int \frac{d^3p}{(2\pi)^3} \frac{\vec{p}}{E^2} a_{\vec{p}}^{(+)^*}(t) a_{\vec{p}}^{(-)}(t) \right], \tag{4.126}$$

where $\Re$ represents taking the real part of the quantity in square brackets. The first term in equation 4.126 represents the expected uniform motion of the wave packet. In the second term

$$a_{\vec{p}}^{(+)^*}(t) a_{\vec{p}}^{(-)}(t) \sim e^{2i|E|t} \tag{4.127}$$

represents a rapid wiggling of the position of the particle about its central location due to the interference of positive- and negative-energy components of the wave. This "jitter" motion is referred to as Zitterbewegung and arises when attempting to localize the particle or from interactions of the particle with a potential.

The positive- and negative-energy components of the wave packet travel in opposite directions. Thus wave packets damp out after a time $\Delta t \sim 1/m$ once the interactions with the potential have ceased. An exception is when the potential is very strong. This problem is called the Klein's paradox for reasons which will become apparent in the next section.

4.10   Klein Paradox for Spin-0 Particles

To localize a particle, a strong external potential must be used to confine it to the desired region. If a free particle of energy E is not to be found more than a distance d outside the confined region, the confining potential V must rise sharply within an interval less than d to a height $V > E$ so that the solution for the particle falls off within a characteristic length less than d. This is as in the Schrödinger theory. But when $(V - E)$ increases beyond mc^2 and the confining length d shrinks to about the Compton wavelength $\hbar/mc$ something different happens. To see what happens, we consider the scattering of a Klein-Gordon particle of energy E and momentum $\vec{p}$ from a electrostatic step-function potential as shown in figure 4.1. This problem is an archetypical problem in nonrelativistic quantum mechanics. For relativistic quantum mechanics, we will find that the solution leads to a paradox – Klein Paradox – when the potential is strong.

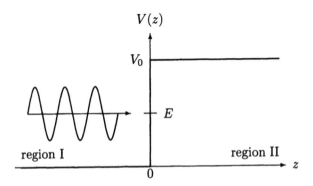

FIGURE 4.1:   Electrostatic potential idealized with a sharp boundary, with an incident free scalar particle of energy E moving to the right in region I.

Mathematically, the step-function electromagnetic potential can be written as

$$\vec{A} = 0 \quad \text{and} \quad qA_0 = V(z), \qquad (4.128)$$

where

$$V(z) = \begin{cases} 0 & \text{for } z < 0, \\ V_0 & \text{for } z > 0. \end{cases} \qquad (4.129)$$

The Klein-Gordon equation in the presence of this potential becomes

$$-\frac{\partial^2 \phi}{\partial t^2} + \nabla^2 \phi - m^2 \phi = 0 \quad \text{for } z < 0, \qquad (4.130)$$

$$\left(i\frac{\partial}{\partial t} - V_0 \right)^2 \phi + \nabla^2 \phi - m^2 \phi = 0 \quad \text{for } z > 0. \qquad (4.131)$$

A positive-energy incoming beam ($z = -\infty$) is a plane wave of the form $e^{-i(Et-pz)}$. We look for solutions of the form

$$
\begin{aligned}
\phi_< &= e^{-iEt}[e^{ipz} + Re^{-ipz}] && \text{for } z < 0, \\
\phi_> &= Te^{-iEt}e^{ip'z} && \text{for } z > 0,
\end{aligned}
\tag{4.132}
$$

where R and T are reflected and transmitted amplitudes, and p' is the momentum in region II. Substitution of these solutions into the Klein-Gordon equation yields

$$
\begin{aligned}
E^2 - p^2 - m^2 = 0 &\Rightarrow E = \sqrt{p^2 + m^2} && \text{for } z < 0, \\
E^2 + V_0^2 - 2EV_0 - p'^2 - m^2 = 0 &\Rightarrow p' = \pm\sqrt{(E - V_0)^2 - m^2} && \text{for } z > 0.
\end{aligned}
\tag{4.133}
$$

The positive sign is chosen for the square-root when $z < 0$, since this is imposed by our initial conditions of solving the problem for a positive-energy incoming beam. The sign for the square-root when $z > 0$ can not yet be specified.

In region II there are three distinct cases, depending on the strength of the potential. This is shown in figure 4.2.

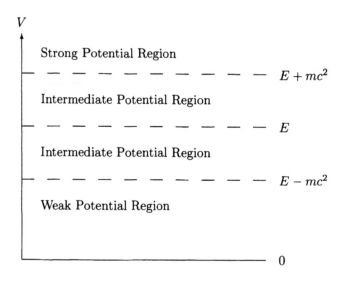

FIGURE 4.2: Energy level diagram for a particle in region II.

weak potential: $V_0 < E - m \quad\Rightarrow\quad E - V_0 > m \quad\Rightarrow\quad p'$ is real, intermediate potential: $(E - m) < V_0 < (E + m)$ $\Rightarrow$ $|E - V_0| < m$ $\Rightarrow$ p' is purely imaginary,

strong potential: $V_0 > E + m$ $\Rightarrow$ $|E - V_0| > m$ $\Rightarrow$ p' is real.

We see that p' can be real or imaginary depending on the strength of the potential V_0. In the region of strong potential, the kinetic energy $(E - V_0)$ is negative, which is forbidden classically. The group velocity is given by $p'/(E - V_0)$ and therefore has the opposite direction to the momentum p' in this region, if we choose $p' > 0$. Since the group velocity is the velocity of the moving wave packet, it looks as if the transmitted wave packet came in from $z = +\infty$; on the other hand this contradicts the initial condition, which only allows an incoming wave packet from $z = -\infty$. We thus have to choose $p' < 0$, i.e. the negative sign of the square root in equation 4.133, for the case of a strong potential. Notice that this condition is not included in the Klein-Gordon equation, but is forced upon us by the physical boundary conditions of our problem.

Imposing the boundary conditions that ϕ and $\partial\phi/\partial z$ be continuous at $z = 0$ gives

$$e^{-iEt}(1 + R) = Te^{-iEt} \quad \Rightarrow \quad 1 + R = T, \tag{4.134}$$

$$e^{-iEt}ip(1 - R) = ip'Te^{-iEt} \quad \Rightarrow \quad 1 - R = \frac{p'}{p}T. \tag{4.135}$$

Solving for R and T we obtain

$$2 = \left(1 + \frac{p'}{p}\right)T \quad \Rightarrow \quad T = \frac{2}{1 + p'/p} = \frac{2p}{p + p'}, \tag{4.136}$$

$$1 - R = \frac{p'}{p}(1 + R) \quad \Rightarrow \quad R = \frac{1 - p'/p}{1 + p'/p} = \frac{p - p'}{p + p'}. \tag{4.137}$$

Recall that the current is defined as

$$\vec{j} = \frac{1}{2im}(\phi^*\vec{\nabla}\phi - \phi\vec{\nabla}\phi^*), \tag{4.138}$$

so the incident current is

$$j_I = \frac{p}{m}. \tag{4.139}$$

The final currents to the left and right of the potential boundary are

$$j_< = \frac{1}{2im}[(e^{-ipz} + R^*e^{ipz})ip(e^{ipz} - Re^{-ipz})$$
$$-(e^{ipz} + Re^{-ipz})ip(-e^{-ipz} + R^*e^{ipz})]$$

$$= \frac{1}{2im} ip[1 + R^* e^{2ipz} - Re^{-2ipz} - |R|^2 + 1 + Re^{-2ipz} - R^* e^{2ipz} - |R|^2]$$

$$= \frac{p}{m}(1 - |R|^2) \tag{4.140}$$

and

$$j_> = \frac{1}{2im}[T^* e^{-ip'^* z} ip' T e^{ip'z} + T e^{ip'z} ip'^* T^* e^{-ip'^* z}]$$

$$= \frac{p' + p'^*}{2m}|T|^2 e^{i(p' - p'^*)z}$$

$$= \begin{cases} \frac{p'}{m}|T|^2 & \text{for } p' \text{ real,} \\ 0 & \text{for } p' \text{ imaginary.} \end{cases} \tag{4.141}$$

The transmission coefficient is

$$T = \frac{j_>}{j_I} = \frac{p' + p'^*}{2p}|T|^2 e^{i(p'-p'^*)z} = \frac{2p(p' + p'^*)}{(p' + p)(p'^* + p)} e^{i(p'-p'^*)z}. \tag{4.142}$$

The reflection coefficient is

$$\mathcal{R} = \frac{j_I - j_<}{j_I} = 1 - (1 - |R|^2) = |R|^2 = \left| \frac{p - p'}{p + p'} \right|^2. \tag{4.143}$$

For the case of the weak potential (p' real), the transmission and reflection coefficients are

$$T = \frac{4pp'}{(p + p')^2}, \quad \mathcal{R} = \left(\frac{p - p'}{p + p'} \right)^2, \quad T + \mathcal{R} = 1. \tag{4.144}$$

Thus the incident beam is partly reflected and partly transmitted. This is similar to the result obtained in nonrelativistic quantum mechanics. The last expression shows that the total probability is conserved.

For the case of the intermediate potential (p' imaginary), the transmission and reflection coefficients are

$$T = 0, \quad \mathcal{R} = 1, \quad T + \mathcal{R} = 1. \tag{4.145}$$

Thus there is no transmission and only reflection. This is again the result obtained in nonrelativistic quantum mechanics, and the probability is conserved.

For the case of the strong potential (p' real and negative), the transmission and reflection coefficients are

$$T = -\frac{4p|p'|}{(p - |p'|)^2}, \quad \mathcal{R} = \left(\frac{p + |p'|}{p - |p'|} \right)^2 > 1, \quad T + \mathcal{R} = 1. \tag{4.146}$$

The probability is still conserved, but only at the cost of a negative transmission coefficient and a reflection coefficient which exceeds unity. The strong potential appears to give rise to a paradox. There is no paradox if we consider that in the strong potential case the potential is strong enough to create particle-antiparticle pairs. The antiparticles are attracted by the potential and create a negative current moving to the right. This is the origin of the negative transmission coefficient. The particles, on the other hand, are reflected from the barrier and combined with the incident particle beam, which is completely reflected, leading to a positive current, moving to the left, and with magnitude greater than that of the incident beam.

4.11   Coulomb Interaction

The study of the structure of hypothetical atomic states with spin-0 constituents is an interesting intellectual problem. In the next chapter, we will compare the results obtained from the Klein-Gordon equation with those of the Dirac equation so that we can tell how much of the observed energy-level structure is due to relativistic effects alone and how much is due to the spin of the electron.

Spin-0 "atoms" can be made in nature when a π^- or K^- meson is captured by a nucleus. The object is called a pionic or kaonic atom. The mass of the pion is 273 times heavier than the electron, and has a half-life of 2.6×10^{-8} s. Since the classical orbital period is about 10^{-21} s, we can think of the pion in a well-defined stationary states, despite its finite half-life.

Consider the Coulomb interaction

$$A_0 = -\frac{Ze}{r} \quad \text{and} \quad \vec{A} = 0, \tag{4.147}$$

and the substitution

$$\hat{p}_\mu \rightarrow \hat{p}_\mu - \frac{e}{c}A_\mu, \tag{4.148}$$

where we are assuming the charge of the Klein-Gordon particle is the same as the electron.

The Klein-Gordon equation with a Coulomb potential is

$$\left[(i\hbar\partial^\mu - \frac{e}{c}A^\mu)^2 - m^2c^2 \right] \Phi = 0$$

$$\left[(i\hbar\partial_0 - \frac{e}{c}A_0)^2 - (-i\hbar\vec{\nabla} - \frac{e}{c}\vec{A})^2 - m^2c^2 \right] \Phi = 0$$

$$\left[\left(i\hbar\frac{\partial}{\partial t} - eA_0 \right)^2 + \hbar^2c^2\nabla^2 - m^2c^4 \right] \Phi = 0. \tag{4.149}$$

For stationary states $\Phi = e^{-iEt/\hbar}\phi(\vec{x})$, we write

$$\left(E + \frac{Ze^2}{r}\right)^2 \phi + (\hbar^2 c^2 \nabla^2 - m^2 c^4)\phi = 0. \tag{4.150}$$

This is a specific case of the more general equation for a stationary state in the presence of a general potential:

$$(E - eV)^2 \phi = c^2 \left(\hat{p} - \frac{e}{c}\vec{A}\right)^2 \phi + m^2 c^4 \phi. \tag{4.151}$$

For spherical coordinates, the Laplacian is

$$\nabla^2 = \frac{1}{r^2}\frac{\partial}{\partial r}\left(r^2\frac{\partial}{\partial r}\right) + \frac{1}{r^2 \sin\theta}\left(\sin\theta\frac{\partial}{\partial\theta}\right) + \frac{1}{r^2 \sin^2\theta}\frac{\partial^2}{\partial\phi^2}. \tag{4.152}$$

For solutions of definite angular momentum l, we have

$$\phi(r,\theta,\phi) = R(r)Y_l^m(\theta,\phi) \tag{4.153}$$

$$\Rightarrow \quad \nabla^2 = \frac{1}{r^2}\frac{d}{dr}\left(r^2\frac{d}{dr}\right) - \frac{l(l+1)}{r^2}, \tag{4.154}$$

where $l = 0, 1, 2, ...$ The Klein-Gordon equation in spherical coordinates is

$$\left[\frac{(E + Ze^2/r)^2 - m^2 c^4}{\hbar^2 c^2}\right]R = \left[-\frac{1}{r^2}\frac{d}{dr}\left(r^2\frac{d}{dr}\right) + \frac{l(l+1)}{r^2}\right]R. \tag{4.155}$$

$$\frac{1}{r^2}\frac{d}{dr}\left(r^2\frac{dR}{dr}\right) + \left[\frac{E^2 - m^2 c^4}{\hbar^2 c^2} + \frac{2Ze^2 E}{\hbar^2 c^2 r} + \frac{Z^2 e^4}{\hbar^2 c^2 r^2} - \frac{l(l+1)}{r^2}\right]R = 0. \tag{4.156}$$

Defining $\frac{E^2 - m^2 c^4}{\hbar^2 c^2} \equiv -\frac{\alpha^2}{4}$, we write

$$\frac{1}{r^2}\frac{d}{dr}\left(r^2\frac{dR}{dr}\right) - \frac{\alpha^2}{4}\left[1 - \frac{8Ze^2 E}{\hbar^2 c^2 \alpha^2 r} - \frac{4Z^2 e^4}{\hbar^2 c^2 \alpha^2 r^2} + \frac{4l(l+1)}{\alpha^2 r^2}\right]R = 0. \tag{4.157}$$

Defining $\gamma \equiv \frac{Ze^2}{\hbar c}$, we write

$$\frac{1}{r^2}\frac{d}{dr}\left(r^2\frac{dR}{dr}\right) - \frac{\alpha^2}{4}\left[1 - \frac{8\gamma E}{\hbar c\alpha^2 r} - 4\frac{\gamma^2 - l(l+1)}{\alpha^2 r^2}\right]R = 0. \tag{4.158}$$

Defining $\lambda \equiv \frac{2E\gamma}{\hbar c\alpha}$, we write

$$\frac{1}{r^2}\frac{d}{dr}\left(r^2\frac{dR}{dr}\right) - \frac{\alpha^2}{4}\left[1 - \frac{4\lambda}{\alpha r} - 4\frac{\gamma^2 - l(l+1)}{(\alpha r)^2}\right]R = 0$$

$$\frac{1}{r^2}\frac{d}{dr}\left(r^2\frac{dR}{dr}\right) + \alpha^2\left[\frac{\lambda}{\alpha r} - \frac{1}{4} - \frac{l(l+1) - \gamma^2}{(\alpha r)^2}\right]R = 0. \quad (4.159)$$

Defining $\rho \equiv \alpha r$ and $\frac{d}{dr} = \frac{d\rho}{dr}\frac{d}{d\rho} = \alpha\frac{d}{d\rho}$, we have

$$\frac{\alpha^2}{\rho^2}\alpha\frac{d}{d\rho}\left(\frac{\rho^2\alpha}{\alpha^2}\frac{dR}{d\rho}\right) + \alpha^2\left[\frac{\lambda}{\rho} - \frac{1}{4} - \frac{l(l+1) - \gamma^2}{\rho^2}\right]R = 0$$

$$\frac{1}{\rho^2}\frac{d}{d\rho}\left(\rho^2\frac{dR}{d\rho}\right) + \left[\frac{\lambda}{\rho} - \frac{1}{4} - \frac{l(l+1) - \gamma^2}{\rho^2}\right]R = 0. \quad (4.160)$$

This radial equation is the same as in the nonrelativistic case if $l(l+1) \rightarrow l(l+1) - \gamma^2$. Solving for E, we have

$$\frac{E^2 - m^2c^4}{\hbar^2c^2} = -\frac{1}{4}\left(\frac{2E\gamma}{\hbar c\lambda}\right)^2,$$

$$E^2 - m^2c^4 = -\frac{E^2\gamma^2}{\lambda^2},$$

$$E^2\left(1 + \frac{\gamma^2}{\lambda^2}\right) = m^2c^4,$$

$$E = mc^2\left(1 + \frac{\gamma^2}{\lambda^2}\right)^{-1/2}. \quad (4.161)$$

We have chosen the positive square-root because in the case of no field, $Z\alpha \rightarrow 0$, and the free-particle solution must be obtained. The parameter λ is determined by the boundary condition on R when $\rho = \infty$.

The remainder of this derivation is left as an exercise for the reader (see problem 4.16). The comparison with the nonrelativistic case is discussed in the problems. In general, the method of solving bound-state problems is the same as in nonrelativistic quantum mechanics but the partial differential equations are different.

4.12   Summary

By combining relativity and quantum mechanics in this chapter, we have uncovered two new phenomena:

1. Relativistic invariant wave packets cannot be localized to less than their Compton wavelength (m^{-1}).

2. For each charged particle there is an antiparticle with the same mass but opposite charge.

There is nothing that prevents us from trying to construct a theory for scalar (or pseudoscalar) bosons interacting with the electromagnetic field. This would enable us to calculate cross sections for the scattering of spinless particles from spinless particles, and the decay of spinless particles to spinless particles. The greatest difficulty in this development would be moving from a description for single-particle scattering in a potential to two-particle scattering. The latter can be described by treating one particle as moving in the field caused by the other particle. The particle field would be calculated from Maxwell's equation $\Box A^\mu = j^\mu$, where the current j^μ is that inferred from single-particle scattering in a potential. The theory would be called scalar electrodynamics. The importance of this theory would be limited because there are no elementary charged scalar particles in nature. We will postpone the development of interactions until later.

The best candidates for the role of pseudoscalar mesons are the π and K. They are unstable and decay by weak interactions. Since this lifetime is very long on a natural time scale, the pion can be considered stable to a good approximation. The more basic problem is that pions have an internal structure. It is well known that mesons are regarded as being composed of two quarks with spin 1/2. Scalar electrodynamics is completely inadequate for describing the coupling of mesons with each other, because the coupling is dominated by the strong interaction.

Many problems arise from attempting to apply a simple single-particle wave function picture to what is obviously a many-body situation. The correct way in which to handle all the subtlety of these problems is to use the formalism of quantum field theory. Nevertheless, the elementary wave function paradigm has allowed us to obtain an accurate sketch of the physics of spinless particles within the limitations of a one-particle theory.

4.13   Problems

1. Write down the wave equation for the nonrelativistic Hamiltonian developed from

$$E^2 = \left(\frac{\vec{p}^{\,2}}{2m} \right)^2,$$

and show that you get positive- and negative-energy solutions.

2. The square-root operator in equation 4.7 is well defined if we write the solutions ψ as a Fourier transform.

(a) Show that

$$i\frac{\partial}{\partial t}\psi(\vec{x},t) = \int d^3x' K(\vec{x},\vec{x}')\psi(\vec{x}',t),$$

where

$$K(\vec{x},\vec{x}') = \int \frac{d^3p}{(2\pi)^3}e^{i\vec{p}\cdot(\vec{x}-\vec{x}')}\sqrt{\vec{p}^2+m^2}.$$

(b) For large $|\vec{x}-\vec{x}'|$ most values of $|\vec{p}|$ except for those with $|\vec{p}| < 1/|\vec{x}-\vec{x}'|$ will lead to rapid oscillations of the exponential and consequently a very small value for the integral. In fact, the integral will be sizable only for $|\vec{x}-\vec{x}'| < 1/m$. Show that the result just obtained may be used via a Taylor series expansion to relate $\psi(x,t+\delta t)$ to values of $\psi(x' \sim x \pm 1/m, t)$.

(c) Show that values of $\psi(x \pm 1/m, t)$ are affecting $\psi(x, t+\delta t)$ even though these two space-time points are outside the forward light cone, and thus causality is violated.

3. Show that in the nonrelativistic limit, the free-particle Klein-Gordon equation becomes the free-particle Schrödinger equation.

4. The Lorentz-invariant step function $\theta(p) \equiv \theta(p_0)$ is defined by

$$\theta(p) = \begin{cases} 1 & \text{for } p_0 > 0, \\ 0 & \text{for } p_0 < 0. \end{cases}$$

It is said that this step function is Lorentz invariant because it only distinguishes between the past and the future, which is a Lorentz invariant concept. Show that this step function is Lorentz invariant.

Hint: I believe this is true only if p is restricted to be a time-like vector.

5. Show that

$$\rho = \frac{i\hbar}{2mc^2}\left(\phi^*\frac{\partial\phi}{\partial t} - \phi\frac{\partial\phi^*}{\partial t}\right)$$

reduces to the nonrelativistic expression in the nonrelativistic limit.

6. Show that

$$q = \int d^3x\phi^{(-)*}(x)i\overleftrightarrow{\partial_0}\phi^{(-)}(x) < 0$$

is satisfied for a solution to the Klein-Gordon equation $\phi^{(-)}(x)$ that is a superposition of negative energy plane-wave solutions.

7. [26] Show that the expectation values of E^2 and $\vec{p}^2$ for a general wave-packet solution of the free-particle Klein-Gordon equation satisfy the equation $\langle E^2 \rangle = c^2 \langle \vec{p}^2 \rangle + (mc^2)^2$. Discuss the connection between this result and the classical equation $E^2 = (c\vec{p})^2 + (mc^2)^2$.

8. Use explicit plane-wave solutions $\phi_{\vec{p}}^{(\pm)}(x)$ to establish the following normalization and orthogonality relationships:

$$\int d^3x \, \phi_{\vec{p},}^{(\pm)^*}(x) i \overleftrightarrow{\partial}_0 \, \phi_{\vec{p}}^{(\pm)}(x) = \pm \delta^3(\vec{p} - \vec{p}'),$$

$$\int d^3x \, \phi_{\vec{p},}^{(\pm)^*}(x) i \overleftrightarrow{\partial}_0 \, \phi_{\vec{p}}^{(\mp)}(x) = 0.$$

9. Derive the conserved current for a scalar wave interacting with an electromagnetic field with minimal coupling.

10. The positive- and negative-energy two-component solutions of the Klein-Gordon equation in the Schrödinger form were defined as

$$\chi^{(\pm)}(\vec{p}) e^{\mp iEt + i\vec{p}\cdot\vec{x}} \equiv \frac{1}{2\sqrt{mE}} \begin{pmatrix} m \pm E \\ m \mp E \end{pmatrix} e^{\mp iEt + i\vec{p}\cdot\vec{x}}.$$

Show that the $\chi^{(\pm)}(\vec{p})$ are orthonormalized.

11. Show that in the nonrelativistic limit

$$\chi^{(-)}(\vec{p}) \approx \begin{pmatrix} 0 \\ 1 \end{pmatrix}.$$

12. By completeness, any wave packet can be expanded in terms of a linear combination of positive- and negative-energy solutions:

$$\phi(\vec{x},t) = \int \frac{d^3p}{(2\pi)^3} \left[a_{\vec{p}}^{(+)}(t) \chi^{(+)}(\vec{p}) e^{i\vec{p}\cdot\vec{x}} + a_{\vec{p}}^{(-)}(t) \chi^{(-)}(\vec{p}) e^{-i\vec{p}\cdot\vec{x}} \right].$$

The $\chi^{(\pm)}(\vec{p})$ are defined as

$$\chi^{(\pm)}(\vec{p}) e^{\mp iEt + i\vec{p}\cdot\vec{x}} \equiv \frac{1}{2\sqrt{mE}} \begin{pmatrix} m \pm E \\ m \mp E \end{pmatrix} e^{\mp iEt + i\vec{p}\cdot\vec{x}}$$

and the $a_{\vec{p}}^{(\pm)}(t) = e^{\mp iEt} f^{(\pm)}(|\vec{p}|)$, where $f^{(\pm)}(|\vec{p}|)$ are general scalar functions of the magnitude of $\vec{p}$. Derive the normalization requirement for $\langle \phi | \phi \rangle = \pm 1$.

13. Show that the energy and momentum expectation values in the two-component Schrödinger representation are given by

$$\hat{E} = \langle \phi | H | \phi \rangle = \int d^3x \phi^\dagger(\vec{x}, t) \sigma_3 H \phi(\vec{x}, t)$$

$$= \int \frac{d^3p}{(2\pi)^3} \sqrt{m^2 + \vec{p}^2} \left(|a_{\vec{p}}^{(+)}(t)|^2 + |a_{\vec{p}}^{(-)}(t)|^2 \right)$$

and

$$\hat{p} = \langle \phi | -i\vec{\nabla} | \phi \rangle = \int d^3x \phi^\dagger(\vec{x}, t) \sigma_3 (-i\vec{\nabla}_x) \phi(\vec{x}, t)$$

$$= \int \frac{d^3p}{(2\pi)^3} \vec{p} \left(|a_{\vec{p}}^{(+)}(t)|^2 + |a_{\vec{p}}^{(-)}(t)|^2 \right).$$

14. Invert

$$\phi(\vec{x}, t) = \int \frac{d^3p}{(2\pi)^3} \left[a_{\vec{p}}^{(+)}(t) \chi^{(+)}(\vec{p}) e^{i\vec{p}\cdot\vec{x}} + a_{\vec{p}}^{(-)}(t) \chi^{(-)}(\vec{p}) e^{-i\vec{p}\cdot\vec{x}} \right]$$

to obtain expressions for $a_{\vec{p}}^{(+)}(t)$ and $a_{\vec{p}}^{(-)}(t)$.

15. Starting with a positive-energy wave packet only, show that there is a minimum width to the wave packet, and that it can not be localized within a distance smaller than the Compton wavelength without creating negative-energy states.

16. Solve the coulomb potential problem. In equation 4.160, we obtained

$$\frac{1}{\rho^2} \frac{d}{d\rho} \left(\rho^2 \frac{dR}{d\rho} \right) + \left[\frac{\lambda}{\rho} - \frac{1}{4} - \frac{l(l+1) - \gamma^2}{\rho^2} \right] R = 0,$$

where $\gamma \equiv \frac{Ze^2}{\hbar c}$, $\lambda \equiv \frac{2E\gamma}{\hbar c \alpha}$, $\alpha^2 \equiv \frac{4(m^2c^4 - E^2)}{\hbar^2 c^2}$, and $\rho \equiv \alpha r$.

Look for solutions that are finite at $\rho = 0$ and ∞, and show that

$$\lambda = n' + s + 1,$$

where n' is 0 or a positive integer, and s is the non-negative solution of

$$s(s+1) = l(l+1) - \gamma^2.$$

Show that the expression for the energy can be expanded in powers of γ^2, and to order γ^4 is

$$E = mc^2 \left[1 - \frac{\gamma^2}{2n^2} - \frac{\gamma^4}{2n^4} \left(\frac{n}{l + \frac{1}{2}} - \frac{3}{4} \right) \right],$$

where $n = n' + l + 1$ is the total quantum number and can take on positive integer values. Identify the rest energy, the energy in the nonrelativistic theory, and the fine-structure energy. Calculate the spread of the fine-structure levels for a given n. *Note:* They are much larger than observed experimentally in the hydrogen spectrum.

17. [14]

(a) Show that the energy levels of a mesonic atom are given by

$$E = m \left[1 + \frac{Z^2 \alpha^2}{\left(n - l - \frac{1}{2} + \sqrt{(l + \frac{1}{2})^2 - Z^2 \alpha^2} \right)^2} \right]^{-\frac{1}{2}}.$$

(b) Show that the ground-state energy for any mesonic atom heavier than $Z = 69$ is complex. Explain what this complex energy means.

(c) Mesonic atoms have been studied at places like Los Alamos Laboratory and it has been found that the ground states of atoms as heavy as lead ($Z = 82$) or uranium ($Z = 92$) are quite stable. How do you reconcile this fact with the result just obtained? Be as quantitative as you can.

18. Solve the Klein-Gordon equation for an attractive square-well potential of depth V_0 and radius a, after determining the continuity conditions at $r = a$. Obtain an explicit expression for the minimum V_0 with given a that just binds a particle of mass m.

19. Solve the Klein-Gordon equation for an exponential potential of the form

$$V(r) = -Z\alpha e^{-r/\alpha},$$

with $\alpha = mc^2 e^2 / \hbar c$; α characterizes the range of the potential. Restrict yourself to s-states ($l = 0$) only.

20. Consider a free Klein-Gordon particle of mass m and charge e immersed in a uniform magnetic field B in the z-direction. Using the gauge $\vec{A} = 1/2(\vec{B} \times \vec{r})$ show that motion is quantized with energy

$$E_n = \sqrt{m^2 + p_z^2 + eB(2n + 1)} \quad \text{for } n = 0, 1, 2, \ldots$$

21. [14] A rapidly varying electric field can lead to the creation of particle-antiparticle pairs. Calculate to lowest order in α the probability per unit volume per unit time for producing such pairs in the presence of an external electric field,

$$\vec{E}(t) = \hat{x}a\cos\omega t,$$

and show that the probability is

$$\text{prob} = VT\frac{\alpha a^2}{6}\left(1 - \frac{4m^2}{\omega^2}\right)^{\frac{3}{2}}\theta(\omega - 2m).$$

Suggestion: Use as an interaction potential the usual form

$$H_{\text{int}} = e\int d^3x j_\mu A^\mu,$$

where

$$j_\mu = \frac{i}{2m}(\phi^*\partial_\mu\phi - \phi\partial_\mu\phi^*)$$

and

$$\vec{A}(t) = -\hat{x}\frac{a}{\omega}\sin\omega t.$$

Utilize normalized plane-waves solutions of the Klein-Gordon equation,

$$\phi(x) = \sqrt{\frac{m}{E}}e^{-i(Et-\vec{p}\cdot\vec{x})}, \quad \text{with} \quad E = \sqrt{\vec{p}^2 + m^2},$$

and simple first-order perturbation theory

$$\text{amp} = -i\int_{-T/2}^{T/2}\langle f|H_{\text{int}}(t)|0\rangle dt.$$

22. [14] Suppose a pionic atom is placed in a uniform magnetic field described by the vector potential $\vec{A} = 1/2(\vec{B}\times\vec{r})$.

(a) Neglecting the quadratic term (justify this) show that this problem can be exactly solved to yield the energy levels

$$E = E_{B=0}\left(1 - 2\omega_L\frac{m}{m_\pi}\right)^{\frac{1}{2}},$$

where $\omega_L = eB/2m_\pi$ is the Larmor frequency and m is the eigenvalue of L along the direction of the magnetic field.

(b) Evaluate the nonrelativistic limit of the Klein-Gordon equation in this case and show that the effective Hamiltonian is

$$H = H_{B=0} - \frac{e}{2m_\pi} \vec{L} \cdot \vec{B} \left(1 - \frac{\vec{p}^2}{2m_\pi^2} + \cdots \right).$$

Thus the usual Bohr magneton $e/2m_\pi$ is reduced by relativistic effects to

$$\frac{e}{2m_\pi} \left(1 - \frac{\vec{p}^2}{2m_\pi^2} \right).$$

(c) Calculate the energy shift induced by the magnetic field using perturbation theory and show that this result agrees with the exact answer to first order in $\vec{B}$.

Chapter 5

Dirac Equation

We have seen in the previous chapter that the Klein-Gordon equation can describe the motion of a particle with spin 0. In this chapter, we will develop the Dirac equation which can be used to describe the motion of a particle with spin 1/2. It is conceptually easier to investigate the properties of the Dirac equation in its single-particle wave equation form, as we will do here. A characteristic feature of the Dirac relativistic wave equation[1] is that the spin of the particle is built into the theory from the beginning, and is not added afterwards as Pauli added the electron spin to the nonrelativistic Schrödinger equation. Built-in features, such as spin, provide a useful measure of the applicability of a particular wave equation to the description of a particular spin of a particle.

5.1 Wave Equation for a Spin-1/2 Particle

Since most of the problems with the Klein-Gordon equation were because it was second order in the time derivative, we seek a relativistic covariant equation of the form

$$i\hbar \frac{\partial \psi}{\partial t} = \hat{H}\psi, \tag{5.1}$$

which is first order in the time derivative, and should have positive definite probability density. Since the Schrödinger and Klein-Gordon equations are both second order in the space derivative, we try an equation that is first order in the space derivatives. We write

$$i\hbar \frac{\partial \psi}{\partial t} = \frac{\hbar c}{i}\left(\alpha_1 \frac{\partial \psi}{\partial x^1} + \alpha_2 \frac{\partial \psi}{\partial x^2} + \alpha_3 \frac{\partial \psi}{\partial x^3}\right) + \beta mc^2\psi, \tag{5.2}$$

where, for the moment, α_i and β are constants of unspecified structure. Using operator notation we have

---

[1]P.A.M. Dirac, "The Quantum Theory of the Electron", Proc. Roy. Soc. **117** (1928) 610-624.

$$\boxed{i\hbar\frac{\partial\psi}{\partial t} = \left(c\vec{\alpha}\cdot\hat{p} + \beta mc^2\right)\psi}\,, \tag{5.3}$$

where $\hat{p}$ is the three-vector momentum operator.

For invariance under spatial rotations, $\vec{\alpha}$ can not be numbers and ψ can not be a scalar (see problem 5.1). In analogy with the spin wave function of nonrelativistic quantum mechanics, we choose ψ to be a column vector, and α_i and β to be matrices. Explicitly,

$$
\begin{aligned}
i\hbar\frac{\partial\psi_\sigma}{\partial t} &= \sum_{\tau=1}^{N}\hat{H}_{\sigma\tau}\psi_\tau \\
&= \frac{\hbar c}{i}\sum_{\tau=1}^{N}\left(\alpha_1\frac{\partial}{\partial x^1} + \alpha_2\frac{\partial}{\partial x^2} + \alpha_3\frac{\partial}{\partial x^3}\right)_{\sigma\tau}\psi_\tau + \sum_{\tau=1}^{N}\beta_{\sigma\tau}mc^2\psi_\tau.
\end{aligned}
\tag{5.4}
$$

We thus have N coupled first-order equations.

These equations must

1. have free-particle solutions that satisfy $E^2 = \vec{p}^2c^2 + m^2c^4$,

2. yield a continuity equation and probability interpretation of ψ, and

3. be Lorentz covariant.

For the first condition to be satisfied, each component of ψ_σ must satisfy the Klein-Gordon equation; all components of a wave function of definite mass must satisfy the Klein-Gordon equation. Some wave functions satisfy other wave equations as well, such as the Dirac equation, depending on whether, or not, there are more components than independent particle states. Applying the operator in equation 5.1 twice gives

$$
\begin{aligned}
\left(i\hbar\frac{\partial}{\partial t}\right)\left(i\hbar\frac{\partial}{\partial t}\right)\psi &= (\hat{H})(\hat{H})\psi, \\
-\hbar^2\frac{\partial^2\psi}{\partial t^2} &= -\hbar^2c^2\left(\sum_{i=1}^{3}\alpha_i\frac{\partial}{\partial x^i}\right)\left(\sum_{j=1}^{3}\alpha_j\frac{\partial}{\partial x^j}\right)\psi \\
&\quad + \frac{\hbar mc^3}{i}\sum_{i=1}^{3}(\alpha_i\beta + \beta\alpha_i)\frac{\partial\psi}{\partial x^i} + \beta^2 m^2c^4\psi \\
&= -\hbar^2c^2\sum_{i,j=1}^{3}\frac{\alpha_i\alpha_j + \alpha_j\alpha_i}{2}\frac{\partial^2\psi}{\partial x^i\partial x^j}
\end{aligned}
$$

$$+\frac{\hbar m c^3}{i}\sum_{i=1}^{3}(\alpha_i\beta+\beta\alpha_i)\frac{\partial\psi}{\partial x^i}+\beta^2 m^2 c^4\psi. \quad (5.5)$$

To obtain the Klein-Gordon equation, the following must be satisfied

$$\alpha_i\alpha_j+\alpha_j\alpha_i=2\delta_{ij}, \quad (5.6)$$
$$\alpha_i\beta+\beta\alpha_i=0, \quad (5.7)$$
$$\alpha_i^2=\beta^2=I, \quad (5.8)$$

where I represents an $N\times N$ unit matrix. Normally we will not write the unit matrix explicitly in an equation unless it is required for clarity. This should not create any confusion since matrices can only equal matrices.

No terms in the Hamiltonian can have any space or time coordinates. Such terms would have the property of space-time dependent energies, and would give rise to forces and noninertial reference frames. Space and time derivatives can only appear in $\hat{p}$ and $\hat{E}$, but not in α_i or β, since the equation is to be linear in the partial derivatives. Thus α_i and β are independent of $\vec{x}$, t, $\vec{p}$, and E, and hence commute with them.

Since the Hamiltonian must be hermitian, α_i and β must be hermitian matrices. Since the matrices are hermitian they must be square.

Since α_i and β anticommute according to equation 5.7, they are traceless. This can be seen as follows:

$$\beta\alpha_i=-\alpha_i\beta,$$
$$\alpha_i=-\beta\alpha_i\beta,$$
$$\mathrm{Tr}[\alpha_i]=-\mathrm{Tr}[\beta\alpha_i\beta]=-\mathrm{Tr}[\beta^2\alpha_i]=-\mathrm{Tr}[\alpha_i]=0, \quad (5.9)$$

where the last line follows from the cyclic property of the trace.

Since $\alpha_i^2=\beta^2=I$, the eigenvalues of α_i and β are ± 1. Since the trace is the sum of eigenvalues, α_i and β must be of even dimensions. For $N=2$, only three anticommuting matrices exist – the Pauli matrices (equation 2.22). Thus the smallest dimension allowed is $N=4$.

If one matrix is diagonal, the others can not be diagonal or they would commute with the diagonal matrix. We can write a representation that is hermitian, traceless, and has eigenvalues of ± 1 (see problem 5.2):

$$\alpha_i=\begin{pmatrix} 0 & \sigma_i \\ \sigma_i & 0 \end{pmatrix} \quad \text{and} \quad \beta=\begin{pmatrix} I & 0 \\ 0 & -I \end{pmatrix}, \quad (5.10)$$

where σ_i are the 2×2 Pauli matrices (equation 2.22), I is the 2×2 unit matrix, and 0 is the 2×2 null matrix. This choice of α_i and β is not unique. All matrices related to these by any unitary $N\times N$ matrix U, which preserves the anticommutation relationships (equations 5.6 and 5.7), are allowed: $\alpha_i'=$

$U\alpha_i U^{-1}$ and $\beta' = U\beta U^{-1}$ (see problem 5.3). We will continue to study the properties of these matrices as needed in subsequent sections.

Now what can we say about the Dirac state function? It is a function of the continuous space-time coordinates $(\vec{x}, t)$. It also has four extra degrees of freedom, which we can label by $r = 1, 2, 3, 4$. We can now ask what these degrees of freedom, represented by the discrete index r, represent? We will see that the Dirac equation describes particles of spin $1/2$. However, since spin-$1/2$ means two new degrees of freedom and ψ has four components, we can still ask what is the reason for this seemingly redundant duplication in the number of states? This question will be addressed in a subsequent section of this chapter.

5.2   Current Conservation

To study the Dirac current, we will use an identical approach to section 4.5. The hermitian conjugate wave function is a row vector $\psi^\dagger = (\psi_1^*, \psi_2^*, \psi_3^*, \psi_4^*)$. Multiplying the Dirac equation 5.2 by the conjugate wave function from the left gives

$$i\hbar\psi^\dagger \frac{\partial\psi}{\partial t} = \frac{\hbar c}{i} \sum_{k=1}^{3} \psi^\dagger \alpha_k \frac{\partial\psi}{\partial x^k} + mc^2\psi^\dagger\beta\psi. \tag{5.11}$$

Forming the Dirac equation for the conjugate wave function and multiplying by the wave function from the right gives

$$-i\hbar\frac{\partial\psi^\dagger}{\partial t}\psi = -\frac{\hbar c}{i} \sum_{k=1}^{3} \frac{\partial\psi^\dagger}{\partial x^k} \alpha_k\psi + mc^2\psi^\dagger\beta\psi, \tag{5.12}$$

where $\alpha_i^\dagger = \alpha_i$ and $\beta^\dagger = \beta$. Subtracting equation 5.11 from equation 5.12 gives

$$i\hbar\frac{\partial}{\partial t}\psi^\dagger\psi = \sum_{k=1}^{3} \frac{\hbar c}{i} \frac{\partial}{\partial x^k}(\psi^\dagger \alpha_k\psi). \tag{5.13}$$

Writing the result as a continuity equation, $\frac{\partial\rho}{\partial t} + \vec{\nabla}\cdot\vec{j} = 0$, gives the probability density

$$\rho = \psi^\dagger\psi, \tag{5.14}$$

and the probability current density

$$j^k = c\psi^\dagger\alpha_k\psi. \tag{5.15}$$

Equation 5.14 would lead us to believe that the problem with not obtaining a positive-definite current, that we encountered for the Klein-Gordon wave functions, seems to have gone away for Dirac wave functions. The form of the probability density is identical to the nonrelativistic form. The operator $c\alpha^k$ in the probability current (equation 5.15) looks like a velocity operator. We will follow up on these two ideas in subsequent sections.

Integrating the continuity equation over all space and using the divergence theorem gives

$$\int d^3x \frac{\partial \rho}{\partial t} + \int d^3x \vec{\nabla} \cdot \vec{j} = 0$$

$$\frac{\partial}{\partial t} \int d^3x \rho + \int \vec{j} \cdot d\vec{S} = 0$$

$$\frac{\partial}{\partial t} \int d^3x \psi^\dagger \psi = 0. \tag{5.16}$$

We have now proven the first two conditions required of the Dirac equation on page 74. We still need to show $(c\rho, \vec{j})$ forms a four-vector under a Lorentz transformation and that the Dirac equation is Lorentz covariant.

5.3 Dirac Particle at Rest

We search for a solution to the Dirac equation for a particle at rest. At rest, a particle has an infinitely large de Broglie wavelength and the wave function is uniform over all space: $\hat{p}\psi = 0$. Therefore we drop the momentum operator terms and the Dirac equation reduces to

$$i\hbar \frac{\partial \psi}{\partial t} = \beta mc^2 \psi. \tag{5.17}$$

For our representation of β (equation 5.10), the solutions are $\psi(\vec{p} = 0)$:

$$\psi^1(0) = e^{-(imc^2/\hbar)t} \begin{pmatrix} 1 \\ 0 \\ 0 \\ 0 \end{pmatrix}, \quad \psi^2(0) = e^{-(imc^2/\hbar)t} \begin{pmatrix} 0 \\ 1 \\ 0 \\ 0 \end{pmatrix}, \tag{5.18}$$

$$\psi^3(0) = e^{+(imc^2/\hbar)t} \begin{pmatrix} 0 \\ 0 \\ 1 \\ 0 \end{pmatrix}, \quad \psi^4(0) = e^{+(imc^2/\hbar)t} \begin{pmatrix} 0 \\ 0 \\ 0 \\ 1 \end{pmatrix}. \tag{5.19}$$

ψ^1 and ψ^2 are positive-energy solutions, while ψ^3 and ψ^4 are negative-energy solutions. We may have expected positive- and negative-energy solutions since we required ψ to also satisfy the Klein-Gordon equation. However, we now see that we have four solutions rather than two solutions as in the Klein-Gordon case. At first sight the Dirac equation seems to have worsened our situation with extra unwanted solutions. We will see soon enough that these extra solutions have desirable physical significance.

5.4   Electromagnetic Interaction

We now introduce the interaction of the electromagnetic field with the solutions of the Dirac equation. Let us assume, for now, the existence of single-particle charged solutions of the Dirac equation. The coupling to an electromagnetic field will subsequently be used to obtain an interpretation of the internal structure of Dirac particles.

Consider the interaction of a point particle of charge e with an external electromagnetic field[2]. We make the usual gauge-invariant minimal substitution introduced in equation 4.53

$$\hat{p}^{\mu} \to \hat{p}^{\mu} - \frac{e}{c}A^{\mu}, \tag{5.20}$$

where e is the magnitude of the charge of the electron. The Dirac equation becomes

$$c\left[i\hbar\frac{\partial}{\partial(ct)} - \frac{e}{c}A_0\right]\psi = \left[c\vec{\alpha}\cdot\left(\hat{p} - \frac{e}{c}\vec{A}\right) + \beta mc^2\right]\psi \tag{5.21}$$

or

$$i\hbar\frac{\partial\psi}{\partial t} = \left[c\vec{\alpha}\cdot\left(\hat{p} - \frac{e}{c}\vec{A}\right) + eA_0 + \beta mc^2\right]\psi. \tag{5.22}$$

This equation contains the interaction with the electromagnetic field:

$$i\hbar\frac{\partial\psi}{\partial t} = \left(\hat{H} + \hat{H}'\right)\psi, \tag{5.23}$$

where $\hat{H}$ is the original free-particle Hamiltonian and the piece of the Hamiltonian due to interactions with the electromagnetic potential is

$$\hat{H}' = -\frac{e}{c}c\vec{\alpha}\cdot\vec{A} + eA_0 = -\frac{e}{c}\hat{v}\cdot\vec{A} + eA_0, \tag{5.24}$$

where we have defined

$$\hat{v} = c\vec{\alpha} \tag{5.25}$$

---

[2]We immediately assume the charge of the particle described by the Dirac equation is $q = e$.

to be a velocity operator based on its appearance in the probability current density (equation 5.15).

We can further convince ourselves of the velocity operator by examining the relativistic extension of the Ehrenfest Theorem. The Ehrenfest Theorem gives the law of motion of the mean values of the coordinates and the conjugate momenta of a quantum system. It stipulates that the equations of motion of these mean values are formally identical to the Hamilton equations of classical mechanics, except that the quantities which occur on both sides of the classical equations must be replaced by their average values. We can use the theorem by looking at the relativistic extension of the Ehrenfest relationship:

$$\frac{d}{dt}\langle \hat{x} \rangle = \frac{i}{\hbar}\langle [\hat{H}, \hat{x}] \rangle = \frac{i}{\hbar}\langle [c\vec{\alpha} \cdot \hat{p}, \hat{x}] \rangle = \frac{ci}{\hbar}\langle \alpha_i [\hat{p}_i, \hat{x}] \rangle = c\langle \vec{\alpha} \rangle, \qquad (5.26)$$

where we have used the commutator relationship $[\hat{x}_i, \hat{p}_j] = i\hbar\delta_{ij}$.

5.5 Nonrelativistic Limit of the Dirac Equation

For consistency, the Dirac equation should reduce to the Schrödinger wave equation for nonrelativistic quantum mechanics in the nonrelativistic limit. We examine the nonrelativistic limit for the case of a positive-energy Dirac particle in the presence of an electromagnetic potential. The Dirac equation with the electromagnetic potential can be written as equation 5.22.

Consider a two-component representation of $\psi(\vec{x}, t) = \begin{bmatrix} \phi(\vec{x}, t) \\ \chi(\vec{x}, t) \end{bmatrix}$, where the four-component spinor[3] ψ is decomposed into a pair of two-component spinors, ϕ and χ. Substitution of this form into the Dirac equation 5.22 gives

$$i\hbar\frac{\partial}{\partial t}\begin{pmatrix} \phi \\ \chi \end{pmatrix} = \begin{pmatrix} c\vec{\sigma} \cdot \left[\hat{p} - \frac{e}{c}\vec{A}\right]\chi \\ c\vec{\sigma} \cdot \left[\hat{p} - \frac{e}{c}\vec{A}\right]\phi \end{pmatrix} + eA_0\begin{pmatrix} \phi \\ \chi \end{pmatrix} + mc^2\begin{pmatrix} \phi \\ -\chi \end{pmatrix}. \qquad (5.27)$$

If the rest energy mc^2 is the largest occurring energy, our two-component solution is approximately

$$\begin{bmatrix} \phi(\vec{x}, t) \\ \chi(\vec{x}, t) \end{bmatrix} = \begin{bmatrix} \phi_0(\vec{x}, t) \\ \chi_0(\vec{x}, t) \end{bmatrix} e^{-imc^2t/\hbar}, \qquad (5.28)$$

where $\phi_0(\vec{x}, t)$ and $\chi_0(\vec{x}, t)$ are slowly varying functions of time. Substitution of this nonrelativistic solution into the Dirac equation now gives

---

[3] A spinor will be defined more formally in section 5.9.

$$i\hbar\frac{\partial}{\partial t}\begin{pmatrix}\phi_0\\\chi_0\end{pmatrix}=\begin{pmatrix}c\vec{\sigma}\cdot\left[\hat{p}-\frac{e}{c}\vec{A}\right]\chi_0\\c\vec{\sigma}\cdot\left[\hat{p}-\frac{e}{c}\vec{A}\right]\phi_0\end{pmatrix}+eA_0\begin{pmatrix}\phi_0\\\chi_0\end{pmatrix}-2mc^2\begin{pmatrix}0\\\chi_0\end{pmatrix}. \quad (5.29)$$

If the kinetic energy is small compared to the rest energy, χ_0 is a slowly varying function of time and

$$\left|i\hbar\frac{\partial\chi_0}{\partial t}\right|\ll|mc^2\chi_0|. \quad (5.30)$$

If the electrostatic potential is weak, the potential energy is small compared to the rest energy:

$$|eA_0\chi_0|\ll|mc^2\chi_0|. \quad (5.31)$$

With the last two approximations, the lower component in equation 5.29 becomes

$$0\approx c\vec{\sigma}\cdot\left(\hat{p}-\frac{e}{c}\vec{A}\right)\phi_0-2mc^2\chi_0, \quad (5.32)$$

$$\chi_0=\frac{\vec{\sigma}\cdot\left(\hat{p}-\frac{e}{c}\vec{A}\right)}{2mc}\phi_0. \quad (5.33)$$

The lower component χ is often referred to as the "small" component of the wave function ψ, relative to the "large" component ϕ. The small component is approximately a factor of v/c less than the large component in the nonrelativistic limit.

The upper component of equation 5.29 becomes

$$i\hbar\frac{\partial\phi_0}{\partial t}=\frac{\vec{\sigma}\cdot\left(\hat{p}-\frac{e}{c}\vec{A}\right)\vec{\sigma}\cdot\left(\hat{p}-\frac{e}{c}\vec{A}\right)}{2m}\phi_0+eA_0\phi_0. \quad (5.34)$$

Using identity 2.25, we have

$$\vec{\sigma}\cdot\left(\hat{p}-\frac{e}{c}\vec{A}\right)\vec{\sigma}\cdot\left(\hat{p}-\frac{e}{c}\vec{A}\right)$$
$$=\left(\hat{p}-\frac{e}{c}\vec{A}\right)^2+i\vec{\sigma}\cdot\left[\left(-i\hbar\vec{\nabla}-\frac{e}{c}\vec{A}\right)\times\left(-i\hbar\vec{\nabla}-\frac{e}{c}\vec{A}\right)\right]$$
$$=\left(\hat{p}-\frac{e}{c}\vec{A}\right)^2-\frac{e\hbar}{c}\vec{\sigma}\cdot(\vec{\nabla}\times\vec{A}), \quad (5.35)$$

wherein obtaining the last line we remember that we are operating on a wave function ϕ_0, and have used the identity

$$\vec{\nabla}\times\vec{A}\phi_0+\vec{A}\times(\vec{\nabla}\phi_0)=(\vec{\nabla}\times\vec{A})\phi_0. \quad (5.36)$$

From the definition of the electromagnetic vector potential,

$$\vec{\sigma} \cdot \left(\hat{p} - \frac{e}{c} \vec{A} \right) \vec{\sigma} \cdot \left(\hat{p} - \frac{e}{c} \vec{A} \right) = \left(\hat{p} - \frac{e}{c} \vec{A} \right)^2 - \frac{e\hbar}{c} \vec{\sigma} \cdot \vec{B}. \qquad (5.37)$$

Therefore the Dirac equation for the large component becomes

$$i\hbar \frac{\partial \phi_0}{\partial t} = \left[\frac{\left(\hat{p} - \frac{e}{c} \vec{A} \right)^2}{2m} - \frac{e\hbar}{2mc} \vec{\sigma} \cdot \vec{B} + eA_0 \right] \phi_0. \qquad (5.38)$$

This is the two-component Pauli equation for the theory of spin in nonrelativistic quantum mechanics. The two components of ϕ_0 describe the spin degrees of freedom.

The Pauli equation, and thus the Dirac equation, yields the correct gyromagnetic factor of $g = 2$ for a free electron. To see this, we turn on a *weak* field so that we can neglect the term quadratic in $e\vec{A}$. The square of the momentum operator becomes

$$\left(\hat{p} - \frac{e}{c} \vec{A} \right)^2 \approx \hat{p}^2 - \frac{e}{c} \hat{p} \cdot \vec{A} - \frac{e}{c} \vec{A} \cdot \hat{p} = \hat{p}^2 - 2\frac{e}{c} \vec{A} \cdot \hat{p}, \qquad (5.39)$$

where we have used $\hat{p} \cdot \vec{A} = -i\hbar \nabla \cdot \vec{A} = 0$ by the Coulomb gauge[4].

Choosing a homogeneous magnetic field

$$\vec{A} = \frac{1}{2} \vec{B} \times \vec{x}, \qquad (5.40)$$

we have

$$\left(\hat{p} - \frac{e}{c} \vec{A} \right)^2 \approx \hat{p}^2 - \frac{e}{c} (\vec{B} \times \vec{x}) \cdot \hat{p} = \hat{p}^2 - \frac{e}{c} \vec{B} \cdot \hat{L}, \qquad (5.41)$$

where $\hat{L} = \vec{x} \times \hat{p}$ is the operator of orbital angular momentum.

Defining $\hat{s} = \frac{1}{2} \hbar \vec{\sigma}$ as the spin operator, the Pauli equation now becomes

$$
\begin{aligned}
i\hbar \frac{\partial \phi_0}{\partial t} &= \left[\frac{\hat{p}^2}{2m} - \frac{e}{2mc} \hat{L} \cdot \vec{B} - \frac{e\hbar}{2mc} \vec{\sigma} \cdot \vec{B} + eA_0 \right] \phi_0 \\
&= \left[\frac{\hat{p}^2}{2m} - \frac{e}{2mc} \hat{L} \cdot \vec{B} - \frac{e}{mc} \hat{s} \cdot \vec{B} + eA_0 \right] \phi_0 \\
&= \left[\frac{\hat{p}^2}{2m} - \frac{e}{2mc} (\hat{L} + 2\hat{s}) \cdot \vec{B} + eA_0 \right] \phi_0. \qquad (5.42)
\end{aligned}
$$

Generally, the intrinsic magnetic moment $\vec{\mu}$ is related to the spin vector $\vec{s}$ by

---

[4]Specific gauges, such as the Coulomb gauge, will be discussed in section 7.2.

$$\vec{\mu} = g\mu_B \vec{s}, \tag{5.43}$$

where $\mu_B = e\hbar/2mc$ is the Bohr magneton and g is called the Lande g-factor. The gyromagnetic ratio is $g\mu_B = \mu/s$, i.e. the ratio of the magnetic to mechanical moment. Thus the Dirac theory predicts for particles of spin 1/2 that

$$g = 2. \tag{5.44}$$

Predicting the correct gyromagnetic ratio of the electron is one of the great triumphs of the Dirac equation.

Experimental values for the g-factors for real spin-1/2 particles have been measured:

$$\text{electron} \quad g_{\exp} = 2\left(1 + \frac{\alpha}{2\pi} + \cdots\right), \tag{5.45}$$

$$\text{proton} \quad g_{\exp} = 2(1 + 1.79), \tag{5.46}$$

$$\text{neutron} \quad g_{\exp} = 2(0 - 1.91). \tag{5.47}$$

In the case of the proton and neutron, a detailed look at these systems reveals that they are far from being simple point-like spin-1/2 structures. In the case of the electron, there exists no substructure. As far as we know, the electron really is a point particle. However, the electron can fragment into an electron-photon system which yields a modification to the g-factor, but only at order α. From experiment, $g = 2.00232$ and the difference from $g = 2$ can be accounted for by higher-order contributions (see section 8.13.3.2). This is a great triumph for quantum electrodynamics and is one of the reasons it is the "best" theory we have.

The Pauli equation is a nonrelativistic wave equation for a spin-1/2 particle. The Dirac equation reduces to the Pauli equation at low velocities. We are thus led to believe the Dirac equation describes a particle with spin 1/2 at both low and high velocities. The Dirac equation has the spin-1/2 property of the solution built into the theory from the beginning. The spin comes into the Dirac theory when the second-order differential Klein-Gordon equation is made first order.

We will discuss the anomalous magnetic moment of the electron further in section 8.13.3.2.

5.6   Constants of the Motion

Constants of the motion $\hat{C}$ are those dynamic variables that commute with the Hamiltonian. The mean value of the variable remains constant in time.

If at time $t = 0$ the wave function is an eigenfunction of $\hat{C}$ with eigenvalue c, this will be the case for all time. One says that c is a good quantum number. It is important in quantum mechanics to identify which variables $\hat{C}$ are constants of the motion and then to seek out their possible eigenfunctions and eigenvalues.

For a free particle ($\hbar = c = 1$)

$$H = \vec{\alpha} \cdot \vec{p} + \beta m. \tag{5.48}$$

With the usual definition of the orbital angular momentum

$$\vec{L} = \vec{x} \times \vec{p}, \tag{5.49}$$

we will find (equation 5.53) that none of the components of $\vec{L}$ commute with H. Thus the orbital angular momentum is not a constant of the motion. On the other hand, we find that

$$\vec{J} = \vec{L} + \frac{1}{2}\vec{\Sigma} \tag{5.50}$$

with

$$\vec{\Sigma} = \begin{pmatrix} \vec{\sigma} & 0 \\ 0 & \vec{\sigma} \end{pmatrix} \tag{5.51}$$

does commute with H and thus is a constant of the motion. The 2×2 identity matrix has not been explicitly written in the $\vec{J}$ and $\vec{L}$ terms of the three-component matrix equation 5.50. $1/2\vec{\Sigma}$ appears to be a generalized spin operator in the four-component representation. For solutions at rest (equations 5.18 and 5.19) the eigenvalues of this operator are $\pm 1/2$.

We now prove $[J, H] = 0$.[5] To do this, we simply verify that this result is true for J_1. We write

$$[J_1, H] = [J_1, \vec{\alpha} \cdot \vec{p}] = [L_1, \vec{\alpha} \cdot \vec{p}] + \frac{1}{2}[\Sigma_1, \vec{\alpha} \cdot \vec{p}]. \tag{5.52}$$

For the first term in equation 5.52,

$$
\begin{aligned}
[L_1, \vec{\alpha} \cdot \vec{p}] &= \alpha_i[x_2 p_3 - x_3 p_2, p_i] \\
&= \alpha_i([x_2, p_i]p_3 - [x_3, p_i]p_2) \\
&= i(\alpha_2 p_3 - \alpha_3 p_2) \\
&= -i(\vec{p} \times \vec{\alpha})_1, \tag{5.53}
\end{aligned}
$$

where we have used $[x_i, p_j] = i\delta_{ij}$. For the second term in equation 5.52, we make use of the relationship

---

[5]We use the square bracket to represent the commutator of two operators: $[A, B] = AB - BA$.

$$\vec{\Sigma} = \gamma^5 \vec{\alpha} = \vec{\alpha}\gamma^5, \tag{5.54}$$

where $\gamma^5 = \begin{pmatrix} 0 & 1 \\ 1 & 0 \end{pmatrix}$. Using γ^5, we have

$$[\Sigma_1, \vec{\alpha} \cdot \vec{p}] = [\gamma^5\alpha_1, \vec{\alpha} \cdot \vec{p}] = \gamma^5[\alpha_1, \alpha_i]p_i. \tag{5.55}$$

Since (see problem 5.6)

$$[\alpha_i, \alpha_j] = 2i\epsilon_{ijk}\Sigma_k, \tag{5.56}$$

we obtain

$$\frac{1}{2}[\Sigma_1, \vec{\alpha} \cdot \vec{p}] = i\gamma^5\epsilon_{1ij}\Sigma_jp_i = i(\vec{p} \times \vec{\alpha})_1. \tag{5.57}$$

Thus, we have shown that

$$[J_1, \vec{\alpha} \cdot \vec{p}] = 0. \tag{5.58}$$

By symmetry it now follows that quite generally

$$[\vec{J}, \vec{\alpha} \cdot \vec{p}] = 0. \tag{5.59}$$

As a consequence, J^2, J_3, and H can be simultaneously diagonalized, as in nonrelativistic quantum mechanics. The components of $\vec{J}$ do not commute with each other.

While the rest-frame 3rd-component of spin can be used to parameterize the spin states, it is not Lorentz invariant. If spin is an intrinsic property of a particle, it seems reasonable to expect the magnitude of the spin not to change under a Lorentz transformation. For $\vec{p} \neq 0$, the operator $1/2\vec{\Sigma}$ is no longer a suitable spin operator, since it fails to commute with the energy operator $\vec{\alpha} \cdot \hat{p} + \beta m$, and hence it cannot be given a definite value at the same time as the energy. Another drawback to the rest-frame spin eigenstates is that they cannot be used to describe massless-particle wave functions, since it is not possible to transform to a rest frame for massless particles.

Since there are still two independent states for each energy, there must be some other operator which commutes with the energy operator H and momentum operator $\hat{p}$ whose eigenvalues can be used to distinguish the states. Such an operator is not unique, but a common choice is helicity.

The component of spin along (or against) a particle's momentum has a simple meaning in any frame and is a valid concept even for a massless particle. We see that the operator $\vec{\Sigma} \cdot \hat{p}$ commutes with H (see problem 5.7) and $\vec{p}$, where $\hat{p}$ is now the unit vector pointing in the direction of the momentum, $\vec{p}/|\vec{p}|$, not the momentum operator. The "spin" component in the direction of the motion, $\frac{1}{2}\vec{\sigma} \cdot \hat{p}$, is therefore a "good" quantum number and can be used to label the states. We call this quantum number the helicity of the state. Helicity is manifestly rotationally invariant (see problem 5.8); helicity states also transform simply under Lorentz transformations (see problem 5.9). We will return to the idea of helicity at the end of the next section.

5.7 Free Motion of a Dirac Particle

Consider the Dirac equation 5.3 without potentials – the free particle equation. We seek positive-energy $\psi^{(+)}(x)$ and negative-energy $\psi^{(-)}(x)$ plane-wave solutions of definite momentum, of the form

$$
\begin{aligned}
\psi^{(+)}(x) &= u(p)e^{-i(Et-\vec{p}\cdot\vec{x})/\hbar} \\
&= u(p)e^{-ip\cdot x/\hbar}
\end{aligned}
\tag{5.60}
$$

and

$$
\begin{aligned}
\psi^{(-)}(x) &= v(p)e^{i(Et-\vec{p}\cdot\vec{x})/\hbar} \\
&= v(p)e^{ip\cdot x/\hbar},
\end{aligned}
\tag{5.61}
$$

where $p_0 = E > 0$ and we have chosen $\vec{p}$ to have the opposite sign for the two different energy solutions. $\psi^{(\pm)}(x)$ are eigenfunctions of energy and momentum with eigenvalues $\pm E$ and $\pm \vec{p}$, respectively. $u(p)$ and $v(p)$ are four-component spinors, which depend on the magnitude of the energy and momentum, and the mass. We call $u(p)$ and $v(p)$ four-spinors or spinors for short. Sometimes they are called bispinors. The latter terminology is most often used when $u(p)$ and $v(p)$ are represented by a pair of two-component spinors.

We split up the four-component spinors into a pair of two-component spinors giving

$$
u(p) = \begin{pmatrix} u_1 \\ u_2 \end{pmatrix} \quad \text{and} \quad v(p) = \begin{pmatrix} v_1 \\ v_2 \end{pmatrix}.
\tag{5.62}
$$

For these solutions, the Dirac equation becomes a set of algebraic relationships – four linear homogeneous equations. With our representation for $\vec{\alpha}$ and β, the positive-energy solution gives

$$
E \begin{pmatrix} u_1 \\ u_2 \end{pmatrix} = c \begin{pmatrix} 0 & \vec{\sigma} \\ \vec{\sigma} & 0 \end{pmatrix} \cdot \hat{p} \begin{pmatrix} u_1 \\ u_2 \end{pmatrix} + mc^2 \begin{pmatrix} 1 & 0 \\ 0 & -1 \end{pmatrix} \begin{pmatrix} u_1 \\ u_2 \end{pmatrix},
\tag{5.63}
$$

$$
(E - mc^2)u_1 - (c\vec{\sigma}\cdot\hat{p})u_2 = 0,
\tag{5.64}
$$

$$
(E + mc^2)u_2 - (c\vec{\sigma}\cdot\hat{p})u_1 = 0.
\tag{5.65}
$$

Similarly for the negative-energy solution

$$(E + mc^2)v_1 - (c\vec{\sigma} \cdot \hat{p})v_2 = 0, \tag{5.66}$$

$$(E - mc^2)v_2 - (c\vec{\sigma} \cdot \hat{p})v_1 = 0. \tag{5.67}$$

These linear homogenous systems of equations have solutions only if the determinate of the coefficients vanishes. Using identity 2.25 $(\vec{\sigma} \cdot \vec{p})^2 = \vec{p}^2$, the requirement of a vanishing determinant of the coefficients leads to $E^2 = (pc)^2 + (mc^2)^2$, which is automatically satisfied since we require solutions of the Dirac equation to also be solutions of the Klein-Gordon equation.

Since $E + mc^2$ is never zero, we can rewrite equation 5.65 for u_2 and equation 5.66 for v_1:

$$u_2 = \frac{c\vec{\sigma} \cdot \vec{p}}{E + mc^2} u_1, \tag{5.68}$$

$$v_1 = \frac{c\vec{\sigma} \cdot \vec{p}}{E + mc^2} v_2. \tag{5.69}$$

Equations 5.64 and 5.67 are not independent, but are identically satisfied by substituting solutions 5.68 and 5.69, respectively. There are therefore two linearly independent positive-energy solutions for every momentum $\vec{p}$, and two linearly independent negative-energy solutions for every momentum $-\vec{p}$. In the nonrelativistic limit, u_2 and v_1 are smaller than u_1 and v_2 by an amount of order v/c.

We define

$$u_1 = N\chi, \tag{5.70}$$

$$v_2 = N'\chi', \tag{5.71}$$

where χ and χ' are two-component spinors, and N and N' are normalization constants. There exists two linearly independent spinors $u(p)$ corresponding to the two linearly independent values for χ; call these χ_r, where $r = 1, 2$. Similarly for $v(p)$ we require χ'_r, where $r = 1, 2$. χ_r and χ'_r are arbitrary two-component quantities subject only to the usual normalization conditions $\chi_r^\dagger \chi_s = \delta_{rs}$ and $\chi_r'^\dagger \chi_s' = \delta_{rs}$.

The normalization constants N and N' are determined from the normalizations of $u(p)$ and $v(p)$. One's first thought might be to place normalization conditions upon $u^\dagger(p)u(p)$ and $v^\dagger(p)v(p)$. However, as we shall see later on (equation 5.213), this is a nonrelativistic way of thinking and would not be Lorentz invariant. The four solutions would also not be orthogonal. We will find that a standard normalization[6] is

---

[6]Another popular normalization is $u_r^\dagger(p)\beta u_s(p) = 2m\delta_{rs}$ and $v_r^\dagger(p)\beta v_s(p) = -2m\delta_{rs}$.

$$u_r^\dagger(p)\beta u_s(p) = \delta_{rs}, \tag{5.72}$$
$$v_r^\dagger(p)\beta v_s(p) = -\delta_{rs}. \tag{5.73}$$

We now proceed to calculate the normalization constants N and N'.

$$N^2\left(\chi^\dagger\chi - \chi^\dagger\frac{c^2(\vec{\sigma}\cdot\vec{p})(\vec{\sigma}\cdot\vec{p})}{(E+mc^2)^2}\chi\right) = 1$$

$$N^2\left(1 - \frac{c^2\vec{p}^2}{(E+mc^2)^2}\right) = 1, \tag{5.74}$$

or

$$N = \sqrt{\frac{(E+mc^2)^2}{(E+mc^2)^2 - c^2\vec{p}^{\,2}}}$$

$$= \sqrt{\frac{(E+mc^2)^2}{E^2 + 2Emc^2 + m^2c^4 - c^2\vec{p}^2}}$$

$$= \sqrt{\frac{(E+mc^2)^2}{2mc^2(E+mc^2)}}$$

$$= \sqrt{\frac{E+mc^2}{2mc^2}}. \tag{5.75}$$

Similarly, for the negative-energy normalization constant, we obtain

$$N' = \sqrt{\frac{E+mc^2}{2mc^2}}. \tag{5.76}$$

The complete set of positive- and negative-energy free-particle solutions is

$$\psi_r^{(+)}(x) = \sqrt{\frac{E+mc^2}{2mc^2}}\left(\begin{array}{c}\chi_r \\ \frac{c(\vec{\sigma}\cdot\vec{p})}{mc^2+E}\chi_r\end{array}\right)e^{-ip\cdot x/\hbar}, \tag{5.77}$$

$$\psi_r^{(-)}(x) = \sqrt{\frac{E+mc^2}{2mc^2}}\left(\begin{array}{c}\frac{c(\vec{\sigma}\cdot\vec{p})}{mc^2+E}\chi_r' \\ \chi_r'\end{array}\right)e^{ip\cdot x/\hbar}. \tag{5.78}$$

The amplitude of the plane waves contains one arbitrary two-component quantity for each energy. Thus, for a given sign of the energy and momentum, there are two different independent states. We will see that these states correspond to the two possible values of the spin component.

In the rest frame $(\vec{p} = 0)$, we have

$$\psi_r^{(+)} = \begin{pmatrix} \chi_r \\ 0 \end{pmatrix} e^{-imc^2 t/\hbar}, \tag{5.79}$$

$$\psi_r^{(-)} = \begin{pmatrix} 0 \\ \chi_r' \end{pmatrix} e^{imc^2 t/\hbar}. \tag{5.80}$$

Comparing with the previously obtained rest solutions (equations 5.18 and 5.19), we identify

$$\chi_1 = \chi_1' = \begin{pmatrix} 1 \\ 0 \end{pmatrix}, \tag{5.81}$$

$$\chi_2 = \chi_2' = \begin{pmatrix} 0 \\ 1 \end{pmatrix}. \tag{5.82}$$

We saw in section 5.6 that the free-particle Hamiltonian does not commute with the four-dimensional generalization of the spin-vector operator,

$$\hat{s} = \frac{\hbar}{2}\vec{\Sigma} = \frac{\hbar}{2}\begin{pmatrix} \vec{\sigma} & 0 \\ 0 & \vec{\sigma} \end{pmatrix}. \tag{5.83}$$

On the other hand, it was mentioned in section 5.6 that helicity is a good quantum number. We now show that helicity can be used to classify free one-particle states.

We form

$$\vec{\Sigma} \cdot \hat{p} = \begin{pmatrix} \vec{\sigma} & 0 \\ 0 & \vec{\sigma} \end{pmatrix} \cdot \hat{p} \tag{5.84}$$

and show that it commutes with the free Dirac Hamiltonian operator $\hat{H}_f$,

$$[\hat{H}_f, \vec{\Sigma} \cdot \hat{p}] = [c\vec{\alpha} \cdot \hat{p} + mc^2\beta, \vec{\Sigma} \cdot \hat{p}]. \tag{5.85}$$

Since $[\beta, \vec{\Sigma}] = 0$, we need only consider $[\vec{\alpha} \cdot \hat{p}, \vec{\Sigma} \cdot \hat{p}]$. In our representation (equation 5.10) for the α matrices

$$\begin{aligned}
[\hat{H}_f, \vec{\Sigma} \cdot \hat{p}] &= c[\vec{\alpha} \cdot \hat{p}, \vec{\Sigma} \cdot \hat{p}] \\
&= c\vec{\alpha} \cdot \hat{p}\vec{\Sigma} \cdot \hat{p} - c\vec{\Sigma} \cdot \hat{p}\vec{\alpha} \cdot \hat{p} \\
&= c\begin{pmatrix} 0 & \vec{\sigma} \cdot \hat{p} \\ \vec{\sigma} \cdot \hat{p} & 0 \end{pmatrix}\begin{pmatrix} \vec{\sigma} \cdot \hat{p} & 0 \\ 0 & \vec{\sigma} \cdot \hat{p} \end{pmatrix} - c\begin{pmatrix} \vec{\sigma} \cdot \hat{p} & 0 \\ 0 & \vec{\sigma} \cdot \hat{p} \end{pmatrix}\begin{pmatrix} 0 & \vec{\sigma} \cdot \hat{p} \\ \vec{\sigma} \cdot \hat{p} & 0 \end{pmatrix} \\
&= c\begin{pmatrix} 0 & (\vec{\sigma} \cdot \hat{p})^2 \\ (\vec{\sigma} \cdot \hat{p})^2 & 0 \end{pmatrix} - c\begin{pmatrix} 0 & (\vec{\sigma} \cdot \hat{p})^2 \\ (\vec{\sigma} \cdot \hat{p})^2 & 0 \end{pmatrix} \\
&= 0. \tag{5.86}
\end{aligned}$$

Also $[\hat{p}, \vec{\Sigma} \cdot \hat{p}] = 0$, therefore $\hat{p}$, $\hat{H}_f$, and $\vec{\Sigma} \cdot \hat{p}$ can all be diagonalized simultaneously.

The term $\vec{\Sigma} \cdot \hat{p}$ can be rewritten as the helicity operator $\hat{\Lambda}_S$

$$\hat{\Lambda}_S \equiv \frac{\hbar}{2} \vec{\Sigma} \cdot \frac{\hat{p}}{|\vec{p}|} = \hat{s} \cdot \frac{\hat{p}}{|\vec{p}|}. \tag{5.87}$$

We see that helicity is the projection of the spin onto the direction of the momentum.

For an electron propagating in the z-direction, $\vec{p} = (0, 0, p)$,

$$\hat{\Lambda}_S = \frac{\hbar}{2} \vec{\Sigma}_z \cdot \frac{p_z}{|\vec{p}|} \equiv \hat{s}_z = \frac{\hbar}{2} \begin{pmatrix} 1 & 0 & 0 & 0 \\ 0 & -1 & 0 & 0 \\ 0 & 0 & 1 & 0 \\ 0 & 0 & 0 & -1 \end{pmatrix}. \tag{5.88}$$

The eigenvalues are $\pm\hbar/2$ and the eigenvectors of $\hat{\Lambda}_S$ are

$$\begin{pmatrix} \chi_1 \\ 0 \end{pmatrix}, \quad \begin{pmatrix} \chi_2 \\ 0 \end{pmatrix}, \quad \begin{pmatrix} 0 \\ \chi_1 \end{pmatrix}, \quad \begin{pmatrix} 0 \\ \chi_2 \end{pmatrix}, \tag{5.89}$$

with

$$\chi_1 = \begin{pmatrix} 1 \\ 0 \end{pmatrix} \quad \text{and} \quad \chi_1 = \begin{pmatrix} 0 \\ 1 \end{pmatrix}. \tag{5.90}$$

Therefore the complete set of solutions with momentum along the z-axis is

$$\psi_1^{(+)} = \sqrt{\frac{E + mc^2}{2mc^2}} \begin{pmatrix} \begin{pmatrix} 1 \\ 0 \end{pmatrix} \\ \frac{cp}{E+mc^2} \begin{pmatrix} 1 \\ 0 \end{pmatrix} \end{pmatrix} e^{-i(Et-pz)/\hbar}, \tag{5.91}$$

$$\psi_2^{(+)} = \sqrt{\frac{E + mc^2}{2mc^2}} \begin{pmatrix} \begin{pmatrix} 0 \\ 1 \end{pmatrix} \\ \frac{-cp}{E+mc^2} \begin{pmatrix} 0 \\ 1 \end{pmatrix} \end{pmatrix} e^{-i(Et-pz)/\hbar}, \tag{5.92}$$

$$\psi_1^{(-)} = \sqrt{\frac{E + mc^2}{2mc^2}} \begin{pmatrix} \frac{cp}{E+mc^2} \begin{pmatrix} 1 \\ 0 \end{pmatrix} \\ \begin{pmatrix} 1 \\ 0 \end{pmatrix} \end{pmatrix} e^{i(Et-pz)/\hbar}, \tag{5.93}$$

$$\psi_2^{(-)} = \sqrt{\frac{E + mc^2}{2mc^2}} \begin{pmatrix} \frac{-cp}{E+mc^2} \begin{pmatrix} 0 \\ 1 \end{pmatrix} \\ \begin{pmatrix} 0 \\ 1 \end{pmatrix} \end{pmatrix} e^{i(Et-pz)/\hbar}. \tag{5.94}$$

5.8   Covariant Form of the Dirac Equation

We now cast the Dirac equation into a more apparent covariant form and thus satisfy the last requirement of the Dirac equation stated on page 74. Multiplying equation 5.3 by β/c from the left and defining

$$\gamma^0 \equiv \beta \quad \text{and} \quad \gamma^i \equiv \beta\alpha_i, \tag{5.95}$$

where $i = 1, 2, 3$, we have

$$i\hbar \left(\gamma^0 \frac{\partial}{\partial x^0} + \gamma^1 \frac{\partial}{\partial x^1} + \gamma^2 \frac{\partial}{\partial x^2} + \gamma^3 \frac{\partial}{\partial x^3} \right) \psi - mc\psi = 0$$

$$\left(i\hbar\gamma^\mu \frac{\partial}{\partial x^\mu} \right) \psi - mc\psi = 0$$

$$(i\hbar\gamma \cdot \partial) \psi - mc\psi = 0, \tag{5.96}$$

where

$$\gamma \cdot \partial = \gamma^\mu \frac{\partial}{\partial x^\mu} = \frac{\gamma^0}{c} \frac{\partial}{\partial t} + \vec{\gamma} \cdot \vec{\nabla}. \tag{5.97}$$

In terms of the four-momentum operator, we write[7]

$$(\gamma^\mu \hat{p}_\mu - mc)\psi = 0. \tag{5.98}$$

Introducing the Feynman dagger, or slash notation, for an arbitrary four-vector a_μ, we define

$$\rlap{/}{a} \equiv \gamma^\mu a_\mu = g_{\mu\nu}\gamma^\mu a^\nu = \gamma^0 a^0 - \vec{\gamma} \cdot \vec{a}. \tag{5.99}$$

We write

$$\boxed{(\rlap{/}{\hat{p}} - mc)\psi = 0} \,, \tag{5.100}$$

the covariant form of the Dirac equation.

We introduce the electromagnetic interaction by the usual minimal substitution for gauge invariance

$$\left(\rlap{/}{\hat{p}} - \frac{e}{c} \rlap{/}{A} - mc \right) \psi = 0. \tag{5.101}$$

Let us now study the properties of the γ^μ matrices; we use the properties given by equations 5.6, 5.7, and 5.8. For the space components

---

[7]Notice that we are now using $\hat{p}$ to represent the four-component operator of momentum. It previously was used to represent the three-momentum operator.

$$\alpha_i \alpha_k + \alpha_k \alpha_i = 2\delta_{ik},$$
$$\beta \alpha_i \alpha_k + \beta \alpha_k \alpha_i = 2\beta \delta_{ik},$$
$$-\alpha_i \beta \alpha_k - \alpha_k \beta \alpha_i = 2\beta \delta_{ik},$$
$$-\beta \alpha_i \beta \alpha_k - \beta \alpha_k \beta \alpha_i = 2\delta_{ik},$$
$$\gamma^i \gamma^k + \gamma^k \gamma^i = -2\delta_{ik}. \tag{5.102}$$

For the mixed time and space components

$$\alpha_i \beta + \beta \alpha_i = 0,$$
$$\beta \alpha_i \beta + \beta \beta \alpha_i = 0,$$
$$\gamma^i \gamma^0 + \gamma^0 \gamma^i = 0. \tag{5.103}$$

For the time component

$$\beta^2 = 1 \quad \Rightarrow \quad \gamma^0 \gamma^0 + \gamma^0 \gamma^0 = 2. \tag{5.104}$$

Putting together equations 5.102, 5.103, and 5.104, we write

$$\boxed{\gamma^\mu \gamma^\nu + \gamma^\nu \gamma^\mu = 2g^{\mu\nu}}, \tag{5.105}$$

where the matrices are 4×4 and $\mu, \nu = 0, 1, 2, 3$. Although the Dirac matrices γ^μ are written with Greek indices, they are not four-vectors. They have the same value in every reference frame and do not change under a Lorentz transformation. They do have the index lowering and raising properties discussed in section 2.5.

The γ^i matrices are antihermitian and γ^0 is hermitian:

$$(\gamma^i)^\dagger = (\beta \alpha_i)^\dagger = \alpha_i^\dagger \beta^\dagger = \alpha_i \beta = -\beta \alpha_i = -\gamma^i, \tag{5.106}$$
$$(\gamma^0)^\dagger = \beta^\dagger = \beta = \gamma^0. \tag{5.107}$$

This can be summarized by writing

$$\boxed{(\gamma^\mu)^\dagger = \gamma^0 \gamma^\mu \gamma^0}. \tag{5.108}$$

Up until now everything is representation independent. Using our previous representation (equation 5.10),

$$\gamma^i \equiv \beta \alpha_i = \begin{pmatrix} I & 0 \\ 0 & -I \end{pmatrix} \begin{pmatrix} 0 & \sigma_i \\ \sigma_i & 0 \end{pmatrix} = \begin{pmatrix} 0 & \sigma_i \\ -\sigma_i & 0 \end{pmatrix}, \tag{5.109}$$

$$\gamma^0 \equiv \beta = \begin{pmatrix} I & 0 \\ 0 & -I \end{pmatrix}. \tag{5.110}$$

Other frequently encountered representations of the γ-matrices are the Majorana representation and the chiral representation. In the Majorana representation, all elements of all the γ-matrices are imaginary. One possible Majorana representation is

$$\gamma^0 \equiv \begin{pmatrix} 0 & \sigma_2 \\ \sigma_2 & 0 \end{pmatrix}, \gamma^1 \equiv i\begin{pmatrix} 0 & \sigma_1 \\ \sigma_1 & 0 \end{pmatrix}, \gamma^2 \equiv i\begin{pmatrix} I & 0 \\ 0 & -I \end{pmatrix}, \gamma^3 \equiv i\begin{pmatrix} 0 & \sigma_3 \\ \sigma_3 & 0 \end{pmatrix}. \quad (5.111)$$

In the chiral representation, all of the γ-matrices have diagonal elements which are zero. One possible chiral representation is

$$\gamma^0 \equiv \begin{pmatrix} 0 & -I \\ -I & 0 \end{pmatrix} \quad \text{and} \quad \gamma^i \equiv \begin{pmatrix} 0 & \sigma^i \\ -\sigma_i & 0 \end{pmatrix}. \quad (5.112)$$

Both of these representations are not unique and several different sets of γ-matrices with the same Majorana and chiral properties are possible.

5.9 Proof of Covariance

The physics expressed by any relativistic equation, and by the Dirac equation in particular, must be independent of the Lorentz frame that is used. Hence to be a true description of the physics, the equation itself must display this same invariance with respect to the choice of coordinates. We have to verify that, if we write down the wave equation in a different Lorentz frame, the solutions of the new wave equation may be put into one to one correspondence with those of the original equation in such a way that corresponding solutions may be assumed to represent the same state. In actual fact, the relativity principle requires this invariance of form only with respect to proper Lorentz transformations, and also with respect to space and time translations, but it happens that the free-particle Dirac equation is formally invariant with respect to the complete Poincaré group.

The Dirac space is a four dimensional abstract space unrelated to physical space-time. To discuss the Lorentz transformation of a Dirac wave function, Dirac equation, or Dirac matrix element, requires that we first construct a representation of each Lorentz transformation on the Dirac space. In general, a representation of a group is a mapping of each element of the group Λ onto a matrix $S(\Lambda)$ which preserves the group multiplication law (see section 2.10).

To prove Lorentz covariance two conditions must be satisfied:

1. If $(i\hbar\gamma^\mu\partial_\mu - mc)\psi(x) = 0$ then $(i\hbar\tilde{\gamma}^\mu\partial'_\mu - mc)\psi'(x') = 0$, where $\partial'_\mu \equiv \frac{\partial}{\partial x'^\mu}$.

2. Given $\psi(x)$ of observer O, there must be a prescription for observer O' to compute $\psi'(x')$, which describes to O' the same physical state.

It can be shown that all 4×4 matrices (with γ^0 hermitian and γ^i antihermitian) are equivalent up to a unitary transformation[8]:

$$\tilde{\gamma}_\mu = U^\dagger \gamma_\mu U, \tag{5.113}$$

where $U^\dagger = U^{-1}$ and $\tilde{\gamma}_\mu$ is just another representation of γ_μ. We drop the distinction between $\tilde{\gamma}^\mu$ and γ^μ, and write

$$(\not{p}' - mc)\psi'(x') = 0, \tag{5.114}$$

where $\not{p}' \equiv i\hbar\gamma^\mu\partial'_\mu$.

The wave function ψ is a four-component column matrix – a four-component spinor – not a four component vector. Therefore if we wish to transform ψ, we must use some transformation S other than Λ developed in chapter 3. We require that the transformation between ψ and ψ' be linear since the Dirac equation and Lorentz transformation are linear:

$$\psi'(x') = \psi'(\Lambda x + d) = S(\Lambda)\psi(x) = S(\Lambda)\psi(\Lambda^{-1}(x' - d)), \tag{5.115}$$

where $S(\Lambda)$ is a 4×4 matrix which depends only on the relative velocities of O and O'. $S(\Lambda)$ has an inverse if $O \to O'$ and also $O' \to O$. The inverse is

$$\psi(x) = S^{-1}(\Lambda)\psi'(x') = S^{-1}(\Lambda)\psi'(\Lambda x + d), \tag{5.116}$$

or we may write

$$\psi(x) = S(\Lambda^{-1})\psi'(x') = S(\Lambda^{-1})\psi'(\Lambda x + d) \tag{5.117}$$

$$\Rightarrow \quad S(\Lambda^{-1}) = S^{-1}(\Lambda). \tag{5.118}$$

We now write

$$(i\hbar\gamma^\mu\partial_\mu - mc)\psi(x) = 0$$
$$(i\hbar\gamma^\mu\partial_\mu - mc)S^{-1}(\Lambda)\psi'(x') = 0$$
$$S(\Lambda)(i\hbar\gamma^\mu\partial_\mu - mc)S^{-1}(\Lambda)\psi'(x') = 0$$
$$(i\hbar S(\Lambda)\gamma^\mu S^{-1}(\Lambda)\partial_\mu - mc)\psi'(x') = 0. \tag{5.119}$$

Using $\frac{\partial}{\partial x^\mu} = \frac{\partial x'^\nu}{\partial x^\mu}\frac{\partial}{\partial x'^\nu} = \Lambda^\nu{}_\mu\frac{\partial}{\partial x'^\nu}$, we have

---

[8]R.H. Good, Rev. Mod. Phys. 27 (1955) 187.

$$\left(i\hbar S(\Lambda)\gamma^\mu S^{-1}(\Lambda)\Lambda^\nu{}_\mu \frac{\partial}{\partial x'^\nu} - mc \right) \psi'(x') = 0. \tag{5.120}$$

Therefore we require

$$S(\Lambda)\gamma^\mu S^{-1}(\Lambda)\Lambda^\nu{}_\mu = \gamma^\nu, \tag{5.121}$$

or

$$\boxed{S^{-1}(\Lambda)\gamma^\nu S(\Lambda) = \Lambda^\nu{}_\mu \gamma^\mu} . \tag{5.122}$$

This relationship defines $S(\Lambda)$ only up to an arbitrary phase factor. For Lorentz invariance of the Dirac equation, this phase factor is further restricted to a $\pm$ sign if we require that the $S(\Lambda)$ form a representation of the Lorentz group. A wave function transforming according to equation 5.115 and 5.116 by means of equation 5.122 is a four-component Lorentz spinor.

To determine S we first specify S for an infinitesimal Lorentz transformation. We then build up a finite S by applying the infinitesimal transformation an infinite number of times. Consider an infinitesimal proper Lorentz transformation

$$\Lambda^\nu{}_\mu = g^\nu{}_\mu + \Delta\omega^\nu{}_\mu, \tag{5.123}$$

where $\Delta\omega^{\nu\mu}$ is antisymmetric for an invariant proper time interval (see section 3.1.2). Each of the six independent non-vanishing components of $\Delta\omega^{\mu\nu}$ generates an infinitesimal Lorentz transformation;

$$\Delta\omega^0{}_1 = -\Delta\omega^{01} = -\Delta\beta, \tag{5.124}$$

for a transformation to a coordinate system moving with velocity $c\Delta\beta$ along the x-direction;

$$\Delta\omega^1{}_2 = -\Delta\omega^{12} = \Delta\phi, \tag{5.125}$$

for a rotation through an angle $\Delta\phi$ about the z-axis.

We expand S in powers of $\Delta\omega^{\mu\nu}$. To first order,

$$S = I - \frac{i}{4}\sigma_{\mu\nu}\Delta\omega^{\mu\nu}, \tag{5.126}$$

$$S^{-1} = I + \frac{i}{4}\sigma_{\mu\nu}\Delta\omega^{\mu\nu}, \tag{5.127}$$

with $\sigma_{\mu\nu} = -\sigma_{\nu\mu}$, or S is trivially the identity. The factor $\pm i/4$ is a convention that will prove to be convent later on. The operators $\sigma_{\mu\nu}$ and $\Delta\omega^{\mu\nu}$ act in different spaces, such that $\Delta\omega^{\mu\nu}$ acts in coordinate space on x_μ, while $\sigma_{\mu\nu}$ acts in wave-function – spinor space – on $\psi(x)$.

We now solve for $\sigma_{\mu\nu}$. Equation 5.122 becomes

$$(g^\nu{}_\mu + \Delta\omega^\nu{}_\mu)\gamma^\mu = \left(I + \frac{i}{4}\sigma_{\alpha\beta}\Delta\omega^{\alpha\beta}\right)\gamma^\nu\left(I - \frac{i}{4}\sigma_{\alpha\beta}\Delta\omega^{\alpha\beta}\right),$$

$$\Delta\omega^\nu{}_\mu\gamma^\mu = \frac{i}{4}\sigma_{\alpha\beta}\Delta\omega^{\alpha\beta}\gamma^\nu - \gamma^\nu\frac{i}{4}\sigma_{\alpha\beta}\Delta\omega^{\alpha\beta}$$

$$= -\frac{i}{4}\Delta\omega^{\alpha\beta}(\gamma^\nu\sigma_{\alpha\beta} - \sigma_{\alpha\beta}\gamma^\nu)$$

$$= -\frac{i}{4}\Delta\omega^{\alpha\beta}[\gamma^\nu, \sigma_{\alpha\beta}]. \tag{5.128}$$

Also,

$$\Delta\omega^\nu{}_\mu\gamma^\mu = 1/2(\Delta\omega^\nu{}_\mu\gamma^\mu + \Delta\omega^\nu{}_\mu\gamma^\mu)$$

$$= 1/2(\Delta\omega^\alpha{}_\mu g^\nu{}_\alpha\gamma^\mu + \Delta\omega^\beta{}_\mu g^\nu{}_\beta\gamma^\mu)$$

$$= 1/2(\Delta\omega^\alpha{}_\beta g^\nu{}_\alpha\gamma^\beta + \Delta\omega^\beta{}_\alpha g^\nu{}_\beta\gamma^\alpha)$$

$$= 1/2(\Delta\omega^{\alpha\beta}g^\nu{}_\alpha\gamma_\beta + \Delta\omega^{\beta\alpha}g^\nu{}_\beta\gamma_\alpha)$$

$$= 1/2\Delta\omega^{\alpha\beta}(g^\nu{}_\alpha\gamma_\beta - g^\nu{}_\beta\gamma_\alpha)$$

$$= -i/4\Delta\omega^{\alpha\beta}2i(g^\nu{}_\alpha\gamma_\beta - g^\nu{}_\beta\gamma_\alpha). \tag{5.129}$$

Combining equations 5.128 and 5.129 gives,

$$2i(g^\nu{}_\alpha\gamma_\beta - g^\nu{}_\beta\gamma_\alpha) = [\gamma^\nu, \sigma_{\alpha\beta}]. \tag{5.130}$$

We must find six matrices $\sigma_{\alpha\beta}$ which satisfy the above equation. We try the antisymmetric product of two gamma matrices

$$\boxed{\sigma_{\mu\nu} = \frac{i}{2}[\gamma_\mu, \gamma_\nu]}, \tag{5.131}$$

where $i/2$ is a convention that will prove convenient later on. Substituting equation 5.131 into the right-hand side of equation 5.130 gives

$$[\gamma^\nu, \sigma_{\alpha\beta}] = \frac{i}{2}[\gamma^\nu, [\gamma_\alpha, \gamma_\beta]]$$

$$= \frac{i}{2}([\gamma^\nu, \gamma_\alpha\gamma_\beta] - [\gamma^\nu, \gamma_\beta\gamma_\alpha])$$

$$= \frac{i}{2}([\gamma^\nu, \gamma_\alpha\gamma_\beta] + [\gamma^\nu, \gamma_\alpha\gamma_\beta] - 2[\gamma^\nu, g_{\beta\alpha}])$$

$$= i[\gamma^\nu, \gamma_\alpha\gamma_\beta]$$

$$= i(\gamma^\nu\gamma_\alpha\gamma_\beta - \gamma_\alpha\gamma_\beta\gamma^\nu)$$

$$= i(\gamma^\nu \gamma_\alpha \gamma_\beta + \gamma_\alpha \gamma^\nu \gamma_\beta - 2\gamma_\alpha g^\nu_\beta)$$
$$= i(\gamma^\nu \gamma_\alpha \gamma_\beta - \gamma^\nu \gamma_\alpha \gamma_\beta + 2g^\nu_\alpha \gamma_\beta - 2\gamma_\alpha g^\nu_\beta)$$
$$= 2i[g^\nu_\alpha \gamma_\beta - g^\nu_\beta \gamma_\alpha], \tag{5.132}$$

which is the left-hand side of equation 5.130. Therefore equation 5.131 is a solution to equation 5.130. Thus equation 5.126 becomes

$$\boxed{S = I - \frac{i}{4}\sigma_{\mu\nu}\Delta\omega^{\mu\nu} = I + \frac{1}{8}[\gamma_\mu, \gamma_\nu]\Delta\omega^{\mu\nu}} . \tag{5.133}$$

We now construct finite proper Lorentz transformations. We define

$$\Delta\omega^\nu{}_\mu = \Delta\omega(I_n)^\nu{}_\mu, \tag{5.134}$$

where $\Delta\omega$ is an infinitesimal parameter of the Lorentz group. I_n is a 4×4 matrix for a general unit space-time rotation around an axis in the direction labeled by n; ν labels the rows and μ labels the columns of the matrix.

We write the finite transformation using $\Delta\omega \to \omega/N$ as

$$x'^\nu = \lim_{N\to\infty} \left(g + \frac{\omega}{N}I_n\right)^\nu{}_{\alpha_1} \left(g + \frac{\omega}{N}I_n\right)^{\alpha_1}{}_{\alpha_2} \cdots \left(g + \frac{\omega}{N}I_n\right)^{\alpha_{N-1}}{}_{\alpha_N} x^{\alpha_N}$$
$$= (e^{\omega I_n})^\nu{}_\mu x^\mu. \tag{5.135}$$

In hyperbolic notation

$$x'^\nu = (\cosh\omega I_n + \sinh\omega I_n)^\nu{}_\mu x^\mu$$
$$= \left[\left(1 + \frac{(\omega I_n)^2}{2!} + \frac{(\omega I_n)^4}{4!} + \cdots\right) + \left(\frac{\omega I_n}{1!} + \frac{(\omega I_n)^3}{3!} + \cdots\right)\right]^\nu{}_\mu x^\mu. \tag{5.136}$$

For Lorentz translations (boosts), $(I_n)^3 = I_n$ and we write

$$x'^\nu = \left[\left(1 + (I_n)^2\frac{\omega^2}{2!} + (I_n)^2\frac{\omega^4}{4!} + \cdots\right) + I_n\left(\frac{\omega}{1!} + \frac{\omega^3}{3!} + \cdots\right)\right]^\nu{}_\mu x^\mu$$
$$= (1 - I_n^2 + I_n^2\cosh\omega + I_n\sinh\omega)^\nu{}_\mu x^\mu. \tag{5.137}$$

For space rotations, $(I_n)^3 = -I_n$ and we write

$$x'^\nu = \left[\left(1 + (I_n)^2\frac{\omega^2}{2!} - (I_n)^2\frac{\omega^4}{4!} + \cdots\right) + I_n\left(\frac{\omega}{1!} - \frac{\omega^3}{3!} + \cdots\right)\right]^\nu{}_\mu x^\mu$$
$$= (1 + (I_n)^2 - (I_n)^2\cos\omega + I_n\sin\omega)^\nu{}_\mu x^\mu. \tag{5.138}$$

Turning now to the construction of a finite spinor transformation S, we have

$$S = \lim_{N \to \infty} \left(1 - \frac{i}{4}\frac{\omega}{N}\sigma_{\mu\nu}I_n^{\mu\nu}\right)^N = \exp\left(-\frac{i}{4}\omega\sigma_{\mu\nu}I_n^{\mu\nu}\right). \qquad (5.139)$$

The following sections consider finite transformations for a rotation in three-space, a general Lorentz boost, spatial inversion, and time reflection.

5.9.1 Rotations

Consider a rotation through an angle ϕ about the z-axis:

$$(I_3)^\nu_{\ \mu} = \begin{pmatrix} 0 & 0 & 0 & 0 \\ 0 & 0 & 1 & 0 \\ 0 & -1 & 0 & 0 \\ 0 & 0 & 0 & 0 \end{pmatrix} \qquad (5.140)$$

and

$$(I_3)^2 = \begin{pmatrix} 0 & 0 & 0 & 0 \\ 0 & -1 & 0 & 0 \\ 0 & 0 & -1 & 0 \\ 0 & 0 & 0 & 0 \end{pmatrix}. \qquad (5.141)$$

Explicitly, equation 5.138 becomes

$$\begin{pmatrix} x^{0'} \\ x^{1'} \\ x^{2'} \\ x^{3'} \end{pmatrix} = \begin{pmatrix} 1 & 0 & 0 & 0 \\ 0 & \cos\omega & \sin\omega & 0 \\ 0 & -\sin\omega & \cos\omega & 0 \\ 0 & 0 & 0 & 1 \end{pmatrix} \begin{pmatrix} x^0 \\ x^1 \\ x^2 \\ x^3 \end{pmatrix}, \qquad (5.142)$$

where ω is the finite rotation angle ϕ.

The matrix I_3 has all zero elements except for $I^1_{\ 2} = -I^2_{\ 1} = 1$, or $I^{12} = -I^{21} = -1$. Thus,

$$\sigma_{\mu\nu}I^{\mu\nu} = 2\sigma_{12}I^{12} = -2i\gamma_1\gamma_2 = -2\begin{pmatrix} \sigma_3 & 0 \\ 0 & \sigma_3 \end{pmatrix} = -2\Sigma_3. \qquad (5.143)$$

Therefore,

$$S = \exp\left(-\frac{i}{4}\phi\sigma_{\mu\nu}I_n^{\mu\nu}\right) = \exp\left(\frac{i}{2}\phi\Sigma_3\right). \qquad (5.144)$$

For a rotation about an arbitrary axis $\hat{n}$, we write

$$S = \exp\left(\frac{i}{2}\phi\vec{\Sigma}\cdot\hat{n}\right). \qquad (5.145)$$

Since $\vec{\Sigma}$ is hermitian, $S^\dagger = S^{-1}$ for spatial rotations.

By applying the rotation operator on the solution for the Dirac particle at rest with spin in the z-direction, it is possible to form spin states in any arbitrary direction.

The appearance of the half-angle in equation 5.145 tells us that $S(\phi+2\pi) = -S(\phi)$ and $S(\phi+4\pi) = S(\phi)$. It thus takes a rotation of 4π to return $\psi(x)$ to its original value. This is a characteristic of half-integer spin particles. Thus the Dirac wave function transforms under rotations like the wave function of a particle of spin $1/2$. We now feel justified in previously referring to the wave function of the Dirac equation as a spinor. Because of the half-angle result, the arbitrary sign of S mentioned after equation 5.122 cannot be removed with S still forming a representation of the Lorentz group. Because of the double valueness of $\psi(x)$ under a rotation of 2π, physical observables in the Dirac theory must be bilinear, or an even power in $\psi(x)$. Only in this case do observables become identical under a rotation of 2π; a property of observables we know from experience.

5.9.2   Lorentz Boosts

Consider a Lorentz boost with a velocity β along the x-axis:

$$(I_x)^\nu{}_\mu = \begin{pmatrix} 0 & -1 & 0 & 0 \\ -1 & 0 & 0 & 0 \\ 0 & 0 & 0 & 0 \\ 0 & 0 & 0 & 0 \end{pmatrix} \qquad (5.146)$$

and

$$(I_x)^2 = \begin{pmatrix} 1 & 0 & 0 & 0 \\ 0 & 1 & 0 & 0 \\ 0 & 0 & 0 & 0 \\ 0 & 0 & 0 & 0 \end{pmatrix}. \qquad (5.147)$$

Explicitly, equation 5.137 becomes

$$\begin{pmatrix} x^{0'} \\ x^{1'} \\ x^{2'} \\ x^{3'} \end{pmatrix} = \begin{pmatrix} \cosh\omega & -\sinh\omega & 0 & 0 \\ -\sinh\omega & \cosh\omega & 0 & 0 \\ 0 & 0 & 0 & 0 \\ 0 & 0 & 0 & 0 \end{pmatrix} \begin{pmatrix} x^0 \\ x^1 \\ x^2 \\ x^3 \end{pmatrix}, \qquad (5.148)$$

where ω is the rapidity parameter given by $\beta = \tanh\omega$.

For an arbitrary Lorentz boost direction $\hat{v} = \vec{p}/|\vec{p}|$, we may write in terms of the direction[9],

---

[9]We are using $I^\mu{}_\nu$ a little differently than previously defined, but the important property of $(I_n)^3 = I_n$ is still obeyed.

$$I^\mu_{\ \nu} = \begin{pmatrix} 0 & -\cos\alpha & -\cos\beta & -\cos\gamma \\ -\cos\alpha & 0 & 0 & 0 \\ -\cos\beta & 0 & 0 & 0 \\ -\cos\gamma & 0 & 0 & 0 \end{pmatrix}. \tag{5.149}$$

Raising the ν index makes $I^{\mu\nu}$ antisymmetric. Also $\sigma_{\mu\nu}$ is antisymmetric. Therefore,

$$\begin{aligned} \sigma_{\mu\nu} I^{\mu\nu} &= 2(\sigma_{01}\cos\alpha + \sigma_{02}\cos\beta + \sigma_{03}\cos\gamma) \\ &= 2i(\gamma_0\gamma_1\cos\alpha + \gamma_0\gamma_2\cos\beta + \gamma_0\gamma_3\cos\gamma) \\ &= -2i(\alpha_1\cos\alpha + \alpha_2\cos\beta + \alpha_3\cos\gamma) \\ &= -2i\vec{\alpha}\cdot\hat{v}. \end{aligned} \tag{5.150}$$

From equation 5.139, the finite spinor transformation for a general Lorentz boost becomes

$$\begin{aligned} S &= \exp\left(-\frac{\omega}{2}\vec{\alpha}\cdot\hat{v}\right) \\ &= \cosh\left(-\frac{\omega}{2}\vec{\alpha}\cdot\hat{v}\right) + \sinh\left(-\frac{\omega}{2}\vec{\alpha}\cdot\hat{v}\right) \\ &= \cosh\left(\frac{\omega}{2}\vec{\alpha}\cdot\hat{v}\right) - \sinh\left(\frac{\omega}{2}\vec{\alpha}\cdot\hat{v}\right). \end{aligned} \tag{5.151}$$

From the properties of the α_i matrices, we have (see problem 5.14)

$$(\vec{\alpha}\cdot\hat{v})^2 = I. \tag{5.152}$$

Therefore, we write

$$\begin{aligned} S &= \cosh\left(\frac{\omega}{2}\right) - \vec{\alpha}\cdot\hat{v}\sinh\left(\frac{\omega}{2}\right) \\ &= \cosh\left(\frac{\omega}{2}\right)\left[I - \vec{\alpha}\cdot\hat{v}\tanh\left(\frac{\omega}{2}\right)\right]. \end{aligned} \tag{5.153}$$

Unlike for rotations, $S^\dagger \neq S^{-1}$ for Lorentz boosts. However, by expanding S in a power series, it has the property (see problem 5.15)

$$\boxed{S^{-1} = \gamma_0 S^\dagger \gamma_0}. \tag{5.154}$$

This can also be generalized to include rotations (see problem 5.15).

5.9.3   Spatial Inversion

We now consider the improper Lorentz transformation of reflection in space or the parity transformation:

$$\vec{x}' = -\vec{x} \quad \text{and} \quad t' = t. \tag{5.155}$$

We need to solve equation 5.122 for

$$\Lambda^{\nu}_{\ \mu} = \begin{pmatrix} 1 & 0 & 0 & 0 \\ 0 & -1 & 0 & 0 \\ 0 & 0 & -1 & 0 \\ 0 & 0 & 0 & -1 \end{pmatrix} = g^{\nu\mu}. \tag{5.156}$$

We denote the Lorentz operator $S(\Lambda)$ for parity by P. Consider the following ansatz:

$$P = e^{i\phi}\gamma_0, \tag{5.157}$$

where ϕ is an arbitrary real phase. Using equation 5.122, we have

$$e^{-i\phi}\gamma_0\gamma^\nu e^{i\phi}\gamma_0 = \gamma^0\gamma^\nu\gamma^0 = g^{\nu\mu}\gamma^\mu, \tag{5.158}$$

as required.

In analogy to the proper Lorentz transformations for which a rotation of 4π reproduces the original spinors, we postulate that four space inversions will reproduce the original spinors,

$$P^4\psi = \psi = (e^{i\phi}\gamma^0)^4\psi = e^{i4\phi}(\gamma^0)^4\psi = e^{i4\phi}\psi. \tag{5.159}$$

Therefore,

$$e^{i4\phi} = 1 \quad \Rightarrow \quad e^{i\phi} = \pm 1 \text{ or } \pm i. \tag{5.160}$$

We see that

$$P^{-1} = e^{-i\phi}\gamma_0^{-1} = e^{-i\phi}\gamma_0, \tag{5.161}$$
$$P^\dagger = e^{-i\phi}\gamma_0^\dagger = e^{-i\phi}\gamma_0, \tag{5.162}$$

and $P^{-1} = P^\dagger \quad \Rightarrow \quad P$ is unitary.
The wave function thus transforms as

$$\psi'(x') = \psi'(-\vec{x},t) = e^{i\phi}\gamma_0\psi(\vec{x},t). \tag{5.163}$$

In the nonrelativistic limit, $\psi = \begin{pmatrix} \phi \\ 0 \end{pmatrix}$ and ψ approaches an eigenstate of P.

The positive- and negative-energy states at rest have opposite eigenvalues, or intrinsic parities:

$$P\psi^1(0) = +e^{i\phi}\psi^1(0),$$
$$P\psi^2(0) = +e^{i\phi}\psi^2(0),$$
$$P\psi^3(0) = -e^{i\phi}\psi^3(0),$$
$$P\psi^4(0) = -e^{i\phi}\psi^4(0). \tag{5.164}$$

The intrinsic parity of a Dirac particle and antiparticle with the same mass are opposite, which has important consequences. This is to be contrasted to the Klein-Gordon case wherein one finds identical parities for the particle and antiparticle solutions.

5.9.4 Time Reflection

Consider the transformation corresponding to a pure time reflection. We need to solve equation 5.122 for

$$\Lambda^\nu{}_\mu = \begin{pmatrix} -1 & 0 & 0 & 0 \\ 0 & 1 & 0 & 0 \\ 0 & 0 & 1 & 0 \\ 0 & 0 & 0 & 1 \end{pmatrix}. \tag{5.165}$$

We denote the Lorentz operator $S(\Lambda)$ for time reflection by T. We see that equation 5.122 is satisfied (see problem 5.16) by

$$T = \gamma^1\gamma^2\gamma^3. \tag{5.166}$$

T defined in this way is known as the Racah time reflection operator.

5.10 Covariance of the Continuity Equation

As mentioned previously on page 77, we still need to show that the conserved current transforms as a four-vector under a Lorentz transformation. Using the gamma matrices and their properties, we have

$$j^0 = c\rho = c\psi^\dagger\psi = c\psi^\dagger\gamma^0\gamma^0\psi, \tag{5.167}$$
$$j^k = c\psi^\dagger\alpha_k\psi = c\psi^\dagger\gamma^0\gamma^0\alpha_k\psi = c\psi^\dagger\gamma^0\gamma^k\psi. \tag{5.168}$$

Therefore,

$$j^\mu(x) = c\psi^\dagger(x)\gamma^0\gamma^\mu\psi(x). \tag{5.169}$$

Under a Lorentz transformation, we have

$$
\begin{aligned}
j'^{\mu}(x') &= c\psi^{\dagger}(x)S^{\dagger}\gamma^{0}\gamma^{\mu}S\psi(x) \\
&= c\psi^{\dagger}(x)\gamma^{0}S^{-1}\gamma^{\mu}S\psi(x) \\
&= c\psi^{\dagger}(x)\gamma^{0}\Lambda^{\mu}_{\ \nu}\gamma^{\nu}\psi(x) \\
&= \Lambda^{\mu}_{\ \nu}j^{\nu}(x),
\end{aligned}
\tag{5.170}
$$

where equations 5.154 and 5.122 have been used in the second and third lines, respectively. Thus the current density $j^{\mu}(x)$ transforms like a four-vector under a Lorentz transformation.

5.11 Adjoint Spinor

Since a spinor is an intrinsically complex entity, the Dirac equation is not sufficient to determine the space-time behavior of ψ without also using the complex conjugate equation, or some equivalent form. Because the combination $\psi^{\dagger}\gamma_{0}$ occurs so often, we define

$$
\overline{\psi}(x) \equiv \psi^{\dagger}\gamma_{0},
\tag{5.171}
$$

where $\overline{\psi}(x)$ is the adjoint Dirac spinor. This definition allows quantities such as the current density to be written in a concise form.

The adjoint spinor Lorentz transformation properties follow from

$$
\begin{aligned}
\psi'(x') &= S\psi(x), \\
{\psi'}^{\dagger}(x') &= \psi^{\dagger}(x)S^{\dagger} = \psi^{\dagger}(x)\gamma_{0}S^{-1}\gamma_{0}, \\
{\psi'}^{\dagger}(x')\gamma_{0} &= \overline{\psi}'(x') = \overline{\psi}(x)S^{-1}.
\end{aligned}
\tag{5.172}
$$

From the definition of the adjoint spinor, the four-current is

$$
j^{\mu} = c\overline{\psi}\gamma^{\mu}\psi,
\tag{5.173}
$$

where $c\gamma^{\mu}$ acts like a current operator.

5.12 Bilinear Covariants

The wave functions themselves do not represent observables directly, but one can construct bilinear expressions of the wave functions. Some of these have a simple physical interpretation. The Dirac matrices are simply constants and have the same value in all Lorentz frames. However, when contracted with $\overline{\psi}$ and ψ, different bilinears have their own distinct tensor transformation properties.

The combination of all the gamma matrices occurs often enough that we define

$$\boxed{\gamma_5 = \gamma^5 = i\gamma^0\gamma^1\gamma^2\gamma^3} \, . \tag{5.174}$$

We see that $(\gamma_5)^2 = I$, $\gamma_5^\dagger = \gamma_5$ (see problem 5.18), and γ_5 anticommutes with γ^μ:

$$\gamma^\mu\gamma_5 = i\gamma^\mu\gamma^0\gamma^1\gamma^2\gamma^3 = -i\gamma^0\gamma^1\gamma^2\gamma^3\gamma^\mu = -\gamma_5\gamma^\mu. \tag{5.175}$$

Also, γ_5 commutes with $\sigma_{\mu\nu}$:

$$[\gamma_5, \sigma_{\mu\nu}] = \frac{i}{2}(\gamma_5\gamma_\mu\gamma_\nu - \gamma_5\gamma_\nu\gamma_\mu - \gamma_\mu\gamma_\nu\gamma_5 + \gamma_\nu\gamma_\mu\gamma_5) = 0. \tag{5.176}$$

Thus γ_5 commutes with S for proper Lorentz transformations but anticommutes with S for improper Lorentz transformations. We write these two observations in the compact form

$$\boxed{S(\Lambda)\gamma_5 = \gamma_5 S(\Lambda)\det|\Lambda|} \, . \tag{5.177}$$

By forming various products of gamma matrices and using the identity matrix, it is possible to construct 16 linearly independent 4×4 matrices Γ_n:

$$\Gamma_S = I; \quad \Gamma_V^\mu = \gamma^\mu; \quad \Gamma_T^{\mu\nu} = \sigma^{\mu\nu}; \quad \Gamma_A^\mu = \gamma^5\gamma^\mu; \quad \Gamma_P = \gamma^5. \tag{5.178}$$

We will see that S, V, T, A, and P represent scalar, vector, tensor, axial vector, and pseudoscalar, respectively. There is one Γ_S which has no gamma matrices, Γ_V^μ has one gamma matrix of which there are four, $\Gamma_T^{\mu\nu}$ has two gamma matrices of which there are six, Γ_A^μ has three gamma matrices of which there are four, and Γ_P has four gamma matrices of which there is one.

We now show how the bilinears $\overline{\psi}(x)\Gamma_n\psi(x)$ transform under a Lorentz transformation. Using equations 5.122, 5.154, and 5.177, we have

$$\overline{\psi}'(x')\psi'(x') = {\psi'}^\dagger(x')\gamma^0\psi'(x') = \psi^\dagger(x)S^\dagger\gamma^0 S\psi(x) = \psi^\dagger(x)\gamma^0 S^{-1}S\psi(x)$$
$$= \overline{\psi}(x)\psi(x). \tag{5.179}$$

$$\overline{\psi}'(x')\gamma_5\psi'(x') = \overline{\psi}(x)S^{-1}\gamma_5 S\psi(x) = \overline{\psi}(x)\det|\Lambda|\gamma_5\psi(x)$$
$$= \det|\Lambda|\overline{\psi}(x)\gamma_5\psi(x). \tag{5.180}$$

$$\overline{\psi}'(x')\gamma^\nu\psi'(x') = \overline{\psi}(x)S^{-1}\gamma^\nu S\psi(x) = \overline{\psi}(x)\Lambda^\nu_{\ \mu}\gamma^\mu\psi(x)$$
$$= \Lambda^\nu_{\ \mu}\overline{\psi}(x)\gamma^\mu\psi(x). \tag{5.181}$$

$$\overline{\psi}'(x')\gamma_5\gamma^\mu\psi'(x') = \overline{\psi}(x)S^{-1}\gamma_5\gamma^\mu S\psi(x) = \overline{\psi}(x)\det|\Lambda|\gamma_5\Lambda^\nu_{\ \mu}\gamma^\mu\psi(x)$$
$$= \det|\Lambda|\Lambda^\nu_{\ \mu}\overline{\psi}(x)\gamma_5\gamma^\mu\psi(x). \tag{5.182}$$

$$\overline{\psi}'(x')\sigma^{\mu\nu}\psi'(x') = i/2\overline{\psi}(x)S^{-1}[\gamma^\mu,\gamma^\nu]S\psi(x) = i/2\overline{\psi}(x)\Lambda^\mu_{\ \alpha}\Lambda^\nu_{\ \beta}[\gamma^\alpha,\gamma^\beta]\psi(x)$$
$$= \Lambda^\mu_{\ \alpha}\Lambda^\nu_{\ \beta}\overline{\psi}(x)\sigma^{\alpha\beta}\psi(x). \tag{5.183}$$

Thus the Lorentz transformation properties of these bilinear forms can be summarized as

$$\text{scalar:} \quad \overline{\psi}'(x')\psi'(x') = \overline{\psi}(x)\psi(x), \tag{5.184}$$

$$\text{pseudoscalar:} \quad \overline{\psi}'(x')\gamma_5\psi'(x') = \det|\Lambda|\overline{\psi}(x)\gamma_5\psi(x), \tag{5.185}$$

$$\text{vector:} \quad \overline{\psi}'(x')\gamma^\mu\psi'(x') = \Lambda^\nu_{\ \mu}\overline{\psi}(x)\gamma^\mu\psi(x), \tag{5.186}$$

$$\text{pseudovector:} \quad \overline{\psi}'(x')\gamma_5\gamma^\nu\psi'(x') = \det|\Lambda|\Lambda^\nu_{\ \mu}\overline{\psi}(x)\gamma_5\gamma^\mu\psi(x), \tag{5.187}$$

$$\text{second-rank tensor:} \quad \overline{\psi}'(x')\sigma^{\mu\nu}\psi'(x') = \Lambda^\mu_{\ \alpha}\Lambda^\nu_{\ \beta}\overline{\psi}(x)\sigma^{\alpha\beta}\psi(x). \tag{5.188}$$

The bilinears should be multiplied by the necessary factors of i to make them hermitian; any new term we added to the Hamiltonian to represent an interaction must be real.

Not all of these covariant forms are realized in electromagnetic scattering theory. In the theory of the weak interaction, and the less established theories involving gravity and super-symmetry, a rich set of bilinear forms are used. Understanding which forms are realized in nature is still an ongoing field of research in particle physics.

5.13 Plane-Wave Solutions

We have already solved the Dirac equation for free particles at rest (section 5.3). The Lorentz transformation may be used to construct the free-particle solutions with an arbitrary velocity. The result should be comparable to the free-particle solutions we obtained in section 5.7.

The exponential part of the solutions is a function of the space-time coordinates but it only involves the dot product of four-vectors. Thus we expect it to be Lorentz invariant. It is easy to see the invariant form of the exponential by writing its form in the rest frame of the particle

$$\exp(-i\epsilon_r p \cdot x/\hbar) \rightarrow \exp(-i\epsilon_r Et/\hbar) \rightarrow \exp(-i\epsilon_r mc^2 t/\hbar), \qquad (5.189)$$

where $r = 1, 2, 3, 4$, and $\epsilon_r = +1$ for $r = 1, 2$ and $\epsilon_r = -1$ for $r = 2, 3$. The positive-energy solutions $(r = 1, 2)$ and negative-energy solutions $(r = 3, 4)$ transform among themselves separately and do not mix with each other under proper Lorentz transformations, as well as, under spatial inversions.

If a particle is at rest in frame S' it will have energy E and momentum $\vec{p}$ in the frame S moving with velocity of magnitude $\beta = v/c$ and direction $\hat{v} = -\vec{p}/|\vec{p}|$ with respect to S'. The Lorentz transformation parameters for this boost are

$$\beta = \frac{pc}{E} \quad \text{and} \quad \gamma = \frac{E}{mc^2}. \qquad (5.190)$$

Using equation 5.153 and recognizing ω as the rapidity parameter (comparing equation 5.148 with 5.153), we write

$$\tanh \frac{\omega}{2} = \frac{\sinh \omega}{1 + \cosh \omega} = \frac{\gamma\beta}{1 + \gamma} = \frac{pc}{E + mc^2}, \qquad (5.191)$$

$$\cosh \frac{\omega}{2} = \sqrt{\frac{1 + \cosh \omega}{2}} = \sqrt{\frac{1 + \gamma}{2}} = \sqrt{\frac{E + mc^2}{2mc^2}}. \qquad (5.192)$$

To calculate the Lorentz transformation matrix S we will need, in our representation,

$$\begin{aligned}
\vec{\sigma} \cdot \vec{p} &= \begin{pmatrix} 0 & p_x \\ p_x & 0 \end{pmatrix} + \begin{pmatrix} 0 & -ip_y \\ ip_y & 0 \end{pmatrix} + \begin{pmatrix} p_z & 0 \\ 0 & -p_z \end{pmatrix} \\
&= \begin{pmatrix} p_z & p_x - ip_y \\ p_x + ip_y & -p_z \end{pmatrix} \\
&= \begin{pmatrix} p_z & p_- \\ p_+ & -p_z \end{pmatrix},
\end{aligned} \qquad (5.193)$$

where $p_\pm = p_x \pm ip_y$. Using these results with equation 5.153 gives

$$S = \sqrt{\frac{E + mc^2}{2mc^2}} \begin{bmatrix} 1 & 0 & \frac{p_z c}{E+mc^2} & \frac{p_- c}{E+mc^2} \\ 0 & 1 & \frac{p_+ c}{E+mc^2} & \frac{-p_z c}{E+mc^2} \\ \frac{p_z c}{E+mc^2} & \frac{p_- c}{E+mc^2} & 1 & 0 \\ \frac{p_+ c}{E+mc^2} & \frac{-p_z c}{E+mc^2} & 0 & 1 \end{bmatrix}. \qquad (5.194)$$

Using equation 5.18 and equation 5.19, we see that each of the spinor pieces of the four general solutions to the Dirac equation correspond to one of the columns of the above transformation matrix. This is identical to the result obtained in section 5.7.

Let us now look at some of the properties of the general solution to the Dirac equation. The general form of the wave function is

$$\psi^r(x) = \omega^r(\vec{p})e^{-i\epsilon_r p \cdot x/\hbar}, \tag{5.195}$$

where ω is a function of the three-momentum $\vec{p}$ and the sign of the energy given by r. Substitution into the Dirac equation gives

$$\left(i\hbar\frac{\partial}{\partial x_\mu}\gamma_\mu - mc\right)\omega^r(\vec{p})e^{-i\epsilon_r p \cdot x/\hbar} = 0,$$

$$\left(i\hbar\frac{-i\epsilon_r p^\mu}{\hbar}\gamma_\mu - mc\right)\omega^r(\vec{p}) = 0,$$

$$(\epsilon_r \slashed{p} - mc)\omega^r(\vec{p}) = 0, \tag{5.196}$$

$$\boxed{(\slashed{p} - \epsilon_r mc)\omega^r(\vec{p}) = 0}, \tag{5.197}$$

which is the Dirac equation in momentum space. It is an algebraic equation without time or space derivatives. The adjoint equation is

$$\psi^{\dagger r}(x)\left(-i\hbar\frac{\overleftarrow{\partial}}{\partial x_\mu}\gamma_\mu^\dagger - mc\right) = 0,$$

$$\omega^r(\vec{p})^\dagger e^{i\epsilon_r p \cdot x/\hbar}\left(-i\hbar\frac{\overleftarrow{\partial}}{\partial x_\mu}\gamma_\mu^\dagger - mc\right) = 0,$$

$$\overline{\omega}^r(\vec{p})\gamma_0\left(-i\hbar\frac{i\epsilon_r p^\mu}{\hbar}\gamma_\mu^\dagger - mc\right) = 0,$$

$$\overline{\omega}^r(\vec{p})\left(\epsilon_r p^\mu\gamma_0\gamma_\mu^\dagger\gamma_0 - mc\right) = 0,$$

$$\overline{\omega}^r(\vec{p})(\epsilon_r \slashed{p} - mc) = 0,$$

$$\overline{\omega}^r(\vec{p})(\slashed{p} - \epsilon_r mc) = 0, \tag{5.198}$$

where equation 5.108 has been used in the third from last line.

We now examine the normalization and completeness relationships for the spinors $\omega^r(\vec{p})$. Since the covariant normalization statement

$$\overline{\omega}^r(\vec{p})\omega^{r'}(\vec{p}) \tag{5.199}$$

is a Lorentz scalar (equation 5.179), we may evaluate it in the rest frame for simplicity. Using equation 5.18 and equation 5.19, we have

$$\overline{w}^r(\vec{p})w^{r\prime}(\vec{p}) = \overline{w}^r(0)w^{r\prime}(0) = w^{r\dagger}(0)\gamma_0 w^{r\prime}(0) = \epsilon_r \delta_{rr\prime}. \qquad (5.200)$$

The completeness relationship is

$$\sum_{r=1}^{4} \epsilon_r w_\alpha^r(\vec{p})\overline{w}_\beta^r(\vec{p}). \qquad (5.201)$$

The adjoint spinor appears in the completeness relationship, rather than the Hermitian conjugate, because the Lorentz transformation of the spinors is not unitary as expressed in equation 5.154. The reason for the factor ϵ_r will become apparent in the next section.

We first evaluate equation 5.201 in the rest frame and then boost it into a general moving frame. For a particle at rest,

$$w^r(0)\overline{w}^r(0) = w^r(0)w^{r\dagger}(0)\gamma_0 = \begin{cases} \begin{pmatrix} 1&0&0&0 \\ 0&0&0&0 \\ 0&0&0&0 \\ 0&0&0&0 \end{pmatrix} & \text{for } r = 1, \\[1.5em] \begin{pmatrix} 0&0&0&0 \\ 0&1&0&0 \\ 0&0&0&0 \\ 0&0&0&0 \end{pmatrix} & \text{for } r = 2, \\[1.5em] \begin{pmatrix} 0&0&0&0 \\ 0&0&0&0 \\ 0&0&-1&0 \\ 0&0&0&0 \end{pmatrix} & \text{for } r = 3, \\[1.5em] \begin{pmatrix} 0&0&0&0 \\ 0&0&0&0 \\ 0&0&0&0 \\ 0&0&0&-1 \end{pmatrix} & \text{for } r = 4. \end{cases} \qquad (5.202)$$

Therefore,

$$\sum_{r=1}^{4} \epsilon_r w^r(0)\overline{w}^r(0) = I \quad \text{or} \quad \sum_{r=1}^{4} \epsilon_r w_\alpha^r(0)\overline{w}_\beta^r(0) = \delta_{\alpha\beta}. \qquad (5.203)$$

For a particle in motion,

$$\sum_{r=1}^{4} \epsilon_r w_\alpha^r(\vec{p})\overline{w}_\beta^r(\vec{p}) = \sum_{r=1}^{4} \epsilon_r S_{\alpha\gamma}\left(\frac{-\vec{p}c}{E}\right) w_\gamma^r(0)\overline{w}_\lambda^r(0) S_{\lambda\beta}^{-1}\left(\frac{-\vec{p}c}{E}\right)$$

$$= S_{\alpha\gamma}\left(\frac{-\vec{p}c}{E}\right)\delta_{\gamma\lambda}S_{\lambda\beta}^{-1}\left(\frac{-\vec{p}c}{E}\right)$$

$$= \delta_{\alpha\beta}. \qquad (5.204)$$

The probability density

$$\omega^{r\dagger}(\epsilon_r \vec{p})\omega^{r'}(\epsilon_{r'}\vec{p}) \tag{5.205}$$

is not Lorentz invariant but transforms as the zeroth component of a vector. Notice the ϵ_r factor in front of the momentum. The three-momentum must change sign for orthogonality between the positive- and negative-energy states. The positive-energy spinor is orthogonal to its hermitian conjugate spinor of the negative energy and reversed momentum.

To determine the value of expression 5.205, we boost the particle from rest:

$$\omega^{r\dagger}(\epsilon_r \vec{p})\omega^{r'}(\epsilon_{r'}\vec{p}) = \omega_\gamma^{r\dagger}(0)S_{\gamma\alpha}^\dagger\left(\frac{-\epsilon_r \vec{p}c}{E}\right)S_{\alpha\delta}\left(\frac{-\epsilon_{r'}\vec{p}c}{E}\right)\omega_\delta^{r'}(0). \tag{5.206}$$

We now calculate $S^\dagger S$. Using equation 5.194, defining $\zeta = (E + mc^2)/c$, and noticing that $p_+^* = p_-$, $p_-^* = p_+$, $p_+ p_+^* = p_x^2 + p_y^2$, $p_- p_-^* = p_x^2 + p_y^2$, we calculate

$$S^\dagger\left(\frac{-\epsilon_r \vec{p}c}{E}\right)S\left(\frac{-\epsilon_{r'}\vec{p}c}{E}\right) = \frac{1}{2mc\zeta}$$

$$\cdot \begin{pmatrix} \zeta & 0 & -\epsilon_r p_z & -\epsilon_r p_+^* \\ 0 & \zeta & -\epsilon_r p_-^* & \epsilon_r p_z \\ -\epsilon_r p_z & -\epsilon_r p_+^* & \zeta & 0 \\ -\epsilon_r p_-^* & \epsilon_r p_z & 0 & \zeta \end{pmatrix} \begin{pmatrix} \zeta & 0 & -\epsilon_{r'} p_z & -\epsilon_{r'} p_- \\ 0 & \zeta & -\epsilon_{r'} p_+ & \epsilon_{r'} p_z \\ -\epsilon_{r'} p_z & -\epsilon_{r'} p_- & \zeta & 0 \\ -\epsilon_{r'} p_+ & \epsilon_{r'} p_z & 0 & \zeta \end{pmatrix}$$

$$= \frac{1}{2mc\zeta} \begin{pmatrix} \zeta^2 + \epsilon_r\epsilon_{r'}(p_z^2 + p_+ p_+^*) & \epsilon_r\epsilon_{r'}(p_-^* - p_+)p_z \\ \epsilon_r\epsilon_{r'}(p_- - p_+^*)p_z & \zeta^2 + \epsilon_r\epsilon_{r'}(p_z^2 + p_- p_-^*) \\ -(\epsilon_r + \epsilon_{r'})p_z\zeta & -(\epsilon_r p_-^* + \epsilon_{r'} p_+)\zeta \\ -(\epsilon_r p_-^* + \epsilon_{r'} p_+)\zeta & (\epsilon_r + \epsilon_{r'})p_z\zeta \\ -(\epsilon_r + \epsilon_{r'})p_z\zeta & -(\epsilon_r p_-^* + \epsilon_{r'} p_+)\zeta \\ -(\epsilon_r p_+^* + \epsilon_{r'} p_-)\zeta & (\epsilon_r + \epsilon_{r'})p_z\zeta \\ \zeta^2 + \epsilon_r\epsilon_{r'}(p_z^2 + p_+ p_+^*) & \epsilon_r\epsilon_{r'}(p_-^* - p_+)p_z \\ \epsilon_r\epsilon_{r'}(p_- - p_+^*)p_z & \zeta^2 + \epsilon_r\epsilon_{r'}(p_z^2 + p_- p_-^*) \end{pmatrix}$$

$$= \frac{1}{2mc\zeta}$$

$$\cdot \begin{pmatrix} \zeta^2 + \epsilon_r\epsilon_{r'}p^2 & 0 & -(\epsilon_r + \epsilon_{r'})p_z\zeta & -(\epsilon_r + \epsilon_{r'})p_+\zeta \\ 0 & \zeta^2 + \epsilon_r\epsilon_{r'}p^2 & -(\epsilon_r + \epsilon_{r'})p_-\zeta & (\epsilon_r + \epsilon_{r'})p_z\zeta \\ -(\epsilon_r + \epsilon_{r'})p_z\zeta & -(\epsilon_r + \epsilon_{r'})p_+\zeta & \zeta^2 + \epsilon_r\epsilon_{r'}p^2 & 0 \\ -(\epsilon_r + \epsilon_{r'})p_+\zeta & (\epsilon_r + \epsilon_{r'})p_z\zeta & 0 & \zeta^2 + \epsilon_r\epsilon_{r'}p^2 \end{pmatrix}. \tag{5.207}$$

If $\epsilon_r = -\epsilon_{r'}$,

$$S^\dagger\left(\frac{-\epsilon_r \vec{p}c}{E}\right)S\left(\frac{-\epsilon_{r'}\vec{p}c}{E}\right) = \frac{\zeta^2 - p^2}{2mc\zeta}I = I \tag{5.208}$$

and

$$w^{r\dagger}(\epsilon_r\vec{p})w^{r'}(\epsilon_{r'}\vec{p}) = w_\gamma^{r\dagger}(0)I_{\gamma\delta}w_\delta^{r'}(0) = 0, \qquad (5.209)$$

since $r \neq r'$.
If $\epsilon_r = \epsilon_{r'}$, then

$$S^\dagger\left(\frac{-\epsilon_r\vec{p}c}{E}\right)S\left(\frac{-\epsilon_{r'}\vec{p}c}{E}\right) = \frac{1}{2mc\zeta}\begin{pmatrix} \zeta^2+p^2 & 0 & -2\epsilon_r p_z\zeta & -2\epsilon_r p_+\zeta \\ 0 & \zeta^2+p^2 & -2\epsilon_r p_-\zeta & 2\epsilon_r p_z\zeta \\ -2\epsilon_r p_z\zeta & -2\epsilon_r p_+\zeta & \zeta^2+p^2 & 0 \\ -2\epsilon_r p_+\zeta & 2\epsilon_r p_z\zeta & 0 & \zeta^2+p^2 \end{pmatrix}.$$
$$(5.210)$$

If $r = r'$,

$$w^{r\dagger}(\epsilon_r\vec{p})w^{r'}(\epsilon_{r'}\vec{p}) = \frac{\zeta^2+p^2}{2mc\zeta} = \frac{E}{mc^2} = \gamma. \qquad (5.211)$$

If $\epsilon_r = \epsilon_{r'}$ and $r \neq r'$,

$$w^{r\dagger}(\epsilon_r\vec{p})w^{r'}(\epsilon_{r'}\vec{p}) = 0. \qquad (5.212)$$

Combining the results of equations 5.209, 5.211, and 5.212, we have

$$w^{r\dagger}(\epsilon_r\vec{p})w^{r'}(\epsilon_{r'}\vec{p}) = \gamma\delta_{rr'}, \qquad (5.213)$$

which is the time component of a four-vector.

5.13.1 Spin

The helicity operator is invariant under rotations (see problem 5.8). It is also invariant for boosts in the direction of $\vec{p}$ (see problem 5.9). Under boosts transverse to $\vec{p}$, however, helicity is not invariant. In general, we may say that the helicity operator is not covariant under arbitrary Lorentz transformations. We would like a covariant description for the intrinsic spin of a particle, so as to be able to transform easily from one frame to another. This description is provided by the spin four-vector.

We have just seen how one can create a state of arbitrary momentum by Lorentz boosting the solution for a particle at rest. In a similar manner, we can create a state of arbitrary spin polarized along the $\vec{s}$-direction by applying the rotation operator,

$$S = \exp\left(\frac{i}{2}\phi\vec{\Sigma}\cdot\vec{s}\right), \qquad (5.214)$$

to the solution for a particle at rest and spin in the z-direction. The defining relationship for such a state $w^r(\vec{p})$ is

$$\vec{\Sigma} \cdot \vec{s}\omega^r(\vec{p}) = \omega^r(\vec{p}), \qquad (5.215)$$

where the spinor $\omega^r(\vec{p})$ corresponds to a particle polarized along the direction of the unit vector $\vec{s}$. Remember that the spin operator is $\frac{\hbar}{2}\vec{\Sigma}$ and thus the eigenvalues of the spin operator in this case are ± 1.

Let $u(p,s)$ denote a spinor of positive energy, momentum p^μ, and spin s^μ. The spin vector is defined as

$$s^\mu = \Lambda^\mu{}_\nu \check{s}^\nu, \qquad (5.216)$$

where $\check{s}^\nu = (0, \hat{s})$ is the spin unit vector in the rest frame and $\Lambda^\mu{}_\nu$ is a Lorentz transformation from the rest frame. Thus we also have $p^\mu = \Lambda^\mu{}_\nu \check{p}^\nu$, where $\check{p}^\nu = (m, 0)$. This tells us that

$$s \cdot s = -1 \qquad (5.217)$$

and

$$p \cdot s = 0, \qquad (5.218)$$

which are true in any frame since they are Lorentz scalars. Thus in the rest frame,

$$\vec{\Sigma} \cdot \check{s}u(\check{p}, \check{s}) = u(\check{p}, \check{s}). \qquad (5.219)$$

Let $v(p,s)$ denote a spinor of negative energy, with spin $-\check{s}$ in the rest frame. In this case,

$$\vec{\Sigma} \cdot \check{s}v(\check{p}, \check{s}) = -v(\check{p}, \check{s}). \qquad (5.220)$$

In an arbitrary Lorentz frame, we define

$$\begin{aligned}
\omega^1(\vec{p}) &= u(p, u_z), \\
\omega^2(\vec{p}) &= u(p, -u_z), \\
\omega^3(\vec{p}) &= v(p, -u_z), \\
\omega^4(\vec{p}) &= v(p, u_z), \qquad (5.221)
\end{aligned}$$

where u_z^μ is a four-vector, which in the rest frame is $\check{u}_z^\mu = (0, \check{u}_z) = (0, 0, 0, 1)$. An arbitrary spinor is thus specified by the sign of the energy, its momentum p^μ, and spin in the rest frame $\check{s}^\mu$.

5.14 Projection Operators for Energy and Spin

5.14.1 Energy Projection Operators

The positive- and negative-energy projection operators can be guessed from the Dirac equation in momentum space (equation 5.197):

$$\Lambda_r(p) = \frac{\epsilon_r \not{p} + mc}{2mc} \quad \text{or} \quad \Lambda_{\pm}(p) = \frac{\pm \not{p} + mc}{2mc}. \qquad (5.222)$$

Applying the "trial" positive-energy projection operator to an arbitrary Dirac spinor and using equation 5.197, we have

$$\Lambda_+(p)w^r(p) = \frac{\not{p} + mc}{2mc} w^r(p) = \frac{\epsilon_r mc + mc}{2mc} w^r(p) = \begin{cases} w^r(p) & \text{for } r = 1,2, \\ 0 & \text{for } r = 3,4, \end{cases} \qquad (5.223)$$

and similarly for the negative-energy operator, as required.

Applying the energy operator twice we have

$$\Lambda_r \Lambda_{r'} = \frac{\epsilon_r \epsilon_{r'} \not{p} \not{p} + (\epsilon_r + \epsilon_{r'}) \not{p} mc + m^2 c^2}{4m^2 c^2}. \qquad (5.224)$$

Since $\not{p} \not{p} = p_\mu p_\nu \gamma^\mu \gamma^\nu = 1/2 p_\mu p_\nu (\gamma^\mu \gamma^\nu + \gamma^\nu \gamma^\mu) = p_\mu p_\nu g^{\mu\nu} = p^2 = m^2 c^2$,

$$\Lambda_r \Lambda_{r'} = \frac{(\epsilon_r + \epsilon_{r'}) \not{p} + (1 + \epsilon_r \epsilon_{r'}) mc}{4mc}$$

$$= \left(\frac{1 + \epsilon_r \epsilon_{r'}}{2} \right) \Lambda_r. \qquad (5.225)$$

Therefore,

$$\Lambda_{\pm}^2(p) = \Lambda_{\pm}(p) \quad \text{and} \quad \Lambda_{\pm}(p) \Lambda_{\mp}(p) = 0 \qquad (5.226)$$

as required of projection operators. Also

$$\Lambda_+(p) + \Lambda_-(p) = \frac{\not{p} + mc}{2mc} + \frac{-\not{p} + mc}{2mc} = 1, \qquad (5.227)$$

as required.

5.14.2 Spin Projection Operators

To deduce the spin-projection operator, we go to the rest frame and try to find a projection operator in a covariant form. A candidate for a spin-up projection operator is $(1 + \sigma_z)/2$, where σ_z is the third Pauli-spin matrix. Removing the explicit z dependence we write

$$\frac{1 + \sigma \cdot \hat{u}}{2}, \tag{5.228}$$

where $\hat{u}$ is a unit three-vector. Extending the operator to four-dimensions in the rest frame, we have

$$\frac{1 + \Sigma_3 u_3}{2} = \frac{1 + i\gamma_1\gamma_2 u_3}{2} = \frac{1 + \gamma_5\gamma_3\gamma_0 u_3}{2} \rightarrow \frac{1 + \gamma_5 \not{u}\gamma_0}{2}, \tag{5.229}$$

where we have used $\Sigma_k = i\epsilon_{ijk}\gamma^i\gamma^j$, which follows from equation 5.56 and the definition of the gamma matrices in terms of the α_i and β matrices.

Because we are in the rest frame, γ_0 acting upon the Dirac spinor becomes ± 1 and we can deal with this overall sign later. The covariant Dirac spin-projection operator has the form

$$\Sigma(u_3) = \frac{1 + \gamma_5 \not{u}_3}{2}. \tag{5.230}$$

Notice that $\Sigma(u_3) \neq \Sigma_3$, and that they are different operators. For a general spin vector s^μ, we have

$$\boxed{\Sigma(s) = \frac{1 + \gamma_5 \not{s}}{2}}. \tag{5.231}$$

In the rest frame,

$$\Sigma(u_3)\omega^1(0) = \frac{1 + \gamma_5\gamma_3}{2}\omega^1(0) = \frac{1 + \Sigma_3\gamma_0}{2}\omega^1(0) = \omega^1(0), \tag{5.232}$$

$$\Sigma(-u_3)\omega^2(0) = \frac{1 - \gamma_5\gamma_3}{2}\omega^2(0) = \frac{1 - \Sigma_3\gamma_0}{2}\omega^2(0) = \omega^2(0), \tag{5.233}$$

$$\Sigma(-u_3)\omega^3(0) = \frac{1 - \gamma_5\gamma_3}{2}\omega^3(0) = \frac{1 - \Sigma_3\gamma_0}{2}\omega^3(0) = \omega^3(0), \tag{5.234}$$

$$\Sigma(u_3)\omega^4(0) = \frac{1 + \gamma_5\gamma_3}{2}\omega^4(0) = \frac{1 + \Sigma_3\gamma_0}{2}\omega^4(0) = \omega^4(0). \tag{5.235}$$

The physical motivation for this seemingly backward association of the eigenvalues for the negative-energy solution will be discussed in section 5.15 when we come to the hole theory.

Using the definition equation 5.221, we write

$$\Sigma(u_z)u(p, u_z) = u(p, u_z),$$
$$\Sigma(u_z)v(p, u_z) = v(p, u_z),$$
$$\Sigma(-u_z)u(p, u_z) = \Sigma(-u_z)v(p, u_z) = 0. \tag{5.236}$$

Because of the covariant form of the projection operator, we may write for any spin vector s^μ

$$\Sigma(s)u(p, s) = u(p, s),$$
$$\Sigma(s)v(p, s) = v(p, s),$$
$$\Sigma(-s)u(p, s) = \Sigma(-s)v(p, s) = 0. \tag{5.237}$$

We now show that the spin projection operators have projection operator properties:

$$\Sigma(\pm s)\Sigma(\pm s) = \frac{1 \pm \gamma_5 \not{s}}{2} \frac{1 \pm \gamma_5 \not{s}}{2} = \frac{1 \pm 2\gamma_5 \not{s} - \gamma_5^2 s^2}{4} = \Sigma(\pm s) \tag{5.238}$$

and

$$\Sigma(\pm s)\Sigma(\mp s) = \frac{1 \pm \gamma_5 \not{s}}{2} \frac{1 \mp \gamma_5 \not{s}}{2} = \frac{1 - \gamma_5 \not{s}\gamma_5 \not{s}}{4} = \frac{1 + \gamma_5^2 s^2}{4} = 0. \tag{5.239}$$

Also,

$$\Sigma(\pm s) + \Sigma(\mp s) = \frac{1 \pm \gamma_5 \not{s}}{2} + \frac{1 \mp \gamma_5 \not{s}}{2} = 1. \tag{5.240}$$

The energy and spin projection operators commute:

$$\begin{aligned}
[\Sigma(s), \Lambda_\pm(p)] &= \frac{1 + \gamma_5 \not{s}}{2} \frac{\pm \not{p} + mc}{2mc} - \Lambda_\pm(p)\Sigma(s) \\
&= \frac{\pm \not{p} + mc \pm \gamma_5 \gamma^\mu \gamma^\nu s_\mu p_\nu + mc\gamma_5 \not{s}}{4mc} - \Lambda_\pm(p)\Sigma(s) \\
&= \frac{\pm \not{p} + mc \pm \gamma_5(2g^{\mu\nu} - \gamma^\nu \gamma^\mu)s_\mu p_\nu + mc\gamma_5 \not{s}}{4mc} - \Lambda_\pm(p)\Sigma(s) \\
&= \frac{\pm \not{p} + mc \pm (2g^{\mu\nu}\gamma_5 + \gamma^\nu \gamma_5 \gamma^\mu)s_\mu p_\nu + mc\gamma_5 \not{s}}{4mc} - \Lambda_\pm(p)\Sigma(s) \\
&= \frac{\pm \not{p} + mc \pm (2\gamma_5 s \cdot p + \not{p}\gamma_5 \not{s}) + mc\gamma_5 \not{s}}{4mc} - \Lambda_\pm(p)\Sigma(s) \\
&= \frac{\pm \not{p} + mc + (\pm \not{p} + mc)\gamma_5 \not{s}}{4mc} - \Lambda_\pm(p)\Sigma(s) \\
&= \frac{(\pm \not{p} + mc)(1 + \gamma_5 \not{s})}{4mc} - \Lambda_\pm(p)\Sigma(s) \\
&= 0. \tag{5.241}
\end{aligned}$$

5.14.3 Projection Operators of Energy and Spin

We have found four operators which project from a given plane-wave solution of momentum $\vec{p}$, the four independent solutions corresponding to positive and negative energy, and to spin up and spin down along a given direction. These projection operators are covariant and satisfy

$$P_r(\vec{p})w^{r'}(\vec{p}) = \delta_{rr'}w^{r'}(\vec{p}), \qquad (5.242)$$

or equivalently,

$$P_r(\vec{p})P_{r'}(\vec{p}) = \delta_{rr'}P_r(\vec{p}). \qquad (5.243)$$

The four projection operators are:

$$
\begin{aligned}
P_1(\vec{p}) &= \Lambda_+(p)\Sigma(u_z), \\
P_2(\vec{p}) &= \Lambda_+(p)\Sigma(-u_z), \\
P_3(\vec{p}) &= \Lambda_-(p)\Sigma(-u_z), \\
P_4(\vec{p}) &= \Lambda_-(p)\Sigma(u_z).
\end{aligned}
\qquad (5.244)
$$

These projection operators are specific to the normalization $\bar{u}u = 1$ and $\bar{v}v = -1$ for the spinors (see page 86).

We shall rely upon these projection operators very frequently in developing rapid and efficient calculational techniques. They permit us to use closure methods, thus avoiding the necessity of writing out matrices and spinor solutions component by component.

5.15 Hole Theory

The energy-level diagram for solutions of the free-particle Dirac equation is shown in figure 5.1. At first sight, it may appear that the negative-energy solutions to the Dirac equation are a major problem with the theory. What stops a positive-energy electron from radiating a photon and falling into a negative-energy (lower-energy) state? This would create a "radiation catastrophe" since nature prefers the lowest energy state. Dirac provided a first solution to this problem by reinterpreting the vacuum state[10].

For our purposes, we define the vacuum to be the lowest energy state or ground state of the system. The vacuum state is the state with all negative-energy levels filled and all positive-energy levels empty, as show in figure 5.2.

---

[10]P.A.M. Dirac, "A Theory of Electrons and Protons", Proc. Roy. Soc. **126** (1930) 360-365.

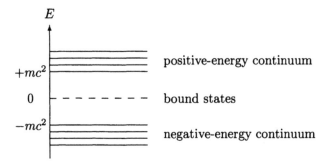

FIGURE 5.1: Spectrum of energy eigenvalues of the free-particle Dirac equation.

The vacuum is required to have all the negative-energy states occupied or the positive-energy electrons could fall into the lower-energy unoccupied states. The negative-energy levels are filled according to the Pauli exclusion principle. The stability of the system is assured since no more electrons can be accommodated in the negative-energy sea.

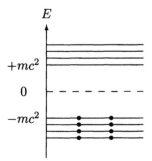

FIGURE 5.2: Negative-energy states are occupied (Dirac sea). This represents the vacuum state.

The most important result of the hole theory is that it was the first theory which introduced a model for the vacuum. The vacuum should be unobservable by having zero energy (mass) and no charge. However, it is clear that the model in this simple form does not have these properties. The states occupied with electrons of negative energy together have infinitely large negative energy and infinitely large negative charge.

The infinite negative energy and infinite negative charge have to be renormalized to zero, i.e. the zero point of energy and charge are chosen in such a way that the Dirac sea has no mass and no charge. The net charge of a given state must be defined with respect to the vacuum. Since a positive-energy electron has charge $-e$, a hole state behaves as if it has charge

$$Q_{\text{hole}} = [Q_{\text{vacuum}} - (-e)] - Q_{\text{vacuum}} = +e, \qquad (5.245)$$

i.e. the negative of the electron charge. In the above expression, Q_{vacuum} is infinite but subtraction of infinite quantities is quite normal for renormalization procedures. Similarly if the momentum of this (negative energy) state is $\vec{p}$, the hole, upon renormalization, will behave as if it has momentum

$$\vec{p}_{\text{hole}} = [\vec{p}_{\text{vacuum}} - \vec{p}] - \vec{p}_{\text{vacuum}} = -\vec{p}. \qquad (5.246)$$

In this case, we expect $\vec{p}_{\text{vacuum}} = 0$ since for each negative-energy state with momentum $\vec{p}$ there is another with momentum $-\vec{p}$. For the energy and spin we have

$$E_{\text{hole}} = [E_{\text{vacuum}} - (-E)] - E_{\text{vacuum}} = +E \qquad (5.247)$$

and

$$1/2\vec{\Sigma}_{\text{hole}} = [1/2\vec{\Sigma}_{\text{vacuum}} - 1/2\vec{\Sigma}] - 1/2\vec{\Sigma}_{\text{vacuum}} = -1/2\vec{\Sigma}, \qquad (5.248)$$

so that a hole state behaves as a positive-energy, positive-charge state of momentum $-\vec{p}$ and spin $-1/2\vec{\Sigma}$.

We now point out that the vacuum can be modified, for instance, by the influence of external fields. If the fields have energy less than $2mc^2$, the fields can cause a redistribution of the charge of the occupied negative-energy states. Hence the fields can produce a measurable vacuum polarization with respect to the state without external fields.

For strong fields with energy greater than $2mc^2$, it is possible to change the vacuum state by lifting a negative-energy electron from the Dirac sea into a positive-energy state. This situation corresponds to a negative-energy electron absorbing radiation and being exited into a positive-energy state (figure 5.3). If this occurs, we observe an electron of charge $-e$ and energy $+E$, and in addition a hole in the negative-energy sea. The hole registers the absence of an electron of charge $-e$ and energy $-E$, and would be interpreted by an observer relative to the vacuum as the presence of a particle of charge $+e$ and energy $+E$, that is, the positron. This is the hole-theory interpretation of pair production.

Correspondingly, a hole in the negative-energy sea, or a positron, is a trap for a positive-energy electron. The positive-energy electron will lower its energy state by emitting radiation and falling into the lower negative-energy hole. This appears to an observer relative to the vacuum as electron-positron pair annihilation with the emission of radiation (figure 5.4).

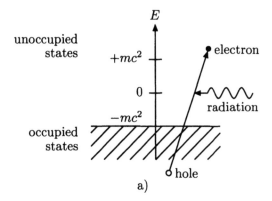

FIGURE 5.3: Pair production in the hole theory.

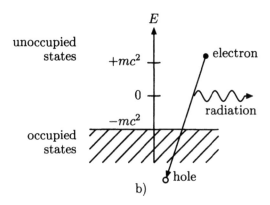

FIGURE 5.4: Pair annihilation in the hole theory.

We recognize that with the hole theory, we move to a many-particle theory describing particles of both signs of charge. No longer does the wave function have the simple probability interpretation of the one-particle theory, since it must now also account for the production and annihilation of electron-positron pairs.

But what about the negative-energy solution of the Klein-Gordon equation? Despite the success of the hole theory for spin-1/2 particles, the Dirac reinterpretation of the vacuum cannot be applied to spin-0 particles, since bosons are not subject to the Pauli exclusion principle. For this reason, we shall not use Dirac's hole theory interpretation for spin-1/2 particles. Instead we shall use a prescription for handling the negative-energy solutions which is due to Stückelberg and Feynman.

5.16   Charge Conjugation

To each particle there is an antiparticle and, in particular, the existence of electrons implies the existence of positrons. In the hole theory, the absence of an energy $-E$ and charge $-e$, is equivalent to the presence of a positron of positive energy $+E$ and charge $+e$. The positron must satisfy the Dirac equation with charge $-e$ replaced by $+e$. Thus there must be a transformation connecting the negative-energy solutions of positive charge with the positive-energy solutions of negative charge. This symmetry transformation is referred to as charge conjugation $\mathcal{C}$. Charge conjugation is a symmetry in nature which is independent of the Lorentz group.

For an electron of negative energy,

$$(i\,\partial\!\!\!/ - e\,A\!\!\!/ - m)\psi = 0, \tag{5.249}$$

where we have written the Dirac equation with an electromagnetic potential in a slightly different, but equivalent, form than equation 5.101.

For a positron with positive energy, we denote the wave function by ψ_C. It satisfies the Dirac equation

$$(i\,\partial\!\!\!/ + e\,A\!\!\!/ - m)\psi_C = 0. \tag{5.250}$$

We look for an operator transforming equations 5.249 and 5.250 into each other. Taking the complex conjugate of the Dirac equation 5.249, multiplying by -1, and remembering that A_μ is real, we have

$$\left[\left(i\frac{\partial}{\partial x^\mu} + eA_\mu\right)(\gamma^\mu)^* + m\right]\psi^* = 0. \tag{5.251}$$

We look for a nonsingular matrix $C\gamma^0$ such that

$$(C\gamma^0)(\gamma^\mu)^*(C\gamma^0)^{-1} = -\gamma^\mu. \tag{5.252}$$

This will cause equation 5.251 to be the same as equation 5.250. It is convenient to separate γ^0 from C. If we can find such an operator, we have

$$(i\,\partial\!\!\!/ + e\,A\!\!\!/ - m)(C\gamma^0\psi^*) = 0, \tag{5.253}$$

with

$$\psi_C = C\gamma^0\psi^* = C(\psi^\dagger\gamma^0)^T = C\overline{\psi}^T \tag{5.254}$$

being the positron wave function. In our representation for the gamma matrices,

$$\gamma^0(\gamma^\mu)^*\gamma^0 = (\gamma^\mu)^T \tag{5.255}$$

follows from equation 5.108, or explicitly,

for $\mu = i$, $\quad \gamma^0(\gamma^i)^*\gamma^0 = \gamma^0(\gamma^i\gamma^0)^* = -\gamma^0\gamma^0(\gamma^i)^* = -(\gamma^i)^* = (\gamma^{i\dagger})^* = (\gamma^i)^T$,

for $\mu = 0$, $\quad \gamma^0(\gamma^0)^*\gamma^0 = \gamma^0 = (\gamma^0)^T$.

Therefore, equation 5.252 becomes

$$(C\gamma^0)(\gamma^\mu)^*(C\gamma^0)^{-1} = C(\gamma^0\gamma^{\mu*}\gamma^0)C^{-1} = C(\gamma^\mu)^T C^{-1} = -\gamma^\mu \qquad (5.256)$$

or

$$\boxed{C^{-1}\gamma^\mu C = -(\gamma^\mu)^T} \, . \qquad (5.257)$$

Since in our representation

$$(\gamma^\mu)^T = \begin{cases} \gamma^\mu & \text{for} \quad \mu = 0, 2, \\ -\gamma^\mu & \text{for} \quad \mu = 1, 3, \end{cases} \qquad (5.258)$$

C must commute with γ^1 and γ^3, and anticommute with γ^0 and γ^2. Therefore, we try

$$\boxed{C = i\gamma^2\gamma^0} \, . \qquad (5.259)$$

This form of C is a standard for the charge conjugation matrix but it is not the only possible form. It suffices to be able to construct a matrix C in any given representation since a unitary transformation to any other representation when applied to this C will give a matrix appropriate to the new representation. We again note that there is a phase arbitrariness in our definition of C.

The properties of C are

$$C^{-1} = i\gamma^0\gamma^2 = -i\gamma^2\gamma^0 = -C, \qquad (5.260)$$
$$C^\dagger = -i\gamma^0(\gamma^2)^\dagger = i\gamma^0\gamma^2 = -i\gamma^2\gamma^0 = -C, \qquad (5.261)$$
$$C^T = i(\gamma^0)^T(\gamma^2)^T = i\gamma^0\gamma^2 = -i\gamma^2\gamma^0 = -C. \qquad (5.262)$$

In our representation, the charge conjugate solution is

$$\psi_C = C\overline{\psi}^T = C\gamma^0\psi^* = i\gamma^2\gamma^0\gamma^0\psi^* = i\gamma^2\psi^*. \qquad (5.263)$$

Let's consider an example. For a negative-energy electron at rest with spin down

$$\psi^4 = \frac{1}{(2\pi)^{3/2}} \begin{pmatrix} 0 \\ 0 \\ 0 \\ 1 \end{pmatrix} e^{imt}. \tag{5.264}$$

The charge conjugate solution is

$$i\gamma^2(\psi^4)^* = i \begin{pmatrix} 0 & 0 & 0 & -i \\ 0 & 0 & i & 0 \\ 0 & i & 0 & 0 \\ -i & 0 & 0 & 0 \end{pmatrix} \begin{pmatrix} 0 \\ 0 \\ 0 \\ 1 \end{pmatrix} \frac{e^{-imt}}{(2\pi)^{3/2}} = \frac{1}{(2\pi)^{3/2}} \begin{pmatrix} 1 \\ 0 \\ 0 \\ 0 \end{pmatrix} e^{-imt} = \psi^1, \tag{5.265}$$

which is a positive-energy state at rest with spin up. In terms of the hole theory, the absence of a spin-down negative-energy electron at rest is equivalent to the presence of a spin-up positive-energy positron at rest.

Noticing that $[C, \gamma_5] = i(\gamma^2\gamma^0\gamma^5 - \gamma^5\gamma^2\gamma^0) = 0$ and recalling that $\gamma_5^\dagger = \gamma_5 = \gamma_5^*$, we apply the transformation to an arbitrary spin-momentum eigenstate $\psi_{\epsilon ps}$:

$$\begin{aligned}
(\psi_{\epsilon ps})_C = C\overline{\psi}_{\epsilon ps}^T = C\gamma_0\psi_{\epsilon ps}^* &= C\gamma_0 \left(\frac{\epsilon\not{p}+m}{2m}\right)^* \left(\frac{1+\gamma_5\not{s}}{2}\right)^* \psi^* \\
&= C \left(\frac{\epsilon\gamma_0\not{p}^*\gamma_0 + m}{2m}\right) \left(\frac{1+\gamma_0\gamma_5\not{s}^*\gamma_0}{2}\right) \gamma_0\psi^* \\
&= C \left(\frac{\epsilon\not{p}^T + m}{2m}\right) \left(\frac{1-\gamma_5\not{s}^T}{2}\right) \gamma_0\psi^* \\
&= \left(\frac{-\epsilon\not{p}+m}{2m}\right) \left(\frac{1+\gamma_5\not{s}}{2}\right) C\gamma_0\psi^* \\
&= \left(\frac{-\epsilon\not{p}+m}{2m}\right) \left(\frac{1+\gamma_5\not{s}}{2}\right) \psi_C. \tag{5.266}
\end{aligned}$$

We see that the operation yields, from a negative-energy solution described by four-vector p_μ and spin s_μ, a positive-energy solution with the same p_μ and s_μ.

In terms of free-particle spinors

$$C\overline{u}^T(p, s) = v(p, s), \tag{5.267}$$
$$C\overline{v}^T(p, s) = u(p, s). \tag{5.268}$$

Therefore, $v(p, s)$ and $u(p, s)$ are charge-conjugate spinors of each other, within a phase factor which may depend on p and s. Notice that s does not change sign under charge conjugation but the spin does reverse.

The charge-conjugate operator applied to a negative-energy solution of the Dirac equation describes a positive-energy particle with identical mass and spin but of opposite charge. If ψ describes a Dirac particle with charge $-e$ in the potential A^μ, then ψ_C describes a particle with charge $+e$ in the same potential A^μ. The Dirac equation is thus invariant under the combined transformation of

$$\psi \to \psi_C = C\overline{\psi}^T, \tag{5.269}$$
$$A^\mu \to A^\mu_C = -A^\mu. \tag{5.270}$$

This is a formal symmetry of the Dirac theory. It transforms the Dirac equation for an electron into the same equation for a positron.

Both ψ and ψ_C propagate forward in time. If a spin-1/2 particle is its own antiparticle, it is called a Majorana fermion. Currently the neutrino is the only possible candidate for such a particle from the known elementary particles.

5.17 Time Reversal

Now consider time-reversal: $t \to t' = -t$. This time we start with the Dirac equation in Hamiltonian form,

$$i\frac{\partial \psi(\vec{x},t)}{\partial t} = H\psi = [\vec{\alpha} \cdot (-i\vec{\nabla} - e\vec{A}) + \beta m + e\phi]\psi(\vec{x},t). \tag{5.271}$$

Defining the transformation $\mathcal{T}$ such that $t' = -t$ and $\psi'(t') = \mathcal{T}\psi(t)$, we have

$$\frac{\partial}{\partial t}(\mathcal{T}i\mathcal{T}^{-1})\psi'(t') = \mathcal{T}H\mathcal{T}^{-1}\psi'(t'). \tag{5.272}$$

Since $\vec{A}$ is generated by currents which reverse sign when the sense of time is reversed,

$$\vec{A}(t) \to \vec{A}'(t') = -\vec{A}(-t), \tag{5.273}$$
$$\phi(t) \to \phi'(t') = +\phi(-t). \tag{5.274}$$

Also $\vec{\nabla}' = +\vec{\nabla}$, since $\vec{x}' = +\vec{x}$; time reversal does not effect the three-space coordinates. The transformation must cause $i \to -i$ to get the correct form, therefore $\mathcal{T}$ can be defined as:

1. take complex conjugate,

2. multiply by the 4×4 constant matrix T.

These rules give

$$\psi'(t') = T\psi^*(t). \tag{5.275}$$

Therefore,

$$i\frac{\partial \psi'(t')}{\partial t'} = [(-T\vec{\alpha}^* T^{-1}) \cdot (-i\vec{\nabla}' - e\vec{A}'(t')) + (T\beta^* T^{-1})m + e\phi'(t')]\psi'(t'). \tag{5.276}$$

This implies T must commute with α_2 and β, and anticommute with α_1 and α_3 in our representation. Therefore, we can try

$$\boxed{T = -i\alpha_1\alpha_3 = -i\alpha_1\gamma^0\gamma^0\alpha_3 = i\gamma^0\alpha_1\gamma^0\alpha_3 = i\gamma^1\gamma^3}. \tag{5.277}$$

The phase factor is arbitrary.

We apply T to a plane-wave solution for a free particle of positive energy. Since

$$T \not{p} = T \not{p}^* = i\gamma^1\gamma^3(\gamma^\mu)^* p_\mu = \begin{cases} i\gamma^\mu\gamma^1\gamma^3 p_\mu \text{ for } \mu = 0, \\ -i\gamma^\mu\gamma^1\gamma^3 p_\mu \text{ for } \mu = 1,2,3, \end{cases} \tag{5.278}$$

$$T\left(\frac{\not{p}+m}{2m}\right)\left(\frac{1+\gamma_5\not{s}}{2}\right)\psi(t) = T\left(\frac{\not{p}^*+m}{2m}\right)T^{-1}T\left(\frac{1+\gamma_5\not{s}^*}{2}\right)T^{-1}\psi'(t')$$

$$= \left(\frac{\not{p}'+m}{2m}\right)\left(\frac{1+\gamma_5\not{s}'}{2}\right)\psi'(t'), \tag{5.279}$$

where $p' = (p_0, -\vec{p})$ and $s' = (s_0, -\vec{s})$. Therefore T projects out a free-particle solution with reversed direction of momentum $\vec{p}$ and spin $\vec{s}$. This is known as "Wigner time reversal".

5.18 Combined CPT Symmetry

In the previous sections, we have shown how one can define in a completely general manner the charge conjugation C, parity P, and time reversal T transformations. In section 3.1.1, we have also briefly discussed the conditions a theory of interactions must satisfy in order to be invariant under these transformations. In this section, we examine the combined effect of these three transformations.

Acting on a Dirac spinor wave function, we found that C transforms a fermion with a given spin orientation into an antifermion with the same spin orientation, $\mathcal{P}$ reverses the momentum without flipping the spin, and $\mathcal{T}$ reverses the momentum and also flips the spin:

$$\mathcal{C}\psi(\vec{x}, t) = C\gamma^0\psi^*(\vec{x}, t) = i\gamma^2\psi^*(\vec{x}, t),$$
$$\mathcal{P}\psi(\vec{x}, t) = P\psi(\vec{x}, t) = e^{i\phi}\gamma^0\psi(\vec{x}, t),$$
$$\mathcal{T}\psi(\vec{x}, t) = T\psi^*(\vec{x}, t) = i\gamma^1\gamma^3\psi^*(\vec{x}, t). \tag{5.280}$$

Combine all three symmetries, $\mathcal{C}$, $\mathcal{P}$, and $\mathcal{T}$, we obtain

$$
\begin{aligned}
\psi_{\mathcal{CPT}}(x') &\equiv \mathcal{CPT}\psi(x) \\
&= i\gamma^2(PT\psi^*(x))^* \\
&= i\gamma^2 e^{-i\phi}\gamma^0(i\gamma^1\gamma^3)^*\psi(x) \\
&= e^{-i\phi}\gamma^2\gamma^0\gamma^1\gamma^3\psi(x) \\
&= -ie^{-i\phi}(i\gamma^0\gamma^1\gamma^2\gamma^3)\psi(x) \\
&= -ie^{-i\phi}\gamma_5\psi(x),
\end{aligned}
\tag{5.281}
$$
$$\tag{5.282}$$

with $x'_\mu = -x_\mu$.

For a positive-energy momentum spin eigenstate, we have

$$
\begin{aligned}
\psi_{\mathcal{CPT}}(x') &= -ie^{-i\phi}\gamma_5\left(\frac{\slashed{p}+m}{2m}\right)\left(\frac{1+\gamma_5\slashed{s}}{2}\right)\psi(x) \\
&= \left(\frac{-\slashed{p}+m}{2m}\right)\left(\frac{1-\gamma_5\slashed{s}}{2}\right)(-ie^{-i\phi}\gamma_5)\psi(x). \tag{5.283}
\end{aligned}
$$

This is a negative-energy momentum spin eigenstate running backwards in space-time, and multiplied by the phase factor $-ie^{-i\phi}\gamma_5$. The Dirac wave function $\psi_{\mathcal{CPT}}$ can represent the positron. Thus positrons are negative-energy electrons running backwards in space-time. $\mathcal{CPT}$ turns incoming particles into outgoing antiparticles, while flipping the spin. This is the basis of the Stückelberg-Feynman form of positron theory that we will exploit in chapter 6.

The free-particle Dirac Hamiltonian is invariant under $\mathcal{C}$, $\mathcal{P}$, and $\mathcal{T}$ separately, and thus is also invariant under the combination of $\mathcal{CPT}$. How does $\mathcal{CPT}$ work in the presence of electromagnetic fields? For an arbitrary solution to the Dirac equation in the presence of electromagnetic fields, the negative-energy eigenvalue equation is

$$[\vec{\alpha}\cdot(-i\vec{\nabla} - e\vec{A}) + \beta m + e\phi]\psi = -E\psi. \tag{5.284}$$

Carrying out the $\mathcal{CPT}$ transformation gives

$$[\vec{\alpha}\cdot(-i\vec{\nabla}' + e\vec{A}'(x')) + \beta m - e\phi'(x')]\psi_{\mathcal{CPT}}(x') = +E\psi_{\mathcal{CPT}}(x'). \tag{5.285}$$

Since

$$\phi'(\vec{x}', t') = \mathcal{P}\phi(\vec{x}, t) = \phi(\vec{x}, t),$$
$$\vec{A}'(\vec{x}', t') = \mathcal{P}\vec{A}(\vec{x}, t) = -\vec{A}(\vec{x}, t),$$
$$\phi'(\vec{x}', t') = \mathcal{T}\phi(\vec{x}, t) = \phi(\vec{x}, t),$$
$$\vec{A}'(\vec{x}', t') = \mathcal{T}\vec{A}(\vec{x}, t) = -\vec{A}(\vec{x}, t), \qquad (5.286)$$

$A'_\mu(x') = +A_\mu(x)$ under $\mathcal{PT}$ or $\mathcal{CPT}$. Thus $\mathcal{CPT}$ is a symmetry that changes the Dirac equation with electromagnetic interactions for an electrons of charge $-e$ to an identical equation for positrons of charge $+e$. The mass of an antiparticle is identical to that of the particle in question. For stable particles this is a consequence of the mass being an eigenvalue (in the rest frame) of the simultaneously commuting operator $\hat{H}$ and $\mathcal{CPT}$. A proof for unstable particles follows from the $\mathcal{CPT}$ invariance of the S-matrix[11].

There may be interactions – not electromagnetic – which are not invariant under $\mathcal{P}$ and/or $\mathcal{C}$ and/or $\mathcal{T}$. Consequently, the various wave functions associated with the various kinds of particles need not be definite covariant representations of the corresponding transformations. Nevertheless, the operations $\mathcal{P}$, $\mathcal{C}$, and $\mathcal{T}$ must certainly have a definite meaning. It turns out that one cannot build a Lorentz-invariant quantum theory with a hermitian Hamiltonian that violates combined $\mathcal{CPT}$. The $\mathcal{CPT}$ theorem claims that if a theory of interacting particles is invariant under the proper Lorentz group, then it will be automatically invariant under the combination of the successive applications of particle-antiparticle conjugation, space inversion, and time reflection.

Other consequences of $\mathcal{CPT}$ symmetry are that the lifetime of an unstable antiparticle is identical to that of the corresponding particle, regardless of the dynamical interaction causing the decay. And likewise, the cross section for particle and antiparticle reactions are equal.

5.19   Free-Particle Solutions and Wave Packets

When dealing with wave packets some new features which were absent in the nonrelativistic theory will arise from the fact that there are negative-energy solutions. It is not possible to exclude the negative-energy states simply by arguing that they are not realized in nature. The positive energy states alone do not represent a complete set of functions. Because of the completeness of plane-wave solutions, we may superimpose plane waves to construct localized

---

[11] The concept of the S-matrix will be discussed in section 6.9.

wave packets. The wave packets are also solutions to the Dirac equation since
it is linear.

A positive-energy wave-packet solution can be written as

$$\psi^{(+)}(\vec{x}, t) = \int \frac{d^3p}{(2\pi\hbar)^{3/2}} \sqrt{\frac{mc^2}{E}} \sum_{\pm s} b(p, s)u(p, s)e^{-ip\cdot x/\hbar}, \qquad (5.287)$$

where $b(p, s)$ is a complex scalar function. Normalizing to unit probability
gives

$$\int d^3x \psi^{(+)\dagger}(\vec{x}, t)\psi^{(+)}(\vec{x}, t)$$

$$= \int \frac{d^3x d^3p d^3p'}{(2\pi\hbar)^3} \sqrt{\frac{m^2c^4}{EE'}} \sum_{\pm s, \pm s'} b^*(p, s)b(p', s')u^\dagger(p, s)u(p', s')e^{i(p-p')\cdot x/\hbar}$$

$$= \int d^3p \frac{mc^2}{E} \sum_{\pm s, \pm s'} b^*(p, s)b(p, s')u^\dagger(p, s)u(p, s')$$

$$= \int d^3p \frac{mc^2}{E} \sum_{\pm s, \pm s'} b^*(p, s)b(p, s') \frac{E}{mc^2}\delta_{ss'}$$

$$= \int d^3p \sum_{\pm s} |b(p, s)|^2 = 1. \qquad (5.288)$$

The average current for such a wave packet is given by the expectation
value of the velocity operator

$$\vec{j}^{(+)} = \int d^3x \psi^{(+)\dagger} c\vec{\alpha}\psi^{(+)}$$

$$= \int d^3x \overline{\psi}^{(+)} c\vec{\gamma}\psi^{(+)}$$

$$= \int \frac{d^3x d^3p d^3p'}{(2\pi\hbar)^3} \sqrt{\frac{m^2c^4}{EE'}} \sum_{\pm s, \pm s'} b^*(p, s)b(p', s')e^{i(p-p')\cdot x/\hbar}$$

$$\cdot \bar{u}(p, s)c\vec{\gamma}u(p', s'). \qquad (5.289)$$

To proceed further, the space part of the Gordon decomposition (see problem 5.31),

$$c\overline{\psi}_2\gamma^\mu\psi_1 = \frac{1}{2m}[\overline{\psi}_2\hat{p}^\mu\psi_1 - (\hat{p}^\mu\overline{\psi}_2)\psi_1] - \frac{i}{2m}\hat{p}_\nu(\overline{\psi}_2\sigma^{\mu\nu}\psi_1), \qquad (5.290)$$

can be used. The Gordon decomposition decomposes the Dirac current density
into a convection current-density term (first term in equation 5.290), similar to

the nonrelativistic case, and an additional spin current-density term (second term in equation 5.290). We now write

$$j_k^{(+)} = \int \frac{d^3x\, d^3p\, d^3p'}{(2\pi\hbar)^3} \sqrt{\frac{m^2c^4}{EE'}} \sum_{\pm s, \pm s'} b^*(p,s) b(p',s') e^{i(p-p')\cdot x/\hbar}$$

$$\cdot\, \bar{u}(p,s) \frac{1}{2m} \left[(p_k + p_k') + \sigma_k^{\nu}(p_\nu - p_\nu') \right] u(p',s'),$$

$$\vec{j}^{(+)} = \int d^3p\, \frac{\vec{p}c^2}{E} \sum_{\pm s} |b(p,s)|^2. \tag{5.291}$$

Using the normalization condition, the current can be written

$$\vec{j}^{(+)} = \langle c\vec{\alpha} \rangle_+ = \left\langle \frac{c^2\vec{p}}{E} \right\rangle. \tag{5.292}$$

Thus, the average current for an arbitrary wave packet formed of positive-energy solutions is just the classical group velocity. In the Schrödinger theory, the velocity operator $\hat{v} = \hat{p}/m$ is proportional to the momentum, but this is not the case in the Dirac theory. In the Dirac theory the velocity operator for a free particle $c\hat{\alpha}$ is no longer a constant. We see that wave packets consisting of plane waves with only positive energy have the expectation value of the velocity $|\langle c\hat{\alpha}\rangle| \sim |\langle c^2\vec{p}/E\rangle| < c$, whereas the eigenvalues of $c\hat{\alpha}$ are exactly $\pm c$. This motivates us to consider wave packets containing both positive and negative-energy solutions:

$$\psi(\vec{x},t) = \int \frac{d^3p}{(2\pi\hbar)^{3/2}} \sqrt{\frac{mc^2}{E}} \sum_{\pm s} [b(p,s)u(p,s)e^{-ip\cdot x/\hbar} + d^*(p,s)v(p,s)e^{+ip\cdot x/\hbar}],$$

$$\tag{5.293}$$

where $b(p,s)$ and $d^*(p,s)$ are complex scalar functions. Normalizing to unit probability, we have

$$\int d^3x\, \psi^\dagger(\vec{x},t)\psi(\vec{x},t)$$

$$= \int \frac{d^3x\, d^3p\, d^3p'}{(2\pi\hbar)^3} \sqrt{\frac{m^2c^4}{EE'}} \sum_{\pm s, \pm s'} \Big[b^*(p,s)b(p',s')u^\dagger(p,s)u(p',s')e^{i(p-p')\cdot x/\hbar}$$

$$+\, b^*(p,s)d^*(p',s')u^\dagger(p,s)v(p',s')e^{-i(p+p')\cdot x/\hbar}$$

$$+\, d(p,s)b(p',s')v^\dagger(p,s)u(p',s')e^{i(p+p')\cdot x/\hbar}$$

$$+\, d(p,s)d^*(p',s')v^\dagger(p,s)v(p',s')e^{-i(p-p')\cdot x/\hbar} \Big]$$

$$= \int d^3p\, \frac{mc^2}{E} \sum_{\pm s, \pm s'} [b^*(p,s)b(p,s')u^\dagger(p,s)u(p,s')$$

$$+ \, b^*(p,s)d^*(-p,s')u^\dagger(p,s)v(-p,s')e^{-2ip_0x_0/\hbar}$$
$$+ \, d(p,s)b(-p,s')v^\dagger(p,s)u(-p,s')e^{2ip_0x_0/\hbar}$$
$$+ \, d(p,s)d^*(p,s')v^\dagger(p,s)v(p,s')]$$

$$= \int d^3p \frac{mc^2}{E} \sum_{\pm s, \pm s'} \left[b^*(p,s)b(p,s')\frac{E}{mc^2}\delta_{ss'} + d(p,s)d^*(p,s')\frac{E}{mc^2}\delta_{ss'} \right]$$

$$= \int d^3p \sum_{\pm s} \left[|b(p,s)|^2 + |d(p,s)|^2 \right] = 1. \tag{5.294}$$

A short calculation (see problem 5.32) shows the current of the wave packet is

$$j^k = \int d^3x \psi^\dagger(\vec{x},t) c\hat{\alpha}_k \psi(\vec{x},t)$$

$$= \int d^3p \left\{ \sum_{\pm s} \left[|b(p,s)|^2 + |d(p,s)|^2 \right] \frac{p^k c^2}{E} \right.$$

$$+ c \sum_{\pm s, \pm s'} b^*(-p,s)d^*(p,s')e^{2ix_0p_0/\hbar}\overline{u}(-p,s)\sigma^{k0}v(p,s')$$

$$\left. - ic \sum_{\pm s, \pm s'} b(-p,s)d(p,s')e^{-2ix_0p_0/\hbar}\overline{v}(p,s)\sigma^{k0}u(-p,s') \right\}. \tag{5.295}$$

The first term represents the time-independent group velocity that appeared before in equation 5.291 for positive-energy only wave packets. The second and third terms are the interferences of the solutions with positive and negative energy, which oscillate time-dependently because of the factors $\exp(\pm 2ip_0x_0/\hbar)$. The frequency of this Zitterbewegung motion is

$$\frac{2p_0c}{\hbar} > \frac{2mc^2}{\hbar} \approx 2 \times 10^{21} \text{ s}^{-1} = 2 \times 10^{12} \text{ GHz}, \tag{5.296}$$

which is very large, and its strength is proportional to the amplitude $d(p,s)$ of the waves with negative energy in the wave packet.

5.20   Klein Paradox for Spin-1/2 Particles

Consider the scattering of an electron of energy E and momentum $p = p_z$ by an electrostatic step-potential of the form (figure 5.5)

$$e\phi = \begin{cases} V_0 & \text{for } z > 0, \\ 0 & \text{for } z < 0. \end{cases} \tag{5.297}$$

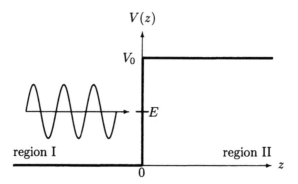

FIGURE 5.5: Electrostatic potential idealized with a sharp boundary, with an incident free-electron wave of energy E moving to the right in region I.

For a free electron, we have $(E/c)^2 = p^2 + m^2c^2$, whereas in the presence of the constant potential

$$\left(\frac{E - V_0}{c}\right)^2 = (p')^2 + m^2c^2, \qquad (5.298)$$

where p' is the momentum of the particle inside the potential.

The Dirac equation for $z < 0$ is

$$(c\alpha\hat{p} + \beta mc^2)\psi = E\psi, \qquad (5.299)$$

while for $z > 0$, we have

$$(c\alpha\hat{p} + \beta mc^2)\psi = (E - V_0)\psi, \qquad (5.300)$$

where it is understood that $\alpha = \alpha_3$ and $\hat{p} = \hat{p}_z$.

The incident wave in region I is

$$\psi_I = a \begin{pmatrix} 1 \\ 0 \\ \frac{pc}{E+mc^2} \\ 0 \end{pmatrix} e^{ipz/\hbar}, \qquad (5.301)$$

with $pc = \sqrt{E^2 - m^2c^4}$. The reflected wave in region I is

$$\psi_R = b \begin{pmatrix} 1 \\ 0 \\ \frac{-pc}{E+mc^2} \\ 0 \end{pmatrix} e^{-ipz/\hbar} + b' \begin{pmatrix} 0 \\ 1 \\ 0 \\ \frac{pc}{E+mc^2} \end{pmatrix} e^{-ipz/\hbar}. \qquad (5.302)$$

The transmitted wave in region II is

$$\psi_T = d \begin{pmatrix} 1 \\ 0 \\ \frac{p'c}{E-V_0+mc^2} \\ 0 \end{pmatrix} e^{ip'z/\hbar} + d' \begin{pmatrix} 0 \\ 1 \\ 0 \\ \frac{-p'c}{E-V_0+mc^2} \end{pmatrix} e^{ip'z/\hbar}, \qquad (5.303)$$

with $p'c = \sqrt{(V_0 - E)^2 - m^2 c^4}$.

From the Klein-Gordon analysis of the Klein Paradox (section 4.10), the interesting energy region is the case of $V_0 > E + mc^2$ (strong field) for which the momentum p' is real and allows the free plane wave to propagate in region II. Continuity at the boundary requires $\psi_I + \psi_R = \psi_T$, and thus

$$a + b = d, \qquad (5.304)$$
$$b' = d', \qquad (5.305)$$
$$\frac{pc}{E+mc^2}a - \frac{pc}{E+mc^2}b = \frac{-p'c}{V_0 - E - mc^2}d, \qquad (5.306)$$
$$\frac{pc}{E+mc^2}b' = \frac{p'c}{V_0 - E - mc^2}d'. \qquad (5.307)$$

The equations for b' and d' can only be satisfied if $b' = d' = 0$. There is thus no spin flip of the electron at the boundary. Also,

$$a - b = -\frac{p'}{p}\frac{E+mc^2}{V_0 - E - mc^2}d$$
$$= -\sqrt{\frac{(V_0 - E + mc^2)(E+mc^2)}{(V_0 - E - mc^2)(E - mc^2)}}d$$
$$\equiv -rd. \qquad (5.308)$$

We thus have

$$a = \frac{d}{2}(1 - r),$$
$$b = \frac{d}{2}(1 + r),$$
$$\frac{b}{a} = \frac{1+r}{1-r}, \qquad (5.309)$$
$$\frac{d}{a} = \frac{2}{1-r}. \qquad (5.310)$$

The particle current is given by

$$j(x) = c\psi^\dagger(x)\alpha_3\psi(x). \qquad (5.311)$$

We calculate the currents:

$$j_I = aa^* \frac{2pc^2}{E + mc^2}, \tag{5.312}$$

$$j_R = -bb^* \frac{2pc^2}{E + mc^2}, \tag{5.313}$$

$$j_T = -dd^* \frac{2p'c^2}{V_0 - E - mc^2}. \tag{5.314}$$

Since r is real, the ratio of currents is

$$\frac{j_R}{j_I} = -\frac{bb^*}{aa^*} = -\frac{(1+r)^2}{(1-r)^2}, \tag{5.315}$$

$$\begin{aligned} \frac{j_T}{j_I} &= -\frac{p'}{p} \frac{E + mc^2}{V_0 - E - mc^2} \frac{dd^*}{aa^*} \\ &= -\frac{4r}{(1-r)^2}. \end{aligned} \tag{5.316}$$

Since $r > 1$, we see that $|j_R| > |j_I|$. This result corresponds to the fact that the flow of j_T is in the $-z$-direction, i.e. the electrons are leaving region II, but according to our assumptions up to now, there are no electrons in region II. A reinterpretation is thus necessary.

To prevent the transition of all electrons to states of negative energy, one has to require that all electron states $E < -mc^2$ are occupied with electrons. The potential $V_0 > mc^2 + E$ raises the electron energy in region II sufficiently for there to be an overlap between the negative-energy continuum for $z > 0$ and the positive-energy continuum for $z < 0$, as shown in figure 5.6. In the case of $V_0 > E + mc^2$, the electrons striking the potential barrier from the left are able to knock additional electrons out of the vacuum on the right, leading to a positron current flowing from left to right in the potential region. It is possible to understand the sign of j_T by assuming that the electrons entering region I are coming from the negative continuum.

$$j_I + j_R = j_I \left[1 - \frac{(1+r)^2}{(1-r)^2} \right] = -\frac{4r}{(1-r)^2} j_I = j_T. \tag{5.317}$$

Since the holes remaining in region II are interpreted as positrons, the phenomena can be understood as electron-positron pair creation at the potential barrier, and is the process responsible for the decay of the vacuum in the presence of a strong field.

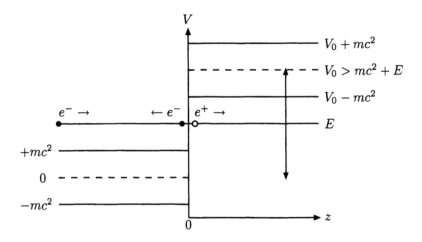

FIGURE 5.6: Klein Paradox interpretation using hole theory.

5.21 Summary

Most of the relativistic effects encountered with the Klein-Gordon equation in chapter 4 have reemerged in the contexts of the Dirac equation in this chapter, such as, the non-localizability of positive-energy-only wave packets and the particle-antiparticle symmetry. The spin-1/2 structure of the solutions to the Dirac equation have given rise to additional new effects.

Nevertheless, problems arise from attempting to apply a simple single-particle wave-function picture to what is obviously a many-body situation. The correct way in which to handle all the subtlety of these problems is to use the formalism of quantum field theory. But the elementary wave function paradigm has allowed us to obtain an accurate sketch of the physics of spin-1/2 particles within the limitations of a one-particle theory.

The hole theory permits the reconciliation of the Dirac theory with the experimental facts: the non-existence of negative-energy states, the existence of positrons, and the creation and annihilation of pairs. It therefore constitutes a considerable step forward. However, it has a number of limitations and difficulties.

First of all, it is incomplete. By postulating the occupation of the quasi-totality of negative-energy states, the theory ceases to be a one-particle theory, even when it sets out to describe a single electron. The formalism of the Dirac theory of a single particle, as set forth, is therefore insufficient for describing such a situation, and it is only in the framework of field theory that one can hope to obtain a self-consistent description.

The hole theory is only a first step in the direction of a correct theory of the quantized electron field. It has the merit of providing a simple picture and can therefore serve as a guide in the elaboration of the correct theory. But pitfalls and contradictions appear when it is pushed too far.

For example, having defined the vacuum as composed of an infinite number of electrons, it is inconsistent to assume that these electrons do not interact. Another weak point of the theory is the apparently very asymmetrical role played by the electrons and the positrons. One can also construct a corresponding charge-conjugate theory where the positrons play the role of the particles and the electrons that of the holes, without any of the physical consequences being changed.

Unlike the situation for fundamental scalar particles, there exists fundamental spin-1/2 particles in nature: the leptons. The three charged leptons: electron, muon, tau, and their antiparticles are believed to be exactly described by the Dirac equation. The corresponding three neutrinos have recently also been added to the set of particles described by the Dirac equation.

The remainder of this book will develop the theory of interacting spin-1/2 particles. This will enable us to calculate cross sections for the scattering of spin-1/2 particles with spin-1/2 particles and with the electromagnetic field. The theory is called quantum electrodynamics. The importance of this theory is far-reaching and has inspired elements of the theories of the strong and weak nuclear forces.

5.22 Problems

1. Show that the Dirac equation can not be invariant under spatial rotations if α_i are numbers and ψ is a scalar wave function.

2. Show that the eigenvalues of $\vec{\alpha}$ are ± 1.

3. Show that the different representations of α_i and β are related by a unitary transformation. Find the transformation and show that it is unitary.

4. Suppose the Dirac wave functions are normalized in the same way as Klein-Gordon wave functions:

$$i \int d^3x \, \overline{\psi}_\alpha(x) \, \overleftrightarrow{\partial_0} \, \psi_\beta(x) = \delta_{\alpha\beta}.$$

Use the Dirac equation to show that these wave functions differ from the ones normalized via

$$\int d^3x \, \psi_\alpha^\dagger(x)\psi_\beta(x) = \delta_{\alpha\beta}$$

by a factor of $\sqrt{2m}$.

5. Show that the current density for a free-particle wave function agrees with the corresponding nonrelativistic expression in the proper limit.

6. Prove that

$$[\alpha_i, \alpha_j] = 2i\epsilon_{ijk}\Sigma_k.$$

7. Show that $\vec{\Sigma} \cdot \hat{p}$ commutes with the Dirac Hamiltonian.

8. Show that helicity is rotationally invariant.

9. Show that the helicity states transform simply under a Lorentz transformation.

10. [12] For a massive fermion, show that handedness is not a good quantum number. That is, show that γ^5 does not commute with the Hamiltonian. However, verify that helicity is conserved but is frame dependent. In particular, show that the helicity is reversed by "overtaking" the particle concerned.

11. [12] For an electron of momentum $\vec{p} = (p\sin\theta, 0, p\cos\theta)$, calculate the $\lambda = +1/2$ helicity eigenspinor(s).

12. Rederive current conservation using the Lorentz invariant form of the Dirac equation.

13. In the Majorana representation all the γ^μ matrices are purely imaginary. Find an explicit form of the Majorana representation.

14. Show that

$$(\vec{\alpha} \cdot \hat{v})^2 = I.$$

15. Show that

$$S^{-1} = \gamma_0 S^\dagger \gamma_0$$

for proper Lorentz transformations, as well as space and time reflections, where S is a 4×4 matrix which is a function of the parameters of the Lorentz transformation and operates upon the four components of the column vector of the wave function satisfying the Dirac equation.

16. Show that equation 5.166 is a solution to equation 5.122.

17. (a) Suppose $\psi(x)$ is the Dirac wave function in a frame $\mathcal{O}$. What is the wave function $\psi'(x')$ in a frame $\mathcal{O}'$ which is obtained from $\mathcal{O}$ by a rotation about the x-axis through $60°$?

(b) If a spin-1/2 particle at rest has its spin in the z-direction, what is the probability that its spin will be observed to be in a direction $60°$ from the z-axis?

(c) Consider a Lorentz transformation to a frame moving in the $+z$-direction with speed v. Obtain the transformation matrix S.

(d) Use the results of part (c) to get the wave function of a spin-1/2 particle whose spin is in the $+z$-direction and whose velocity is v in the $-z$-direction. Assume the wave function in the rest frame is $u = (1, 0, 0, 0)$ and apply the boost of part (c).

(e) Obtain the transformation matrix for the boost of part (c) followed by the rotation of part (a).

(f) In this new frame in which the particle has negative helicity and a velocity $60°$ from the z-axis, calculate the probability that it will be observed to have its spin in the new $+z$-direction.

18. Show that $(\gamma_5)^2 = I$ and $\gamma_5^\dagger = \gamma_5$.

19. Show that γ_5 commutes with S for proper Lorentz transformations Λ but anticommutes with S for improper Lorentz transformations Λ, and hence

$$S(\Lambda)\gamma_5 = \gamma_5 S(\Lambda)\det|\Lambda|.$$

20. Determine how the bilinears

$$\overline{\psi}\psi, \quad \overline{\psi}\gamma_5\psi, \quad \overline{\psi}\gamma^\mu\psi, \quad \overline{\psi}\gamma_5\gamma^\mu\psi, \quad \text{and} \quad \overline{\psi}\sigma^{\mu\nu}\psi$$

transform separately under charge conjugation, space reflection, and time reversal. Using your results, apply the three operators in succession to each bilinear. Comment on the results.

21. Obtain explicit representations for the matrix elements of the operators $\mathcal{O}_i = \gamma^\mu, \gamma^\mu\gamma^5, \gamma^5, \sigma^{\mu\nu}$ between spinors $u(\vec{p}, s)$ and $u(\vec{q}, r)$. Analyze in detail the case of $\vec{p} = (E, 0, 0, p)$ and $\vec{q} = (E, 0, 0, -p)$. That is, compute $\overline{u}(\vec{p}, s)\mathcal{O}_i u(\vec{q}, r)$.

22. Prove

$$\sum_{r=1}^{4} \epsilon_r \omega_\alpha^r(\vec{p}) \overline{\omega}_\beta^r(\vec{p}) = \delta_{\alpha\beta}$$

independently of the specific representation of the Dirac spinors.

23. Prove the completeness relationship

$$\sum_{r=1}^{4} w_\alpha^r(\epsilon^r \vec{p}) w_\beta^{r\dagger}(\epsilon^r \vec{p}) = \frac{E}{m}\delta_{\alpha\beta}.$$

24. Derive equation 5.211 in a representation-free way directly from the Dirac equation.

25. [4] Given a free-particle spinor $u(p)$, construct $u(p+q)$ for $q_\mu \to 0$ and $p \cdot q \to 0$, in terms of $u(p)$ by means of a Lorentz transformation.

26. Derive the following completeness relationships

$$\sum_{\pm s} u_\alpha(p,s)\bar{u}_\beta(p,s) = [\Lambda_+(p)]_{\alpha\beta}$$

and

$$\sum_{\pm s} v_\alpha(p,s)\bar{v}_\beta(p,s) = -[\Lambda_-(p)]_{\alpha\beta}.$$

27. [14] The operators

$$P_L = \frac{1}{2}(1+\gamma_5) \quad \text{and} \quad P_R = \frac{1}{2}(1-\gamma_5)$$

are projection operators which are said to identify states of definite chirality (handedness).

(a) Show that P_L and P_R are legitimate projection operators in that

$$P_L^2 = P_L, \qquad P_R^2 = P_R, \qquad \text{and} \qquad P_L P_R = P_R P_L = 0.$$

(b) Demonstrate that in the limit of high energy $E/m \gg 1$ – or equivalently in the massless limit – that the Dirac spinors for positive helicity (right-handed) and negative helicity (left-handed) states of momentum $\vec{p}$ are given by

$$u_\pm(\vec{p}) = \sqrt{\frac{1}{2}}\left(\begin{array}{c} \chi_{\pm\hat{p}} \\ \pm\chi_{\pm\hat{p}} \end{array}\right),$$

where $\chi_{\pm\hat{p}}$ are spinors such that

$$\vec{\sigma} \cdot \hat{p}\chi_{\pm\hat{p}} = \pm\chi_{\pm\hat{p}}.$$

Note: In order to conveniently deal with massless particles, it is important to use the normalization $u^\dagger(p)u(p) = 1$. The appropriate Dirac spinors can then be found by multiplying the usual forms by the factor $\sqrt{m/E}$. Demonstrate this.

(c) Show that

$$P_L u_-(p) = u_-(p),$$
$$P_R u_+(p) = u_+(p),$$
$$P_L u_+(p) = P_R u_-(p) = 0,$$

so that the chirality operator is equivalent to the helicity operator in this limit.

28. [12] Show that at high energies

$$\gamma^5 u = \begin{pmatrix} \vec{\sigma} \cdot \hat{p} & 0 \\ 0 & \vec{\sigma} \cdot \hat{p} \end{pmatrix} u,$$

where u is the electron spinor. That is, show that in the extreme relativistic limit, the chirality operator γ^5 is equal to the helicity operator; and so, for example, $\frac{1}{2}(1 - \gamma^5)u = u_L$ corresponds to an electron of negative helicity.

29. Examine in detail the influence of the transformation $\psi_C = C\overline{\psi}^T = i\gamma^2\psi^*$ on the eigenfunctions of an electron at rest with negative energy.

30. [4] In order that $\mathcal{T}$ be a symmetry operation of the Dirac theory, the rules of interpretation of the wave function $\psi'(t')$ must be the same as those of $\psi(t)$. This means that observables composed of forms bilinear in ψ' and $\psi'^\dagger$ must have the same interpretation – within a sign, appropriate to the observable – as those of ψ.

(a) Verify that this is so for the current:

$$j'_\mu(x') = j^\mu(x)$$

and also

$$\langle \vec{r} \rangle' = \langle \vec{r} \rangle \quad \text{and} \quad \langle \vec{p} \rangle' = -\langle \vec{p} \rangle.$$

(b) Repeat these calculations for the charge-conjugation transformation C. In particular, show

$$\overline{\psi}_C(x)\gamma_\mu\psi_C(x) = +\overline{\psi}(x)\gamma_\mu\psi(x)$$

and interpret using the hole theory.

31. If ψ_1 and ψ_2 are two arbitrary solutions of the free Dirac equation, prove the Gordon decomposition

$$c\overline{\psi}_2\gamma^\mu\psi_1 = \frac{1}{2m}\left[\overline{\psi}_2\hat{p}^\mu\psi - (\hat{p}^\mu\overline{\psi}_2)\psi_1\right] - \frac{i}{2m}\hat{p}_\nu(\overline{\psi}_2\sigma^{\mu\nu}\psi_1).$$

32. Calculate the current

$$j^k = c \int d^3 x \psi^\dagger(\vec{x}, t) \alpha_k \psi(\vec{x}, t) = c \int d^3 x \bar{\psi}(\vec{x}, t) \gamma^k \psi(\vec{x}, t)$$

for the general wave packet which contains both positive- and negative-energy plane waves.

33. [7] At time $t = 0$, the following wave packet with Gaussian density distribution is defined as

$$\psi'(\vec{x}, 0, s) = \frac{1}{(\pi d^2)^{3/4}} e^{-|x|^2/2d^2} w^1(0).$$

Determine the wave packet at time t developed from the above. Consider the intensity of the negative-energy solutions in the wave packet. What does one learn in general about the applicability of the one-particle interpretation of the Dirac equation?

34. Consider a Dirac electron of mass m in an attractive electrostatic potential

$$V(z) = \begin{cases} 0 & \text{for } z < 0, \\ -V_0 & \text{for } 0 < z < a, \\ 0 & \text{for } z > a. \end{cases}$$

(a) Find an expression for the bound-state energy levels.

(b) Solve the problem of scattering of such an electron with momentum $\vec{p}$ off this potential. Obtain the transmission amplitude and phase shift.

35. Solve the Dirac equation for an attractive square well potential of depth V_0 and radius a, after determining the continuity conditions at $r = a$. Obtain an explicit expression for the minimum V_0, given a that just binds a particle of mass m.

36. Find the exact energy eigenvalues and eigenfunctions for an electron in a uniform magnetic field $\vec{B} = B\hat{z}$, and show the energy eigenvalues can be written as

$$E = \sqrt{m^2 + p_z^2 + 2neB}, \quad \text{for } n = 1, 2, 3 \ldots$$

Compare your results with what is expected nonrelativistically.

37. [14] The Dirac equation describing the interaction of a proton or neutron with an applied external electromagnetic field has an additional term involving the so-called anomalous magnetic moment:

$$\left(i\,\slashed{\nabla} - q_i\,\slashed{A} + \frac{\kappa_i|e|}{4m}\sigma_{\mu\nu}F^{\mu\nu} - m \right)\psi(x) = 0.$$

For the proton, $q_p = e$ and for the neutron $q_n = 0$.

(a) Verify that the choice

$$\kappa_p = 1.79 \quad \text{and} \quad \kappa_n = -1.91$$

corresponds to the observed magnetic moments of these particles, and

(b) Show that the additional interaction disturbs neither the Lorentz covariance of the equation nor the hermiticity of the Hamiltonian.

38. [4] Suppose that the electron had a static electric dipole moment d analogous to its magnetic moment.

(a) Show that this could be accommodated by modifying the Dirac equation to become

$$\left(i\,\slashed{\nabla} - e\,\slashed{A} - i\frac{ed}{4m}\sigma_{\mu\nu}\gamma_5 F^{\mu\nu} - m \right)\psi(x) = 0.$$

Write down the expression for the resulting classical dipole moment.

(b) Demonstrate that the modified Dirac equation is Lorentz covariant under proper Lorentz transformations but not invariant under a parity transformation.

39. [14] Consider a positive-energy spin-1/2 particle at rest. Suppose that at $t = 0$ we apply an external (classical) vector potential

$$\vec{A}(t) = -\hat{x}\frac{a}{\omega}\sin\omega t,$$

which corresponds to an electric field of the form

$$\vec{E}(t) = \hat{x}a\cos\omega t.$$

Show that for $t > 0$ there exists a finite probability of finding the particle in a negative-energy state if such negative-energy states are assumed to be originally empty. In particular, work out quantitatively the two cases: $\omega \ll 2m$ and $\omega \approx 2m$, and comment.

40. [14] As we saw in problem 5.39, a rapidly varying electric field can lead to the creation of particle-antiparticle pairs. Calculate to lowest order the probability per unit volume per unit time of producing fermion pairs in the presence of an external electric field

$$\vec{E}(t) = \hat{x}a\cos\omega t$$

and show that

$$\text{prob} = VT\frac{e^2a^2}{24\pi}\left(1 - \frac{4m^2}{\omega^2}\right)^{\frac{1}{2}}\left(1 + \frac{2m^2}{\omega^2}\right).$$

Suggestion: Utilize normalized plane-wave solutions of the Dirac equation:

$$\psi(x) = \sqrt{\frac{m}{E}}u(p)\exp i(\vec{p}\cdot\vec{x} - Et) \quad \text{with} \quad E = \sqrt{\vec{p}^2 + m^2},$$

and simple first-order perturbation theory

$$\text{amp} = -i\int_{-T/2}^{T/2} dt\langle f|H_{\text{int}}(t)|i\rangle,$$

with $H_{\text{int}} = e\int d^3x\,j_\mu A^\mu$ as in the Klein-Gordon case (see problem 4.21).

Part III

Quantum Electrodynamics

Chapter 6

Propagator Methods

We begin this chapter with a general discussion of the scattering process. To understand the laws governing the interactions of elementary particles, the most popular experiment technique is to scatter a variety of particles by a variety of targets. On the other hand, atoms and molecular structure are largely explored using spectroscopic methods. We can argue that, in some sense, spectroscopy is another form of scattering. The atom in the ground state is excited by some projectile, and then an outgoing photon is observed with the atom going into the ground state again, or possibly another excited state.

Our aim in this book is to calculate transition rates and cross-sections. There are at least two approaches we could take: 1) a systematic approach of quantizing the field, or 2) the propagator formalism, which is more intuitive and leads to being able to perform calculations quickly. The propagator method will be defined in terms of integral equations with boundary conditions incorporating the Stückelberg-Feynman physical interpretation of the positron as a negative-energy electron moving backwards in time.

Starting with scattering processes in the framework of Dirac's theory (relativistic from beginning), the calculations are exact in principle but practically they will be carried out using perturbation theory, i.e. an expansion in terms of small interaction parameters. Although Stückelberg started the relativistic propagator idea, Feynman exploited it in calculations. J. Schwinger and S. Tomonaga developed an alternative formalism, and the latter three physicists were awarded the Nobel Prize in 1965 for the development of quantum electrodynamics.

Quantum electrodynamics (QED) is one of the most successful and most accurate theories known in physics. QED is the archetype for all modern field theories, and important in its own right since it provides the theoretical foundation for atomic physics. The theory also applies to heavy leptons – muon and tau – and in general can be used to describe the electromagnetic interaction of other charged elementary particles.

143

6.1 Nonrelativistic Scattering Theory

In classical mechanics, the collision of two particles is entirely determined by their relative velocity and impact parameter. The impact parameter is the distance at which the particles would pass if they did not interact. In quantum mechanics, motion with a definite velocity along a path is meaningless, and therefore so is the concept of impact parameter. We can however calculate the probability that the particles will deviate, or be scattered, through a given angle as a result of a collision. Throughout this book we mean by collision, an elastic collision in which the internal states of the colliding particles, if they have one, is left unchanged.

We begin by considering the idealized problem of a beam of particles of mass m traveling in the z-direction with velocity v, scattered by a fixed scattering center place at the origin, $r = 0$, and acting primarily in the neighbourhood of the origin. We describe this scattering center by a potential $V(\vec{r})$. As a result of this potential field, the incident beam of particles experiences a force and is deflected. The scattering center in an actual experiment will be an atom, a nucleus, or another particle. The potential which the incident particle experiences is usually more complicated, depending not only on the separation $\vec{r}$ but also on the internal coordinates of the target. Nevertheless, one can in many cases represent the scattering center approximately by such a potential.

A free incident particle moving in the positive direction along the z-axis is described by the plane wave $\psi(z) = Ae^{ikz}$, where $k = p/\hbar = mv/\hbar$. The scattered particles can be described, at a great distance from the scattering center, by an outgoing spherical wave of the form $f(\theta, \phi)e^{ikr}/r$. Its amplitude depends on the scattering angles θ and ϕ, and is inversely proportional to r since the radial flux must fall off as the inverse square of the distance.

The total asymptotic form of the solution of the Schrödinger equation which we require is

$$\psi(r, \theta, \phi) \approx A \left[e^{ikz} + f(\theta, \phi)\frac{e^{ikr}}{r} \right], \quad \text{for } r \to \infty. \tag{6.1}$$

$f(\theta, \phi)$ is called the scattering amplitude of the process. The solution in equation 6.1 is readily verified to satisfy the wave equation asymptotically through terms of order $1/r$ in the region in which $V = 0$, for any form of the function $f(\theta, \phi)$ (see problem 6.1).

To specify the angular distribution of the particles which have been scattered, we consider the passage of these particles through a large sphere of radius R, whose center is at the target (see figure 6.1). Consider an element of surface $dS = R^2 d\Omega$, where $d\Omega$ is the element of solid angle on the sphere subtending the spherical polar angles (θ, ϕ) at the target. The number of scattered particles crossing dS per unit time, or the current of particles scattered across dS, is given by the product of the density of scattered particles at dS times the velocity of scattered particles times the element of area dS, i.e. the current of particles scattered across dS is given by

$$\left| Af(\theta,\phi)\frac{e^{ikR}}{R} \right|^2 vR^2 d\Omega = \left(\frac{\hbar k}{m}\right) |A|^2 |f(\theta,\phi)|^2 d\Omega. \tag{6.2}$$

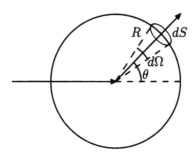

FIGURE 6.1: Definition of the solid angle in scattering. The azimuthal angle ϕ is not shown.

This expression gives the relative probabilities of a particle being scattered into different directions. The absolute probability, however, still depends on the flux of the incident beam of particles striking the target. The current of particles scattered across dS is clearly proportional to the flux of incident particles, i.e. the number of particles incident per unit area per unit time. Since we have taken the density of the incident beam to be $|A|^2$, the flux is simply equal to $J_{\text{inc}} = v|A|^2 = \hbar k |A|^2/m$. We then define the element of cross section $d\sigma(\theta,\phi)$ by

$$d\sigma(\theta,\phi) = \frac{\text{current of particles scatter across } dS}{\text{flux of incident beam}}$$
$$= |f(\theta,\phi)|^2 d\Omega, \tag{6.3}$$

$$\boxed{\frac{d\sigma}{d\Omega}(\theta,\phi) = |f(\theta,\phi)|^2}. \tag{6.4}$$

The differential cross section is equal to the absolute square of the scattering amplitude. The choice of coefficient A is unimportant in the calculation of scattering. We see that the cross section has the dimensions of area. It is the effective area which the target presents to the incident beam of particles for the process considered.

From the angular distribution or differential cross section, we obtain the total cross section by integrating over all angles on the unit sphere:

$$\sigma_{\text{tot}} = \int |f(\theta, \phi)|^2 d\Omega. \tag{6.5}$$

To find the scattering amplitude $f(\theta, \phi)$, i.e. to solve the Schrödinger equation, it is most convenient to convert the Schrödinger equation first to an integral equation.

6.2 Green Functions and Integral Equations

The propagator theory is based on the Green-function method of solving inhomogeneous differential equations. Before we consider wave equations, we explain the method in terms of a simple example that the reader may be familiar with.

Suppose we wish to solve Poisson's equation

$$\nabla^2 \phi(\vec{x}) = -\rho(\vec{x}) \tag{6.6}$$

for a known charge distribution $\rho(\vec{x})$, subject to some boundary conditions. It is easier to first solve the "unit source" problem

$$\nabla_{\vec{x}}^2 G(\vec{x}; \vec{x}') = -\delta^3(\vec{x} - \vec{x}'), \tag{6.7}$$

where $G(\vec{x}; \vec{x}')$ is the potential at $\vec{x}$ due to a unit point source at $\vec{x}'$. We then move this source over the charge distribution and accumulate the total potential at $\vec{x}$ from all possible sources in the volume element d^3x':

$$\phi(\vec{x}) = \int d^3x' G(\vec{x}; \vec{x}') \rho(\vec{x}'). \tag{6.8}$$

We can check directly that ϕ is the desired solution:

$$\nabla_{\vec{x}}^2 \phi(\vec{x}) = \int d^3x' \nabla_{\vec{x}}^2 G(\vec{x}; \vec{x}') \rho(\vec{x}') \tag{6.9}$$

$$= -\int d^3x' \delta^3(\vec{x} - \vec{x}') \rho(\vec{x}')$$

$$= -\rho(\vec{x}).$$

The difficulty now becomes solving equation 6.7 for $G(\vec{x}; \vec{x}')$. In this case, the solution is well known to be the Coulomb potential

$$G(\vec{x}; \vec{x}') = G(\vec{x} - \vec{x}') = \frac{1}{4\pi|\vec{x} - \vec{x}'|}. \tag{6.10}$$

This can be easily verified by substitution into equation 6.7 (see problem 6.2). In the case of quantum mechanics, the equivalent of ϕ appears on both sides of the equation, and so an iterative perturbation-series solution is required.

We now treat the inhomogeneous Schrödinger equation:

$$-\frac{\hbar^2}{2m}\nabla^2\psi + V(\vec{x})\psi = E\psi. \tag{6.11}$$

By defining $k = \sqrt{2mE}/\hbar$ and $\rho(\vec{x}) = 2mV(\vec{x})\psi/\hbar^2$, we can write

$$(\nabla^2 + k^2)\psi = \rho(\vec{x}), \tag{6.12}$$

where the inhomogeneous term ρ depends on ψ.

The Green function equation is

$$(\nabla_{\vec{x}}^2 + k^2)G(\vec{x};\vec{x}') = \delta^3(\vec{x} - \vec{x}'). \tag{6.13}$$

The solution is

$$G(\vec{x};\vec{x}') = G(\vec{x} - \vec{x}') = -\frac{e^{\pm ik|\vec{x}-\vec{x}'|}}{4\pi|x - x'|}, \tag{6.14}$$

which again can be easily verified by substitution into equation 6.13 (see problem 6.3). These solutions correspond to an outgoing spherical wave generated from a unit source at $\vec{x}'$ and to an incoming spherical wave absorbed by a sink at $\vec{x}'$. For scattering problems, we are only interested in the outgoing spherical wave, and shall not consider the solution containing the negative exponential function further.

Equation 6.14 is a solution to the inhomogeneous equation 6.13; to this solution we must add solutions $\psi_0(\vec{x})$ to the homogeneous equation. The general solution to the Schrödinger equation becomes

$$\psi(\vec{x}) = \psi_0(\vec{x}) - \frac{m}{2\pi\hbar^2}\int\frac{e^{ik|\vec{x}-\vec{x}'|}}{|\vec{x} - \vec{x}'|}V(\vec{x}')\psi(\vec{x}')d^3\vec{x}'. \tag{6.15}$$

Equation 6.15 may look like an explicit solution to the Schrödinger equation for any potential. Unfortunately there is a ψ in the integral on the right-hand side. Thus it is not a solution unless we already know the solution. This integral form of the Schrödinger equation is entirely equivalent to the more familiar differential form. It is however particularly well suited to scattering problems.

6.3   The Born Approximation

To calculate the cross section, we need the asymptotic form of equation 6.15 for large distances[1]. A procedure which leads to a very useful approximation, both when the potential is weak and when the kinetic energy is large, is called the Born approximation. Suppose $V(\vec{r}')$ is localized about $r' = 0$ and falls off rapidly for large r'; this is the typical case for scattering problems in quantum electrodynamics.

When calculating $\psi(\vec{r})$ at points far away from the scattering center, $r \gg r'$ for all points that contribute to the integral in equation 6.15. Expanding to lowest order in r'/r, we have

$$|\vec{r} - \vec{r}'| = r\left[1 - 2\frac{\vec{r} \cdot \vec{r}'}{r^2} + \left(\frac{r'}{r}\right)^2\right]^{1/2} \approx r\left(1 - \frac{\vec{r} \cdot \vec{r}'}{r^2}\right) = r - \hat{r} \cdot \vec{r}' \quad (6.16)$$

and

$$\frac{1}{|\vec{r} - \vec{r}'|} = \frac{1}{r}\left[1 - 2\frac{\vec{r} \cdot \vec{r}'}{r^2} + \left(\frac{r'}{r}\right)^2\right]^{-1/2} \approx \frac{1}{r}\left(1 + \frac{\vec{r} \cdot \vec{r}'}{r^2}\right) = \frac{1}{r} + \frac{\hat{r} \cdot \vec{r}'}{r^2},$$
$$(6.17)$$

where $\vec{r} = r\hat{r}$. Defining $\vec{k} = k\hat{r}$, we obtain

$$\frac{e^{ik|\vec{r} - \vec{r}'|}}{|\vec{r} - \vec{r}'|} \approx \frac{e^{ikr}}{r}e^{-i\vec{k} \cdot \vec{r}'} \quad (6.18)$$

to leading order in r'/r.

In the case of scattering, we want $\psi_0(\vec{r}) = Ae^{ikz}$ to represent an incident plane wave. Thus for large r, equation 6.15 becomes

$$\psi(\vec{r}) \approx Ae^{ikz} - \frac{m}{2\pi\hbar^2}\frac{e^{ikr}}{r}\int e^{-i\vec{k} \cdot \vec{r}'}V(\vec{r}')\psi(\vec{r}')d^3\vec{r}'. \quad (6.19)$$

Comparing equation 6.19 with equation 6.1, we read off the scattering amplitude to be

$$f(\theta, \phi) = -\frac{m}{2\pi\hbar^2 A}\int e^{-\vec{k} \cdot \vec{r}}V(\vec{r})\psi(\vec{r})d^3\vec{r}, \quad (6.20)$$

where we have renamed the dummy variable of integration.

---

[1]In this section, we switch our coordinate notation from $\vec{x}$ to $\vec{r}$ since the latter is more common when discussing nonrelativistic scattering, and we will most often work in spherical polar coordinates.

Equation 6.20 is not a useful solution for the scattering amplitude since it still contains the unknown wave function $\psi(\vec{r})$ inside the integral. To proceed further, we consider the case when the scattering potential $V(\vec{r})$ is weak and can be considered as a small perturbation, which slightly distorts the incoming plane wave. As a first approximation, we can use $\psi(\vec{r}) \approx \psi_0(\vec{r}) = A e^{ikz} = A e^{i\vec{k}_0 \cdot \vec{r}}$, where $\vec{k}_0 = k\hat{z}$, inside the integral. This gives the first Born approximation

$$f(\theta, \phi) \approx -\frac{m}{2\pi\hbar^2} \int e^{i(\vec{k}_0 - \vec{k}) \cdot \vec{r}} V(\vec{r}) d^3 \vec{r}, \qquad (6.21)$$

which is simple and therefore of practical use in calculations. Notice that the normalization A has dropped out, as it should.

We may simplify equation 6.21 further by introducing the momentum transfer variable $\hbar\vec{q} = \hbar(\vec{k}_0 - \vec{k})$, as shown in figure 6.2. Since $\vec{k} = k\hat{r}$ and $\vec{k}_0 = k\hat{z}$, $q = 2k\sin(\theta/2)$, and we obtain

$$\boxed{f(\theta, \phi) = -\frac{m}{2\pi\hbar^2} \int e^{i\vec{q} \cdot \vec{r}} V(\vec{r}) d^3 \vec{r}} \ . \qquad (6.22)$$

This shows that the scattering amplitude in a particular direction is determined by the Fourier transform of the potential with the corresponding momentum transfer of the particle during the collision. Equation 6.22 is applicable to the scattering from a potential, which is a function of all the coordinates in general, not only a function of r. We now consider some special cases of equation 6.22.

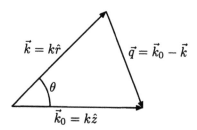

FIGURE 6.2: Relationship between the propagation vectors $\vec{k}_0$ for the incident particle, $\vec{k}$ for the scattered particle, and the scattering angle θ. The momentum transfer in the collision is $\hbar\vec{q}$.

For a spherically symmetrical potential, $V(\vec{r}) = V(r)$. We may integrate equation 6.22 using spherical polar coordinates with $\vec{q}$ taken along the polar axis:

$$f(\theta, \phi) = -\frac{m}{2\pi\hbar^2} \int_0^\infty r^2 V(r) \int_0^{2\pi} d\phi \int_0^\pi e^{iqr\cos\theta} \sin\theta d\theta$$

$$= -\frac{2m}{\hbar^2} \int_0^\infty V(r) \frac{\sin(qr)}{q} r dr. \tag{6.23}$$

The scattering is independent of the angle ϕ; it depends on the speed v and the scattering angle θ only through the combination $q = (2mv/\hbar)\sin(\theta/2)$. If $\theta = 0$, or $q = 0$, the integral diverges if $r^2 V(r)$ does not approach zero as fast as $r \to \infty$.

We now consider the two energy extremes. For low energy (long wavelength) scattering, the exponential factor in equation 6.22 is approximately constant over the scattering region:

$$f(\theta, \phi) \approx -\frac{m}{2\pi\hbar^2} \int V(\vec{r}) d^3\vec{r}. \tag{6.24}$$

Furthermore, if $V(\vec{r}) = V(r)$,

$$f(\theta, \phi) = -\frac{2m}{\hbar^2} \int r^2 V(r) dr. \tag{6.25}$$

In this case, the scattering is isotropic and independent of the velocity.

For high energy (short wavelength) scattering, q is large and the factor $e^{i\vec{q}\cdot\vec{r}}$ in equation 6.22 is a rapidly oscillating function, if the scattering angle is not too small. The integral of its product with the slowly varying function V will be nearly zero. Thus as the velocity increases, the cross section, for not too small θ, tends to zero.

Having calculated the lowest order term of the perturbative expansion of equation 6.20, we now iterate to generate a series of higher-order corrections. We assume the series converges to the exact wave function. Once more, the integral form of the Schrödinger equation is

$$\psi(\vec{r}) = \psi_0(\vec{r}) + \int G(\vec{r} - \vec{r}_0) V(\vec{r}_0) \psi(\vec{r}_0) d^3\vec{r}_0, \tag{6.26}$$

where $G(\vec{r}) \equiv -\frac{m}{2\pi\hbar^2} \frac{e^{ikr}}{r}$. Dropping the arguments of the functions allows us to write the equation in a more schematic form: $\psi = \psi_0 + \int GV\psi$. If we take this expression for ψ and plug it in under the integral of equation 6.26, we obtain

$$\psi = \psi_0 + \int GV\psi_0 + \int \int GVGV\psi. \tag{6.27}$$

Iterating this procedure, we obtain a formal series for ψ:

$$\psi = \psi_0 + \int GV\psi_0 + \int \int GVGV\psi_0 + \int \int \int GVGVGV\psi_0 + \dots. \tag{6.28}$$

In each integral, only the incident wave function ψ_0 appears together with more and more powers of GV. In subsequent sections we will develop this type of series in a covariant manner.

6.4 Propagator Theory

Until now, we have treated scattering using a time-independent approach and looked for stationary states $\Psi(\vec{x},t) = \psi(\vec{x})e^{-iEt/\hbar}$ of the Hamiltonian. In this section, we will follow a time-dependent approach to solve the nonrelativistic scattering problem. We will develop the wave function resulting from the continuous interaction of a free-particle wave function with a potential. This development begins by establishing the formalism for a single scattering off a point potential. These single scatters are then summed to obtain the solution to the scattering from n potentials. The continuum limit is then taken to obtain the scattering solution from a continuous interaction.

We first answer a simple question. Given a wave packet, which in the remote past represented a particle approaching a potential, what does that wave look like in the remote future? We begin by considering the nonrelativistic propagator. Huygens' principle can be written as

$$\psi(\vec{x}',t') = i \int d^3x\, G(\vec{x}',t';\vec{x},t)\psi(\vec{x},t) \quad \text{for} \quad t' > t, \qquad (6.29)$$

where the integral extends over all space. $\psi(\vec{x}',t')$ is the total wave arriving at the point $\vec{x}'$ at time t' and $\psi(\vec{x},t)$ is the original wave. $G(\vec{x}',t';\vec{x},t)$ is the Green function or propagator[2]. In general, one does not consider stationary eigenstates of energy, i.e. stationary waves, in scattering problems. Knowledge of G enables us to construct the physical state which develops in time from any given initial state. This is equivalent to a complete solution to the Schrödinger equation.

Consider a free-particle solution ϕ and its Green function G_0. We introduce a potential $V(\vec{x},t)$ which is "turned on" for a brief interval of time Δt_1 about t_1; $V = 0$ outside of time Δt_1. $V(\vec{x}_1,t_1)$ acts as a source of new waves, and we can write

$$\left(i\frac{\partial}{\partial t_1} - H_0\right)\psi(\vec{x}_1,t_1) = V(\vec{x}_1,t_1)\psi(\vec{x}_1,t_1) \quad \text{for} \quad t = \Delta t_1,$$

$$= 0 \quad \text{for} \quad t \neq \Delta t_1. \qquad (6.30)$$

[2]If one is careful with the factors of $\hbar$ and c, there should be a factor of $1/\hbar$ in front of the Green function.

The new wave function can be written as

$$\psi(\vec{x}_1, t_1) = \phi(\vec{x}_1, t_1) + \Delta\psi(\vec{x}_1, t_1). \qquad (6.31)$$

Substituting this wave function into equation 6.30 and using the free-particle Schrödinger equation gives

$$\left(i\frac{\partial}{\partial t_1} - H_0\right)\Delta\psi(\vec{x}_1, t_1) = V(\vec{x}_1, t_1)[\phi(\vec{x}_1, t_1) + \Delta\psi(\vec{x}_1, t_1)]. \qquad (6.32)$$

The second terms on both sides of the equation are smaller than the first terms on both sides of the equation. This is clear for the right-hand side of the equation. To see it for the left-hand side, we drop the second term on the right-hand side and write

$$i\frac{\partial}{\partial t_1}\Delta\psi(\vec{x}_1, t_1) = V(\vec{x}_1, t_1)\phi(\vec{x}_1, t_1) + H_0\Delta\psi(\vec{x}_1, t_1), \qquad (6.33)$$

$$\Delta\psi(\vec{x}_1, t_1 + \Delta t_1) = -i\int_{t_1}^{t_1+\Delta t_1} dt' \left[V(\vec{x}_1, t')\phi(\vec{x}_1, t') + H_0\Delta\psi(\vec{x}_1, t')\right]. \qquad (6.34)$$

We see that the last term is less than the first term.

After dropping the second term on both sides of equation 6.32, we have

$$i\frac{\partial}{\partial t_1}\Delta\psi(\vec{x}_1, t_1) = V(\vec{x}_1, t_1)\phi(\vec{x}_1, t_1). \qquad (6.35)$$

To first order,

$$\Delta\psi(\vec{x}_1, t_1 + \Delta t_1) = -iV(\vec{x}_1, t_1)\phi(\vec{x}_1, t_1)\Delta t_1. \qquad (6.36)$$

The result shows that the potential produces an additional change in ψ during Δt_1 in addition to that taking place in the absence of V. Since the potential $V(\vec{x}_1, t_1)$ vanishes after the time interval Δt_1, the scattered wave also propagates according to the free propagator G_0. This added wave at a future time t' leads to a new contribution to $\psi(\vec{x}', t')$,

$$\Delta\psi(\vec{x}', t') = i\int d^3x_1 G_0(\vec{x}', t'; \vec{x}_1, t_1)\Delta\psi(\vec{x}_1, t_1)$$

$$= \int d^3x_1 G_0(\vec{x}', t'; \vec{x}_1, t_1)V(\vec{x}_1, t_1)\phi(\vec{x}_1, t_1)\Delta t_1. \qquad (6.37)$$

Here we have replaced $t_1 + \Delta t_1$ by t_1, which is justified in the limit $\Delta t_1 \to 0$. The space-time coordinate $(\vec{x}', t')$ is in the future, $(\vec{x}_1, t_1)$ is the present, and

$(\vec{x}, t)$ is in the past. Thus, the wave ψ developing from an arbitrary wave packet ϕ in the remote past is

$$
\begin{aligned}
\psi(\vec{x}', t') &= \phi(\vec{x}', t') + \Delta\psi(\vec{x}', t') \\
&= \phi(\vec{x}', t') + \int d^3x_1 G_0(\vec{x}', t'; \vec{x}_1, t_1) V(\vec{x}_1, t_1) \phi(\vec{x}_1, t_1) \Delta t_1 \quad (6.38) \\
&= i \int d^3x G_0(\vec{x}', t'; \vec{x}, t) \phi(\vec{x}, t) \\
&\quad + \int d^3x d^3x_1 \Delta t_1 G_0(\vec{x}', t'; \vec{x}_1, t_1) V(\vec{x}_1, t_1) G_0(\vec{x}_1, t_1; \vec{x}, t) \phi(\vec{x}, t) \\
&= i \int d^3x [G_0(\vec{x}', t'; \vec{x}, t) \\
&\quad + \int d^3x_1 \Delta t_1 G_0(\vec{x}', t'; \vec{x}_1, t_1) V(\vec{x}_1, t_1) G_0(\vec{x}_1, t_1; \vec{x}, t)] \phi(\vec{x}, t).
\end{aligned}
$$

$$(6.39)$$

Therefore, this Green function is the integrand

$$
\begin{aligned}
G(\vec{x}', t'; \vec{x}, t) &= G_0(\vec{x}', t'; \vec{x}, t) \\
&\quad + \int d^3x_1 \Delta t_1 G_0(\vec{x}', t'; \vec{x}_1, t_1) V(\vec{x}_1, t_1) G_0(\vec{x}_1, t_1; \vec{x}, t).
\end{aligned}
$$

$$(6.40)$$

The first term represents the propagation from $(\vec{x}, t)$ to $(\vec{x}', t')$ as a free particle. The second term represents propagation from $(\vec{x}, t)$ to $(\vec{x}_1, t_1)$, a scattering at $(\vec{x}_1, t_1)$, and free propagation from $(\vec{x}_1, t_1)$ to $(\vec{x}', t')$.

If we turn on another potential $V(\vec{x}_2, t_2)$ for an interval Δt_2 at time $t_2 > t_1$, the additional contribution to $\psi(\vec{x}', t')$ for $t' > t_2$ is, using equation 6.37,

$$
\begin{aligned}
\Delta\psi(x') &= \int d^3x_2 G_0(x'; 2) V(2) \psi(2) \Delta t_2 \quad (6.41) \\
&= i \int d^3x_2 d^3x \Delta t_2 G_0(x'; 2) V(2) [G_0(2; x) \\
&\quad + \int d^3x_1 \Delta t_1 G_0(2; 1) V(1) G_0(1; x)] \phi(x), \quad (6.42)
\end{aligned}
$$

where we have used the obvious notation: $(x) \equiv (\vec{x}, t)$ and $\psi(2) \equiv \psi(x_2)$. The first term represents a single scattering at time t_2. The second term is a double scattering.

The total wave is obtained by inserting equation 6.38 for $\psi(2)$ into equation 6.41 and adding it to $\phi(x')$:

$$\psi(x') = \phi(x') + \int d^3x_1 \Delta t_1 G_0(x';1)V(1)\phi(1)$$

$$+ \int d^3x_2 \Delta t_2 G_0(x';2)V(2)\phi(2)$$

$$+ \int d^3x_1 \Delta t_1 d^3x_2 \Delta t_2 G_0(x';2)V(2)G_0(2;1)V(1)\phi(1). \tag{6.43}$$

The four terms of the above equation are depicted in figure 6.3. As we proceed through this book, we shall make increasing use of a pictorial representation for amplitudes. These pictures are called Feynman diagrams (or graphs).

If there are n such time intervals when the potential V is turned on,

$$\psi(x') = i \int d^3x\, G(x';x)\phi(x)$$

$$= \phi(x') + \sum_i \int d^3x_i \Delta t_i G_0(x';x_i)V(x_i)\phi(x_i)$$

$$+ \sum_{i,j;t_i>t_j} \int d^3x_i \Delta t_i d^3x_j \Delta t_j G_0(x';x_i)V(x_i)G_0(x_i;x_j)V(x_j)\phi(x_j)$$

$$+ \sum_{i,j,k;t_i>t_j>t_k} \int d^3x_i \Delta t_i d^3x_j \Delta t_j d^3x_k \Delta t_k$$

$$\cdot G_0(x';x_i)V(x_i)G_0(x_i;x_j)V(x_j)G_0(x_j;x_k)V(x_k)\phi(x_k)$$

$$+ \cdots . \tag{6.44}$$

The corresponding Green function is

$$G(x';x) = G_0(x';x) + \sum_i \int d^3x_i \Delta t_i G_0(x';x_i,t_i)V(\vec{x}_i;t_i)G_0(\vec{x}_i,t_i;x)$$

$$+ \sum_{i,j;t_i>t_j} \int d^3x_i \Delta t_i d^3x_j \Delta t_j$$

$$\cdot G_0(x';\vec{x}_i,t_i)V(\vec{x}_i;t_i)G_0(\vec{x}_i,t_i;\vec{x}_j,t_j)V(\vec{x}_j,t_j)G_0(\vec{x}_j,t_j;x)$$

$$+ \cdots , \tag{6.45}$$

which is the probability amplitude per unit space-time volume for a particle wave originating at x to propagate to x', as depicted in figure 6.4. This amplitude is a sum of amplitudes, the nth such term being a product of factors corresponding to figure 6.4. Each line in the figure represents the amplitude $G_0(x_i;x_{i-1})$ that a particle wave originating at x_{i-1} propagates

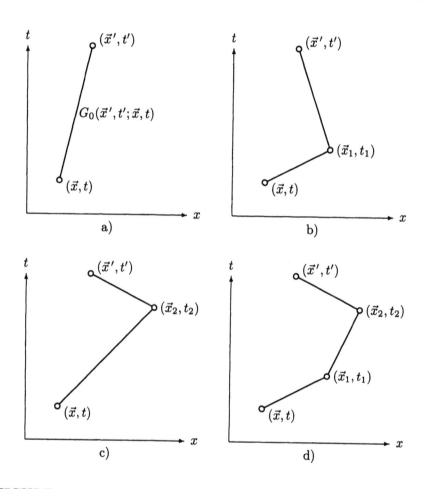

FIGURE 6.3: Space-time diagrams for propagation from (x,t) to (x',t') as
a) a free particle, b) with one scattering at (x_1,t_1), c) with one scattering at
(x_2,t_2), and d) with double scattering.

freely to x_i. At the point x_i it is scattered with probability amplitude per unit space-time volume $V(x_i)$ to a new wave propagating forward in time with amplitude $G_0(x_{i+1}; x_i)$ to the next interaction. This amplitude is then summed over all space-time points in which the interactions can occur.

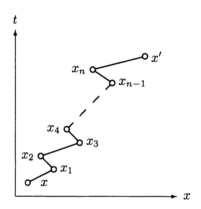

FIGURE 6.4: nth order contribution to $G(x'; x)$.

We may lift the time-ordering restriction $t_i > t_j$, etc., if we define

$$G_0^+(\vec{x}', t'; \vec{x}, t) = \begin{cases} 0 & \text{for } t' < t \\ G_0(\vec{x}', t'; \vec{x}, t) & \text{for } t' > t \end{cases} \qquad (6.46)$$

and

$$G^+(\vec{x}', t'; \vec{x}, t) = \begin{cases} 0 & \text{for } t' < t \\ G(\vec{x}', t'; \vec{x}, t) & \text{for } t' > t \end{cases} . \qquad (6.47)$$

Since it only propagates waves forward in time, G^+ is a retarded propagator. Physically this just means that no Huygens' wavelets $\Delta\psi$ from the ith iteration (at time t_i) appear until after t_i.

If one wants to describe the evolution backwards in time, it is useful to introduce the advanced Green function G^-:

$$G_0^-(\vec{x}', t'; \vec{x}, t) = \begin{cases} -G_0(\vec{x}', t'; \vec{x}, t) & \text{for } t' < t \\ 0 & \text{for } t' > t \end{cases} \qquad (6.48)$$

and

$$G^-(\vec{x}\,',t';\vec{x},t) = \begin{cases} -G(\vec{x}\,',t';\vec{x},t) & \text{for } t' < t \\ 0 & \text{for } t' > t \end{cases}. \tag{6.49}$$

In the limit of a continuous interaction, $\Delta t \to dt$ and $n \to \infty$, so that

$$\sum_i \int d^3x \Delta t_i \to \int d^4x \tag{6.50}$$

and we obtain

$$G^+(x';x) = G_0^+(x';x) + \int d^4x_1 G_0^+(x';x_1)V(x_1)G_0^+(x_1;x)$$

$$+ \int d^4x_1 d^4x_2 G_0^+(x';x_1)V(x_1)G_0^+(x_1,x_2)V(x_2)G_0^+(x_2;x) + \cdots. \tag{6.51}$$

This multiple scattering series is assumed to converge and may be summed to yield

$$G^+(x';x) = G_0^+(x';x) + \int d^4x_1 G_0^+(x';x_1)V(x_1)G^+(x_1;x). \tag{6.52}$$

This is the inhomogeneous Lippmann-Schwinger equation[3]. We have ignored the possibility of bound states in the potential V.

Similarly the series for the wave function $\psi(x')$ can be summed, resulting in

$$\psi(x') = \lim_{t \to -\infty} i \int d^3x\, G^+(x';x)\phi(x)$$

$$= \lim_{t \to -\infty} i \int d^3x \left[G_0^+(x';x) + \int d^4x_1 G_0^+(x';x_1)V(x_1)G^+(x_1;x) \right] \phi(x)$$

$$= \phi(x') + \lim_{t \to -\infty} \int d^4x_1 G_0^+(x';x)V(x_1)i \int d^3x\, G^+(x_1;x)\phi(x)$$

$$= \phi(x') + \int d^4x_1 G_0^+(x';x_1)V(x_1)\psi(x_1). \tag{6.53}$$

This is the integral equation for $\psi(x')$, where the second term is the scattered wave. The integral equations 6.52 and 6.53 are more useful than the original Schrödinger differential equation. They allow a systematic approximation

[3]B.A. Lippmann & J. Schwinger, "Variational Principles for Scattering Processes. I", Phys. Rev. **79** (1950) 469-480.

to be made in the case of weak perturbations, that is, a small perturbation potential V. Moreover, one can easily impose the correct boundary conditions when using integral equations.

It should be noted that not only does $G_0^+(x';x)$ vanish for $t' < t$, but also $G^+(x',x)$. This property of the retarded Green function expresses the principle of causality.

If the infinite series is truncated after a finite number of terms, equation 6.51 allows us to calculate G^+ as a function of V and G_0^+. Given G^+, one can immediately solve the initial value problem. The wave function $\psi(\vec{x}', t')$ is obtained by a simple integration according to equation 6.53 if it is known at some former point in time.

6.5   The Nonrelativistic Propagator

Our goal in this section is to investigate the differential equation which defines G, and in particular to solve for G_0 explicitly, so that the expansion of G can be explicitly carried out. From Huygens' principle,

$$\psi(\vec{x}', t') = i \int d^3x\, G(\vec{x}', t'; \vec{x}, t)\psi(\vec{x}, t) \quad \text{for } t' > t. \tag{6.54}$$

A form valid for all time is

$$\theta(t' - t)\psi(x') = i \int d^3x\, G^+(x'; x)\psi(x). \tag{6.55}$$

Using the chain rule and the fact that $\psi(x')$ satisfies the Schrödinger equation, we have

$$\left[i\frac{\partial}{\partial t'} - H(x')\right]\theta(t' - t)\psi(x') = \left[i\frac{\partial}{\partial t'}\theta(t' - t)\right]\psi(x'). \tag{6.56}$$

Since (see problem 6.8)

$$\frac{d\theta(\tau)}{d\tau} = \delta(\tau), \tag{6.57}$$

we have

$$\left[i\frac{\partial}{\partial t'} - H(x')\right]\theta(t' - t)\psi(x') = i\delta(t' - t)\psi(x'). \tag{6.58}$$

Substituting the right-hand side of equation 6.55 for $\theta(t' - t)\psi(x')$, we have

$$i \int d^3x \left[i \frac{\partial}{\partial t'} - H(x') \right] G^+(x';x)\psi(x)$$

$$= i\delta(t'-t)\psi(x')$$

$$= i \int d^3x \delta(t'-t)\delta^3(\vec{x}'-\vec{x})\psi(x). \qquad (6.59)$$

Equating integrals,

$$\left[i \frac{\partial}{\partial t'} - H(x') \right] G^+(x';x) = \delta(t'-t)\delta^3(\vec{x}'-\vec{x}) = \delta^4(x'-x) \qquad (6.60)$$

is the Green-function equation in the Schrödinger theory. Along with the boundary conditions $G(x';x) = 0$ for $t' < t$ this defines the retarded Green function.

The Green function $G^+(x';x)$ can depend only on the difference of the coordinates (x',t') and (x,t). This is because the wave at $(\vec{x}',t')$ emerging from a unit source at $\vec{x}$, which is turned on at t, depends only on the interval $(\vec{x}'-\vec{x}, t'-t)$.

Consider the Fourier transform of G_0^+,

$$G_0^+(x';x) = G_0^+(x'-x) = \int \frac{d^3p\,d\omega}{(2\pi)^4} e^{i\vec{p}\cdot(\vec{x}'-\vec{x})} e^{-i\omega(t'-t)} G_0^+(\vec{p},\omega) \qquad (6.61)$$

and the free-particle Hamiltonian $H_0(x') = -\nabla_{x'}^2/2m$. The free-particle Schrödinger equation for the Green function can be solved by substituting this Fourier transform into equation 6.60, performing the differentiations, and representing the delta function as a Fourier transform:

$$\left(i \frac{\partial}{\partial t'} + \frac{1}{2m}\nabla_{x'}^2 \right) G_0^+(x';x) = \delta^4(x'-x), \qquad (6.62)$$

$$\int \frac{d^3p\,d\omega}{(2\pi)^4} \left(\omega - \frac{p^2}{2m} \right) G_0^+(\vec{p},\omega) e^{-i\omega(t'-t)+ip\cdot(\vec{x}'-\vec{x})}$$

$$= \int \frac{d^3p\,d\omega}{(2\pi)^4} e^{-i\omega(t'-t)+ip\cdot(\vec{x}'-\vec{x})}$$

$$(6.63)$$

and hence for $\omega \neq p^2/2m$,

$$G_0^+(\vec{p},\omega) = \frac{1}{\omega - p^2/2m}. \qquad (6.64)$$

To complete the expression, we must have a rule for handling the singularity. We will see that the retarded boundary condition

$$G(x'; x) = 0 \quad \text{for} \quad t' < t \qquad (6.65)$$

requires us to add a positive infinitesimal imaginary part $i\epsilon$ (ϵ is a real constant) to the denominator. Substituting equation 6.64 into equation 6.61, we have

$$
\begin{aligned}
G_0^+(x' - x) &= \int \frac{d^3p}{(2\pi)^3} e^{i\vec{p}\cdot(\vec{x}'-\vec{x})} \int_{-\infty}^{\infty} \frac{d\omega}{2\pi} \frac{e^{-i\omega(t'-t)}}{\omega - p^2/2m + i\epsilon} \\
&= \int \frac{d^3p}{(2\pi)^3} e^{i\vec{p}\cdot(\vec{x}'-\vec{x})} \int \frac{d\omega}{2\pi} \frac{e^{-i(\omega+p^2/2m)(t'-t)}}{\omega + i\epsilon} \\
&= \int \frac{d^3p}{(2\pi)^3} e^{i\vec{p}\cdot(\vec{x}'-\vec{x})} e^{-i(p^2/2m)(t'-t)} \int \frac{d\omega}{2\pi} \frac{e^{-i\omega(t'-t)}}{\omega + i\epsilon}. \quad (6.66)
\end{aligned}
$$

The integral over ω is evaluated as a contour integral in the complex ω-plane as shown in figure 6.5. For $t' > t$, the contour may be closed along an infinite semicircle below the real axis in order to ensure exponential damping of the integrand, and the value of the integral is $-i$ by Cauchy's theorem (equation 2.35). For $t > t'$, the contour is closed above the real axis and the integral vanishes because the pole at $-i\epsilon$ lies outside the contour. The integral is thus just $-i$ times the step function.

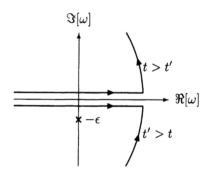

FIGURE 6.5:   Contour in the complex ω-plane for integrating the unit step function.

Continuing our previous integration, equation 6.66, we have

$$G_0^+(x' - x) = \int \frac{d^3p}{(2\pi)^3} e^{i\vec{p}\cdot(\vec{x}'-\vec{x})} e^{-i(p^2/2m)(t'-t)} (-i)\theta(t' - t)$$

$$= -i\theta(t' - t) \int d^3p \frac{e^{i\vec{p}\cdot\vec{x}'-i(p^2/2m)t'}}{(2\pi)^{3/2}} \frac{e^{-i\vec{p}\cdot\vec{x}+i(p^2/2m)t}}{(2\pi)^{3/2}}$$

$$= -i\theta(t' - t) \int d^3p \frac{e^{i(\vec{p}\cdot\vec{x}'-\omega t')}}{(2\pi)^{3/2}} \frac{e^{-i(\vec{p}\cdot\vec{x}-\omega t)}}{(2\pi)^{3/2}}$$

$$= -i\theta(t' - t) \int d^3p \phi_p(\vec{x}',t')\phi_p^*(\vec{x},t), \tag{6.67}$$

where the subscript p stands for plane waves.

The plane-wave solutions arise in equation 6.67 because we are considering the special case of the free-particle propagator. In general,

$$G^+(x';x) = -i\theta(t' - t) \sum_n \psi_n(\vec{x}',t')\psi_n^*(\vec{x},t), \tag{6.68}$$

where $\sum_n$ is a generalized sum and integral over the spectrum of quantum numbers n and $\psi_n(x)$ is a complete set of normalized solutions to the Schrödinger equation, which satisfy the completeness (or closure) statement

$$\sum_n \psi_n(\vec{x}',t)\psi_n^*(\vec{x},t) = \delta^3(\vec{x} - \vec{x}'). \tag{6.69}$$

The Green function $G^+(\vec{x}',t';\vec{x},t)$ defined by equation 6.68 has different times t' and t, yet the completeness relationship 6.69 has equal time t. At first this may appear troublesome, but we can show that equation 6.68 does indeed satisfy equation 6.60.

$$\left(i\hbar\frac{\partial}{\partial t'} - \hat{H}(x')\right) G^+(x';x)$$

$$= \left(i\hbar\frac{\partial}{\partial t'} - \hat{H}(x')\right) \left(-i\theta(t' - t) \sum_n \phi_n^*(\vec{x},t)\phi_n(\vec{x}',t')\right)$$

$$= -i\left(i\hbar\frac{\partial}{\partial t'}\theta(t' - t)\right) \sum_n \phi_n^*(\vec{x},t)\phi_n(\vec{x}',t')$$

$$-i\theta(t' - t) \sum_n \phi_n^*(\vec{x},t) \left(i\hbar\frac{\partial}{\partial t'} - \hat{H}(x')\right) \phi(\vec{x}',t')$$

$$= \hbar\delta(t' - t) \sum_n \phi_n^*(\vec{x},t)\phi_n(\vec{x}',t')$$

$$= \hbar\delta(t' - t) \sum_n \phi_n^*(\vec{x},t)\phi_n(\vec{x}',t)$$

$$= \hbar\delta(t' - t)\delta^3(\vec{x} - \vec{x}') = \hbar\delta^4(x - x'). \tag{6.70}$$

The Schrödinger equation has been used in the second step to eliminate the second term. In the fourth step, t' has been set equal to t because of the delta function.

There is an enormous amount of information contained in $G^+(x'; x)$. All the solutions of the Schrödinger equation, including the bound states, as required in the completeness relationship, appear with equal weight. It is no wonder that G is difficult to compute.

The same Green function which propagates a solution of the Schrödinger equation forward in time also propagates its complex conjugate solution backwards in time. To see this from equation 6.68, we write

$$i \int d^3x G^+(x'; x)\psi_n(x) = \theta(t' - t) \sum_m \psi_m(x') \int d^3x \psi_m^*(x)\psi_n(x)$$

$$= \theta(t' - t)\psi_n(x') \tag{6.71}$$

and

$$i \int d^3x' \psi_m^*(x') G^+(x'; x) = \theta(t' - t) \sum_n \psi_n^*(x) \int d^3x' \psi_n(x')\psi_m^*(x')$$

$$= \theta(t' - t)\psi_m^*(x). \tag{6.72}$$

6.6 Propagator in Relativistic Theory

We now generalize our propagator development of the nonrelativistic theory and apply it to the relativistic theory. Our starting point is provided by the picture of the nonrelativistic $G(x'; x)$ in figure 6.4. One may say the interaction at the ith point, or vertex, destroys the particle propagating up to x_i and creates a particle which propagates on to x_{i+1}, with $t_{i+1} \geq t_i$.

In the relativistic case, there are not only the scattering processes, but also the pair creation and annihilation processes shown in figure 6.6. Diagram 6.6a shows the production of an electron-positron pair by a potential acting at point 1; the two particles of the pair then propagate to points x and x'. Diagram 6.6b shows an electron originating at x and ending up at x'. Along the way, a pair is produced by a potential acting at 1; the positron of the pair annihilates an initial electron in the field at 2; the electron of the pair propagates up to point 3, where it is destroyed by the potential. The potential at 3 creates an electron which propagates to x'. Diagram 6.6c shows a pair produced at 1, both propagating up to 2, and then being destroyed in the field there.

In the relativistic theory, we need not only the amplitude for an electron to be created, to propagate, and to be destroyed as in the nonrelativistic case, but also the amplitude for a positron to be created, to propagate, and to be destroyed. Once the positron amplitude is found, we may then attempt to

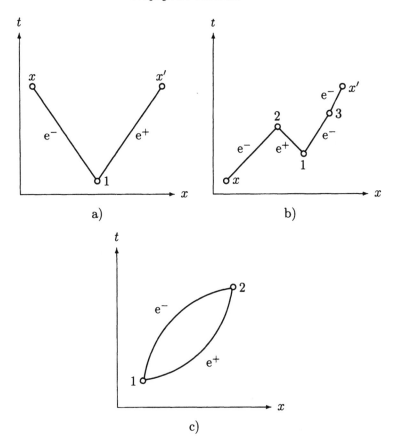

FIGURE 6.6: Examples of space-time diagrams in relativistic theory for
a) pair production, b) scattering, and c) a closed loop.

associate a probability amplitude with each process: pair production, scattering, and annihilation, and to construct the total amplitude for any particular
process, by summing or integrating over all the intermediate paths which can
contribute to the process.

We may determine the positron amplitude in accordance with the hole
theory. Since the existence of a positron is associated with the absence of a
negative-energy electron from the filled sea, we may view the destruction of a
positron as equivalent to the creation of a negative-energy electron.

In addition to electron paths which zigzag forward and backwards in space-
time, there is also the possibility of closed loops (figure 6.6c). Processes
such as these may not simply be ignored. The formalism requires them and
experiments verify their existence.

6.7   Propagator for the Klein-Gordon Equation

Having developed the nonrelativistic propagator, we now turn to the development of the relativistic propagators. To delay the slight complication introduced by the Dirac matrices, we first develop the Klein-Gordon propagator. We will only consider the free-particle Klein-Gordon propagator.

The free-particle Klein-Gordon propagator $\triangle_F(x'-x)$ is defined to satisfy the Green-function equation

$$(\Box_{x'} + m^2)\triangle_F(x'-x) = -\delta^4(x'-x). \tag{6.73}$$

The minus sign on the right-hand side of equation 6.73 is chosen by convention since equation 4.61 also has a minus sign on the right-hand side. In addition to satisfying equation 6.73, the propagator must also only propagate positive-energy solutions forward in time and only propagate negative-energy solutions backwards in time.

Rather than solve the Green-function equation in configuration space, we work in momentum space. The momentum-space representation will be totally adequate for our purposes since we will never require the explicit form of the propagators in configuration space.

The Fourier transform of the momentum-space propagator $\tilde{\triangle}_F(p)$ in configuration space is

$$\triangle_F(x'-x) = \int \frac{d^4p}{(2\pi)^4} e^{-ip\cdot(x'-x)}\tilde{\triangle}_F(p). \tag{6.74}$$

Substituting this Fourier transform into equation 6.73, performing the differentiations, and representing the delta function as a Fourier transform gives

$$\int \frac{d^4p}{(2\pi)^4}(p^2 - m^2)e^{-p\cdot(x'-x)}\tilde{\triangle}_F(p) = \int \frac{d^4p}{(2\pi)^4}e^{-ip\cdot(x'-x)}. \tag{6.75}$$

Equating integrands gives

$$\tilde{\triangle}_F(p) = \frac{1}{p^2 - m^2}, \tag{6.76}$$

provided $p^2 \neq m^2$. There is a singularity at $p^2 - m^2 = p_0^2 - \vec{p}^{\,2} - m^2 = 0$ or in other words, when $p_0 = \pm\sqrt{\vec{p}^{\,2} + m^2} = \pm E$. This means that the particle represented by the Green function is off mass-shell, and p_0 and E are independent variables. The origin of negative-energy waves in relativistic propagator theory is the pole at $p_0 = -E$, which was not present in the nonrelativistic theory.

A prescription for how to handle the singularities is needed to complete the definition of the propagator. The interpretation given to the Green function $\triangle_F(x'-x)$ is that it represents the wave produced at the point x' by a unit source located at x. A necessary physical requirement is that the wave propagating from x into the future consists only of positive-energy components. The boundary conditions on the integration of $\triangle_F(x' - x)$ can provide this requirement.

One can evaluate $\triangle_F(x' - x)$ by Cauchy integration in the p_0-plane. We perform the p_0 integration along the contour in the complex p_0-plane, which is infinitesimally close to the real axis and avoids the poles at $p_0 = \pm E$ in some way. As shown in figure 6.7, we close the contour with a semicircle at infinity: above for $t > t'$ and below for $t' > t$. This causes the exponential to vanish in the limit $\Im[p_0] \to \pm\infty$. Only the positive pole contributes for $t' > t$ and only the negative pole contributes for $t > t'$. After the contour integration has been performed, one must take the limit as the contour approaches the real axis and the enclosed semicircle approaches infinity along the imaginary axis.

For $t' > t$, the contour is closed in the lower half-plane and includes the positive-energy pole at $p_0 = +\sqrt{\vec{p}^2 + m^2}$ only. The factor $e^{-ip_0(t'-t)}$ in the integrand will vanish as $\Im(p_0) \to -\infty$, if $t' > t$. Performing the integral in this case, we have

$$\triangle_F(x' - x) = \int_C \frac{d^4p}{(2\pi)^4} \frac{e^{-ip\cdot(x'-x)}}{p^2 - m^2}$$

$$= \lim_{\epsilon \to 0} \left[\int_{-\infty-i\epsilon}^{0-i\epsilon} \frac{d^4p}{(2\pi)^4} \frac{e^{-ip\cdot(x'-x)}}{p^2 - m^2} + \int_{0+i\epsilon}^{+\infty+i\epsilon} \frac{d^4p}{(2\pi)^4} \frac{e^{ip\cdot(x'-x)}}{p^2 - m^2} \right]$$

$$= \int \frac{d^3p}{(2\pi)^3} e^{i\vec{p}\cdot(\vec{x}'-\vec{x})} \oint_C \frac{dp_0}{2\pi} \frac{e^{-ip_0(t'-t)}}{p^2 - m^2}$$

$$= \int \frac{d^3p}{(2\pi)^3} e^{i\vec{p}\cdot(\vec{x}'-\vec{x})} \oint_C \frac{dp_0}{2\pi} \frac{e^{-ip_0(t'-t)}}{(p_0 + E)(p_0 - E)}. \tag{6.77}$$

Using Cauchy's integral formula (equation 2.35) for the pole at $p_0 = E$ gives

$$\triangle_F(x' - x) = -\int \frac{d^3p}{(2\pi)^3} e^{i\vec{p}\cdot(\vec{x}'-\vec{x})} \frac{ie^{-iE(t'-t)}}{2E}, \tag{6.78}$$

where the minus sign comes from the clockwise direction of the contour enclosing the pole.

For $t > t'$, the contour is closed in the upper half-plane and includes the negative-energy pole at $p_0 = -\sqrt{\vec{p}^2 + m^2}$. The factor $e^{-ip_0(t'-t)}$ in the integrand will vanish as $\Im(p_0) \to +\infty$, if $t > t'$. Performing the integral in this case, we have

$$\triangle_F(x' - x) = -\int \frac{d^3p}{(2\pi)^3} e^{i\vec{p}\cdot(\vec{x}'-\vec{x})} \frac{ie^{iE(t'-t)}}{2E}. \tag{6.79}$$

Notice that the only difference between the result of the two integrations is the sign of E in the exponential.

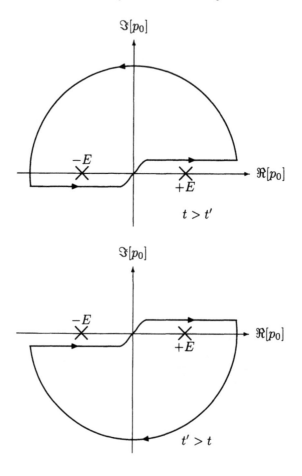

FIGURE 6.7: The causal contours in the complex p_0-plane used for integration of the Klein-Gordon propagator.

We point out that any other choice of the integration contour, for example those shown in figure 6.8, would lead to contributions from negative-energy waves propagating into the future (figure 6.8a) or positive-energy waves propagating into the past (figure 6.8b) (see problem 6.9).

An equivalent procedure for performing the integration along the real axis is to displace the poles slightly away from the axis by a small real number ϵ:

$$p_0 - E \rightarrow p_0 - E + i\epsilon,$$
$$p_0 + E \rightarrow p_0 + E - i\epsilon. \tag{6.80}$$

This procedure is shown in figure 6.9 (see problem 6.10). After the integration one takes the limit as the poles approach the real axis ($\epsilon \rightarrow 0$).

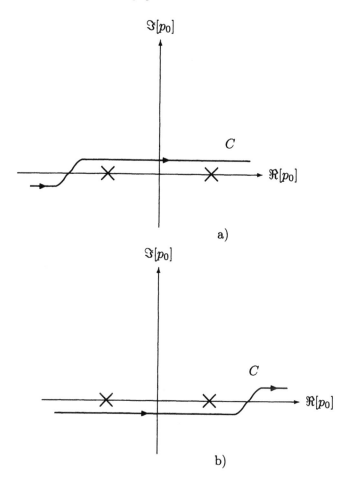

FIGURE 6.8: Integration contours which do not lead to physical solutions.

The procedure of displacing the poles from the real axis is equivalent to adding a small positive imaginary part $+i\epsilon$ to the denominator of the propagator and taking $\epsilon \to 0$ at the end of the calculation:

$$\triangle_F(x' - x) = \int \frac{d^4p}{(2\pi)^4} e^{-ip\cdot(x'-x)} \frac{1}{p^2 - m^2 + i\epsilon}. \tag{6.81}$$

Then the singularity corresponding to positive-energy states is

$$p_0 = +\sqrt{\vec{p}^2 + m^2 - i\epsilon} = +\sqrt{\vec{p}^2 + m^2} - i\epsilon', \tag{6.82}$$

which lies below the real p_0-axis, while the pole corresponding to negative-energy states is

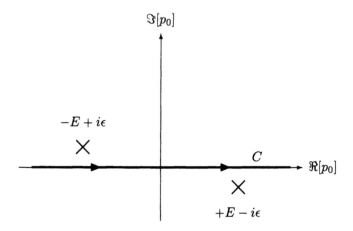

FIGURE 6.9: Displacing the poles away from the real axis allows the integration to be carried out along the real axis in the complex p_0-plane.

$$p_0 = -\sqrt{\vec{p}^2 + m^2 - i\epsilon} = -\sqrt{\vec{p}^2 + m^2} + i\epsilon', \qquad (6.83)$$

which is located above the p_0-axis, as required, ϵ' is another small positive real number.

This prescription is most easily remembered in the form of a rule: to ensure the correct boundary conditions, the mass has to be given a small negative imaginary part $(m \to m - i\epsilon)$. Writing it this way, we have

$$\begin{aligned}
p^2 - m^2 &\to p^2 - (m - i\epsilon)^2 \\
&\approx p^2 - m^2 + 2im\epsilon \\
&= p^2 - m^2 + i\epsilon',
\end{aligned} \qquad (6.84)$$

where $\epsilon' = 2m\epsilon$.

Returning to our two propagator solutions (equations 6.78 and 6.79), we can combine them into one form by using the step functions θ to give

$$\begin{aligned}
\triangle_F(x' - x) = &-i \int \frac{d^3p}{(2\pi)^3 2E} e^{i\vec{p}\cdot(\vec{x}'-\vec{x})} e^{-iE(t'-t)} \theta(t' - t) \\
&-i \int \frac{d^3p}{(2\pi)^3 2E} e^{i\vec{p}\cdot(\vec{x}'-\vec{x})} e^{+iE(t'-t)} \theta(t - t').
\end{aligned} \qquad (6.85)$$

Using the normalized plane-wave solutions (equation 4.22), we can write

$$\Delta_F(x'-x) = -\int d^3p \phi_p^{(+)}(x')\phi_p^{(+)*}(x)\theta(t'-t)$$

$$-\int d^3p \phi_p^{(-)}(x')\phi_p^{(-)*}(x)\theta(t-t'). \qquad (6.86)$$

The propagator now consists of a term (first term) propagating positive-energy solutions and another term (second term) propagating negative-energy solutions. The step functions ensure that the positive-energy solutions are propagated forward in time and the negative-energy solutions backwards in time.

When propagating waves in the Klein-Gordon theory, the operator $i\overleftrightarrow{\partial}_0$ is sandwiched between the propagator and the wave function to allow the orthogonality and normalization conditions (equations 4.46 and 4.47) to be used.

Consider a general wave,

$$\phi(x) = \phi^{(+)}(x)+\phi^{(-)}(x) = \int d^3k a_+(k)\phi_k^{(+)}(x)+\int d^3k a_-(k)\phi_k^{(-)}(x), \quad (6.87)$$

consisting of positive- and negative-energy components. Propagating the positive-energy component gives

$$\int d^3x \Delta_F(x'-x)i\overleftrightarrow{\partial}_0 \phi^{(+)}(x)$$

$$= -i\int d^3x d^3p \left[\phi_p^{(+)}(x')\phi_p^{(+)*}(x)\theta(t'-t)\right.$$

$$\left. + \phi_p^{(-)}(x')\phi_p^{(-)*}(x)\theta(t-t')\right]\overleftrightarrow{\partial}_0 \int d^3k a_+(k)\phi_k^{(+)}(x). \quad (6.88)$$

Using the orthonormal relationships (equations 4.46 and 4.47) gives

$$\int d^3x \Delta_F(x'-x)i\overleftrightarrow{\partial}_0 \phi^{(+)}(x) = -i\int d^3p d^3k a_+(k)\theta(t'-t)\phi_p^{(+)}(x')\delta^3(\vec{p}-\vec{k})$$

$$= -i\theta(t'-t)\int d^3p a_+(p)\phi_p^{(+)}(x')$$

$$= -i\theta(t'-t)\phi^{(+)}(x'). \qquad (6.89)$$

Similarly for the negative-energy component,

$$\int d^3x \Delta_F(x'-x)i\overleftrightarrow{\partial}_0 \phi^{(-)}(x)$$

$$= -i\int d^3x d^3p \left[\phi_p^{(+)}(x')\phi_p^{(+)*}(x)\theta(t'-t)\right.$$

$$+ \phi_p^{(-)}(x')\phi_p^{(-)*}(x)\theta(t - t')\Big] \overleftrightarrow{\partial_0} \int d^3k a_-(k)\phi_k^{(-)}(x)$$

$$= i \int d^3p d^3k a_+(k)\theta(t - t')\phi_p^{(-)}(x')\delta^3(\vec{p} - \vec{k})$$

$$= i\theta(t - t') \int d^3p a_+(p)\phi_p^{(-)}(x')$$

$$= i\theta(t - t')\phi^{(-)}(x'). \tag{6.90}$$

Thus $\Delta_F(x' - x)$ propagates only the positive-energy part of a general wave function forward in time and the negative-energy part backwards in time, as required. The occurrence of a relative minus sign between equations 6.89 and 6.90 results from the difference in the direction of propagation in time.

6.8 Propagator for the Dirac Equation

The relativistic Dirac propagator $S_F(x', x; A)$ is defined to satisfy a Green-function equation

$$\left[\gamma_\mu \left(i\frac{\partial}{\partial x'_\mu} - eA^\mu(x')\right) - m\right]_{\alpha\lambda} S_{F\lambda\beta}(x', x; A) = \delta_{\alpha\beta}\delta^4(x' - x). \tag{6.91}$$

The Dirac propagator is a 4×4 matrix corresponding to the dimensionality of the gamma matrices. Note that the definition of the Dirac propagator differs from the nonrelativistic counterpart; the differential operator $i\partial_{t'} - \hat{H}(x')$ has been multiplied by γ^0 to form the covariant operator $i\hat{\partial}' - e\,\hat{A}' - m$. Suppressing the matrix indices, we have

$$(\hat{\partial}' - e\,\hat{A}' - m)S_F(x', x; A) = \delta^4(x' - x). \tag{6.92}$$

We can compute the free-particle propagator $S_F(x', x)$ using

$$(\hat{\partial}' - m)S_F(x'; x) = \delta^4(x' - x), \tag{6.93}$$

and Fourier transforming to momentum space. $S_F(x'; x)$ depends only on the interval $(x' - x)$. This property is a manifestation of the homogeneity of space and time, and in general would not be valid for the interacting propagator $S_F(x', x; A)$. We try

$$S_F(x'; x) = S_F(x' - x) = \int \frac{d^4p}{(2\pi)^4} e^{-ip\cdot(x' - x)} S_F(p). \tag{6.94}$$

This gives

$$(\not{p} - m)_{\alpha\lambda} S_{F_{\lambda\beta}}(p) = \delta_{\alpha\beta}. \tag{6.95}$$

Solving for the Fourier amplitude $S_F(p)$ and suppressing matrix indices, we find

$$S_F(p) = \frac{1}{\not{p} - m} \equiv \frac{\not{p} + m}{p^2 - m^2} \quad \text{for } p^2 \neq m^2. \tag{6.96}$$

A prescription for how to handle the singularity at $p^2 = m^2$ or $p_0 = \pm E = \pm\sqrt{\vec{p}^2 + m^2}$ is needed. This comes from the boundary conditions put on $S_F(x' - x)$ in the integration.

The interpretation given to the Green function $S_F(x' - x)$ is that it represents the wave produced at the point x' by a unit source located at the point x. The Fourier components of such a localized point source contain many momenta larger than m, the reciprocal of the electron Compton wavelength, and we expect that electrons as well as positrons may be created at x by the source. However, a necessary physical requirement of the hole theory is that the wave propagating from x into the future consist only of positive-energy electron and positron components. Since positive-energy electrons and positrons are represented by wave functions with positive energy-time behavior, $S_F(x' - x)$ can contain in the future $x_0' > x_0$ only positive-energy components.

We perform the p_0 integration along the contour in the complex p_0-plane. For $t' > t$, the contour is closed in the lower half-plane and includes the positive-energy pole at $p_0 = +\sqrt{\vec{p}^2 + m^2} = +E$ only. This gives

$$S_F(x' - x) = \int \frac{d^3p}{(2\pi)^3} e^{i\vec{p}\cdot(\vec{x}'-\vec{x})} \oint \frac{dp_0}{2\pi} \frac{e^{-ip_0(t'-t)}}{p^2 - m^2}(\not{p} + m)$$
$$= -i \int \frac{d^3p}{(2\pi)^3} e^{i\vec{p}\cdot(\vec{x}'-\vec{x})} e^{-iE(t'-t)} \frac{E\gamma_0 - \vec{p}\cdot\vec{\gamma} + m}{2E} \quad \text{for } t' > t,$$

$$\tag{6.97}$$

so that the wave at $(\vec{x}', t')$ contains positive-energy components only. For $t' < t$, the contour can be closed above the real axis, including the pole at $p_0 = -\sqrt{\vec{p}^2 + m^2} = -E$. This gives

$$S_F(x' - x) = -i \int \frac{d^3p}{(2\pi)^3} e^{i\vec{p}\cdot(\vec{x}'-\vec{x})} e^{+iE(t'-t)} \frac{-E\gamma_0 - \vec{p}\cdot\vec{\gamma} + m}{2E} \quad \text{for } t' < t, \tag{6.98}$$

showing that the propagator consists of negative-energy waves for $t' < t$. Any other choice of contours leads to negative-energy waves propagating into the future or positive-energy waves propagating into the past. Moreover, the negative-energy waves propagating into the past that we have just found are welcome; they are the positive-energy positrons. The origin of the negative-energy waves is the pole at $p_0 = -\sqrt{\vec{p}^2 + m^2}$, which was not present in the nonrelativistic theory.

The choice of the contour is summarized by adding a small positive imaginary part to the denominator, or simply taking $m^2 \to m^2 - i\epsilon$, where the limit $\epsilon \to 0$ is understood:

$$S_F(x' - x) = \int \frac{d^4 p}{(2\pi)^4} \frac{e^{-ip \cdot (x'-x)}}{p^2 - m^2 + i\epsilon} (\not{p} + m). \qquad (6.99)$$

The two integrations (equations 6.97 and 6.98) can be combined by changing the dummy variable $\vec{p}$ to $-\vec{p}$ in the negative-energy part, and introducing projection operators:

$$S_F(x' - x) = -i \int \frac{d^3 p}{(2\pi)^3} \frac{m}{E} \left[\Lambda_+(p) e^{-ip \cdot (x'-x)} \theta(t' - t) \right.$$
$$\left. + \Lambda_-(p) e^{ip \cdot (x'-x)} \theta(t - t') \right], \qquad (6.100)$$

with $p_0 = E > 0$.

Equivalently, in terms of normalized plane-wave solutions, we find

$$S_F(x' - x) = -i\theta(t' - t) \int d^3 p \sum_{r=1}^{2} \psi_P^r(x') \overline{\psi}_P^r(x)$$
$$+ i\theta(t - t') \int d^3 p \sum_{r=3}^{4} \psi_P^r(x') \overline{\psi}_P^r(x). \qquad (6.101)$$

We see that $S_F(x' - x)$ carries the positive-energy solutions $\psi^{(+)}$ forward in time and the negative-energy ones $\psi^{(-)}$ backwards in time:

$$\theta(t' - t) \psi^{(+)}(x') = i \int d^3 x S_F(x' - x) \gamma_0 \psi^{(+)}(x), \qquad (6.102)$$

$$\theta(t - t') \psi^{(-)}(x') = -i \int d^3 x S_F(x' - x) \gamma_0 \psi^{(-)}(x). \qquad (6.103)$$

The minus sign in the second equation results from the difference in the direction of propagation in time between equations 6.102 and 6.103.

The propagator $S_F(x' - x)$ is known as the Feynman propagator. As a matter of interest, this spin-1/2 propagator is related to the Klein-Gordon propagator by

$$S_F(x' - x) = (i \not{\partial} - m) \Delta_F(x' - x). \qquad (6.104)$$

From the free propagator $S_F(x' - x)$, we may formally construct the complete Green function and amplitudes for various scattering processes of electrons and positrons in the presence of potentials. The exact Feynman propagator $S_F(x', x; A)$ satisfies equation 6.92 and can be expressed in terms of a superposition of free Feynman propagators:

$$(\hat{p}' - e\,A' - m)S_F(x', x; A) = \delta^4(x' - x), \qquad (6.105)$$

$$(\hat{p}' - m)S_F(x', x; A) = \delta^4(x' - x) + e\,A'S_F(x', x; A) \qquad (6.106)$$
$$= \int d^4 y\,\delta^4(x' - y)[\delta^4(y - x) + e\,A(y)S_F(y, x; A)]$$
$$= \int d^4 y(\hat{p}' - m)S_F(x'; y)$$
$$\cdot [\delta^4(y - x) + e\,A(y)S_F(y, x; A)], \qquad (6.107)$$

$$S_F(x', x; A) = \int d^4 y\,S_F(x' - y)[\delta^4(y - x) + e\,A(y)S_F(y, x; A)]$$
$$= S_F(x' - x) + e\int d^4 y\,S_F(x' - y)\,A(y)S_F(y, x; A). \ (6.108)$$

This is the relativistic counterpart of the Lippmann-Schwinger equation. Another notation for $S_F(x', x; A)$ is $S_F(x'; x; A)$. The integral equation 6.108 determines the complete propagator $S_F(x'; x; A)$ in terms of the free-particle propagator $S_F(x'; x)$.

Proceeding analogously to the nonrelativistic treatment (equation 6.51), the iteration of the integral equation yields the following multiple scattering expansion

$$S_F(x', x; A) = S_F(x' - x) \qquad (6.109)$$
$$+ e\int d^4 x_1 S_F(x' - x_1)\,A(x_1)S_F(x_1 - x)$$
$$+ e^2\int d^4 x_1 d^4 x_2 S_F(x' - x_1)\,A(x_1)S_F(x_1 - x_2)\,A(x_2)S_F(x_2 - x)$$
$$+ \cdots. \qquad (6.110)$$

In analogy to equation 6.53, the exact solution of the Dirac equation

$$(\hat{p} - m)\Psi(x) = e\,A(x)\Psi(x) \qquad (6.111)$$

is completely determined in terms of S_F if one imposes the boundary conditions of Feynman and Stückelberg. If $\psi(x)$ is the solution of the free Dirac equation, i.e. of the homogeneous version, and the potential $V(x)$ occurring in equation 6.53 is replaced by $e\,A(x)$, we have

$$\Psi(x) = \lim_{t' \to -\infty} \int d^4x'\, S_F(x, x'; A)\psi(x')$$

$$= \lim_{t' \to -\infty} \int d^4x'\, [S_F(x - x') + \int d^4y\, S_F(x - y)e\, A(y)S_F(y - x')]\psi(x')$$

$$= \psi(x) + \lim_{t' \to -\infty} \int d^4y\, S_F(x - y)e\, A(y) \int d^4x'\, S_F(y - x')\psi(x'),$$

$$\boxed{\Psi(x) = \psi(x) + e \int d^4y\, S_F(x - y)\, A(y)\Psi(y)}. \qquad (6.112)$$

The scattering wave contains only positive energies in the future and negative energies in the past:

$$\Psi(x) - \psi(x) \to \int d^3p \sum_{r=1}^{2} \psi_P^r(x)[-ie \int d^4y\, \overline{\psi}_P^r(y)\, A(y)\Psi(y)] \quad \text{as} \quad t \to +\infty$$

$$(6.113)$$

and

$$\Psi(x) - \psi(x) \to \int d^3p \sum_{r=3}^{4} \psi_P^r(x)[+ie \int d^4y\, \overline{\psi}_P^r(y)\, A(y)\Psi(y)] \quad \text{as} \quad t \to -\infty.$$

$$(6.114)$$

6.9 S-Matrix

In a scattering process, we may mathematically view the initial incoming state as being transformed into the final outgoing state by a transformation matrix. The S-matrix (or scattering matrix) is the matrix which transforms an incoming state into an outgoing scattered state. We develop the S-matrix by first considering the nonrelativistic case and then the relativistic case.

In order to define properly the scattering problem, there should be no interaction at the initial time,

$$\lim_{t \to -\infty} V(\vec{x}, t) = 0, \qquad (6.115)$$

so that $\phi(\vec{x}, t)$ is a solution of the free-particle wave equation which incorporates the required initial conditions. The exact incoming wave $\psi(\vec{x}, t)$ becomes the incoming wave $\phi(\vec{x}, t)$ in the limit $t \to -\infty$:

$$\lim_{t \to -\infty} \psi(\vec{x}, t) = \phi(\vec{x}, t). \tag{6.116}$$

We are primarily interested in the form of the scattered wave as $t' \to \infty$, and assume the potential vanishes after a certain time:

$$\lim_{t' \to \infty} V(\vec{x}', t') = 0. \tag{6.117}$$

In this limit, the particle emerges from the interaction region, and $\psi(\vec{x}', t')$ becomes a different solution of the free-particle wave equation. All information about the scattered wave may be obtained from the probability amplitude for the particle to arrive in various final free states $\phi_f(\vec{x}', t')$ as $t' \to \infty$, for a given incident wave $\phi_i(\vec{x}, t)$.

Let $\psi_i^{(+)}(\vec{x}, t)$ be a solution of equation 6.38, which reduces to a plane wave of momentum $\vec{k}$ as $t \to -\infty$. The superscript $(+)$ over ψ_i is meant to express the fact that we are dealing with a wave which propagates into the future. The probability amplitude for a given pair of wave functions (f, i) is an element of the S-matrix:

$$
\begin{aligned}
S_{fi} &= \lim_{t' \to \infty} \langle \phi_f(\vec{x}', t') | \psi_i^{(+)}(\vec{x}', t') \rangle \\
&= \lim_{t' \to \infty} \int d^3x' \, \phi_f^*(\vec{x}', t') \psi_i^{(+)}(\vec{x}', t') \\
&= \lim_{t' \to \infty} \lim_{t \to -\infty} i \int d^3x' \, \phi_f^*(\vec{x}', t') \int d^3x \, G^+(\vec{x}', t'; \vec{x}, t) \phi_i(\vec{x}, t) \\
&= \lim_{t' \to \infty} \int d^3x' \, \phi_f^*(\vec{x}', t') \Big[\phi_i(\vec{x}', t') \\
&\qquad + \int d^4x \, G_0^+(\vec{x}', t'; \vec{x}, t) V(\vec{x}, t) \psi_i^{(+)}(\vec{x}, t) \Big] \\
&= \delta^3(\vec{k}_f - \vec{k}_i) \\
&\qquad + \lim_{t' \to \infty} \int d^3x' \, d^4x \, \phi_f^*(\vec{x}', t') G_0^+(\vec{x}', t'; \vec{x}, t) V(\vec{x}, t) \psi_i^{(+)}(\vec{x}, t).
\end{aligned}
\tag{6.118}
$$

We may expand $\psi_i^{(+)}(\vec{x}, t)$ in a multiple scattering series by iteration of equation 6.38, and thus express the S-matrix as a multiple scattering series. If we insert $\psi_i^{(+)}(\vec{x}, t)$ from the iterated solution of equation 6.112, we get an expression for the S-matrix in terms of multiple scattering events:

$$
\begin{aligned}
S_{fi} &= \delta^3(\vec{k}_f - \vec{k}_i) + \lim_{t' \to \infty} \int d^3x' \, d^4x \, \phi_f^*(\vec{x}', t') G_0^+(\vec{x}', t'; \vec{x}, t) V(\vec{x}, t) \phi_i(\vec{x}, t) \\
&\quad + \lim_{t' \to \infty} \int d^3x' \, d^4x_1 \, d^4x \, \phi_f^*(\vec{x}', t') G_0^+(\vec{x}', t'; \vec{x}_1, t_1) V(\vec{x}_1, t_1)
\end{aligned}
$$

$$\cdot \, G_0^+(\vec{x}_1, t_1; \vec{x}, t) V(\vec{x}, t) \phi_i(\vec{x}, t) + \cdots. \tag{6.119}$$

The first term – the delta function – does not describe scattering but characterizes the particle flux without scattering. The second term represents single scattering, the third term double scattering, etc. The terms are coherently summed[4] to give the total S-matrix element.

For the relativistic Dirac case, the S-matrix elements are defined in the same manner as in the nonrelativistic case. Defining $\psi_f(x)$ as the final-state free wave with the quantum numbers f that is observed at the end of the scattering process, we infer

$$S_{fi} = \lim_{t \to \pm\infty} \langle \psi_f(x) | \Psi_i(x) \rangle$$

$$= \lim_{t \to \pm\infty} \langle \psi_f(x) | \psi_i(x) + \int d^4 y \, S_F(x - y) e \, \rlap{/}{A}(y) \Psi_i(y) \rangle. \tag{6.120}$$

Here the limit $t \to +\infty$ is understood if $\psi_f(x)$ describes an electron and $t \to -\infty$ if $\psi_f(x)$ describes a positron, since the latter is considered a negative-energy electron moving backwards in time.

For electron scattering, we have

$$S_{fi} = \delta_{fi} - ie \lim_{t \to +\infty} \langle \psi_f(x) | \int d^3 p \sum_{r=1}^{2} \psi_p^r(x) \int d^4 y \, \overline{\psi}_p^r(y) \, \rlap{/}{A}(y) \Psi_i(y) \rangle, \tag{6.121}$$

while positron scattering is described by

$$S_{fi} = \delta_{fi} + ie \lim_{t \to -\infty} \langle \psi_f(x) | \int d^3 p \sum_{r=3}^{4} \psi_p^r(x) \int d^4 y \, \overline{\psi}_p^r(y) \, \rlap{/}{A}(y) \Psi_i(y) \rangle. \tag{6.122}$$

The $\int d^3 x$ integral, implied by the angular brackets, projects out just that state $\psi_p^r(x)$ whose quantum numbers agree with $\psi_f(x)$. All other terms of the integral-sum $\int d^3 p \sum_r$ do not contribute. For electron scattering (equation 6.121), this yields

$$S_{fi} = \delta_{fi} - ie \int d^4 y \, \overline{\psi}_f(y) \, \rlap{/}{A}(y) \Psi_i(y) \tag{6.123}$$

and a similar expression for positron scattering. Both results can be combined by defining $\epsilon_f = +1$ for positive-energy waves in the future and $\epsilon_f = -1$ for negative energy waves in the past:

---

[4]By coherent we mean that the amplitudes are summed and then squared. By incoherent we mean that the amplitudes are squared and then summed.

$$S_{fi} = \delta_{fi} - ie\epsilon_f \int d^4y \overline{\psi}_f(y)\, A\!\!\!/(y)\Psi_i(y) \quad, \tag{6.124}$$

where $\Psi_i(x)$ stands for the incoming wave, which either reduces at $y_0 \to -\infty$ to an incident positive-energy wave ψ_i carrying quantum number i, or at $y_0 \to +\infty$ to an incident negative-energy wave propagating into the past with quantum number i, according to Feynman-Stückelberg boundary conditions.

Equations 6.112 and 6.124 contain the rules for calculating the pair production and annihilation amplitudes, as well as, the "ordinary" scattering process. In practice, we shall usually calculate only the first non-vanishing contribution – lowest-order in e – to the S-matrix for a given interaction. The validity of this procedure depends on the weakness of the interaction eA and the rapid convergence of the series in powers of the interaction strength.

The application of the theory to specific systems is considerably simplified by the symmetry properties of the S-matrix under charge conjugation. For example, the matrix elements of charge-conjugated processes are equal to those of the non-charge-conjugated processes (see problem 6.17). This means the photon-positron scattering cross section is equal to the photon-electron scattering cross section. The cross section for electron-electron collisions is also valid for positron-positron collisions. These two examples show that the charge symmetry of the theory considerably reduces the number of processes which must be calculated.

The S-matrix is unitary by construction – if the interaction Hamiltonian is hermitian (see problem 6.18). If we start directly with an assumed interaction Hamiltonian then we do not know if the resulting S-matrix is unitary. This is truly important, because unitarity implies conservation of probability, and probability is the link between the formalism and physical reality. If unitarity is violated, then we have nothing that can be interpreted as probability and the link to observed processes disappears. In any case, properties like unitarity, Lorentz invariance, locality, etc. must be the framework within which the theory is formulated.

6.9.1 Electron Scattering

Consider ordinary scattering of electrons. In this case,

$$\Psi_i(y) \overset{y_0 \to -\infty}{\Longrightarrow} \psi_i^{(+E)}(y) = \frac{1}{(2\pi)^{3/2}}\sqrt{\frac{m}{E_-}}u(p_-,s_-)e^{-ip_-\cdot y} \tag{6.125}$$

reduces to an incoming electron wave with positive energy E_-, momentum $\vec{p}_-$, and spin s_-. The minus sign designates the negative charge of the electron.

The nth order contribution to the S-matrix element is

$$S_{fi}^{(n)} = -ie^n \int d^4y_1 \cdots d^4y_n \overline{\Psi}_f^{(+E)}(y_n)\, A\!\!\!/(y_n)S_F(y_n - y_{n-1})\, A\!\!\!/(y_{n-1})$$

$$\cdots \; S_F(y_2 - y_1) \, \slashed{A}(y_1) \psi_i^{(+E)}(y_1). \tag{6.126}$$

This expression contains both types of diagrams as shown in figure 6.10. That is, in addition to ordinary scattering, intermediate pair creation and pair annihilation are included in the series, since the various $d^4 y_i$ integrations also allow for a reverse time ordering, $y_{n+1}^0 < y_n^0$.

6.9.2   Positron Scattering

Consider positron scattering, which, in lowest order in eA, is represented by either of the two equivalent diagrams shown in figure 6.11. The incident wave is an electron of negative energy labeled by quantum numbers $(-\vec{p}_+, -s_+)$ and $\epsilon_f = -1$. The final state (outgoing wave) is also represented as a negative-energy electron.

Arrows on a world line in a Feynman diagram keep track of the entry and exit at each vertex. An arrow forward in time implies positive energy, while an arrow backwards in time signifies negative energy. There is no distinction between particle and antiparticle propagators since the Feynman prescription describes both simultaneously.

6.9.3   Pair Creation

Next, consider pair production as shown in figure 6.12. In this case, $\Psi_i(y)$ at $y_0 \to +\infty$ reduces to a plane wave with negative energy. This particle state propagating backwards in time represents a positron. We use the notation $(\vec{p}_-, s_-)$ for the three-momentum and spin corresponding to the physical electron and $(\vec{p}_+, s_+)$ for the physical positron, where $p_\pm^0 > 0$. The physical positron state at $t \to \infty$ is described by a plane wave of negative energy with quantum numbers $(-\vec{p}_+, -s_+)$ and $\epsilon_f = -1$. The wave propagating backwards in time entering into the vertex is

$$\Psi_i(y) \stackrel{y_0 \to \infty}{\Longrightarrow} \psi_i^{(-E)}(y) = \frac{1}{(2\pi)^{3/2}} \sqrt{\frac{m}{E_+}} v(p_+, s_+) e^{ip_+ \cdot y}. \tag{6.127}$$

This form of the wave function explicitly exhibits the negative energy and negative three-momentum. The fact that the spin direction is reversed, i.e. $-s_+$, is taken into account by the definition of the spinor $v(p_+, s_+)$. Recall (equation 5.221) that the spinors have been defined according to

$$v(p_+, +1/2) = \omega^4(p_+) \quad \text{and} \quad v(p_+, -1/2) = \omega^3(p_+), \tag{6.128}$$

where ω^4 is the spinor corresponding to a negative-energy electron with spin up and ω^3 to a negative-energy electron with spin down.

The final wave function ψ_f in the case of pair production is a positive-energy solution carrying the quantum numbers $(\vec{p}_-, s_-)$ and $\epsilon_f = +1$, and describes an electron.

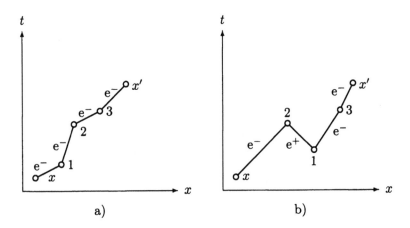

FIGURE 6.10: Two third-order diagrams for electron scattering.

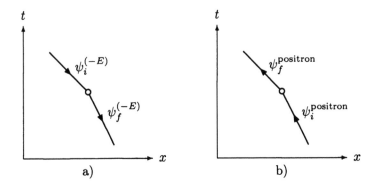

FIGURE 6.11: Positron scattering: a) viewed as a negative-energy electron moving backwards in time and b) viewed as a positron moving forward in time.

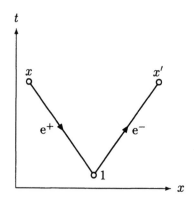

FIGURE 6.12: Pair creation.

The two second-order amplitudes for pair creation are shown in figure 6.13. These two second-order diagrams are said to differ in the time ordering of the two scattering processes. Since the Feynman propagator consists of two parts, there is no need to deal explicity with time orderings when calculating any process; the formula for the S-matrix automatically contains them all. However, we do not work with the general S-matrix, but visualize it one term at a time, and hence one diagram at a time.

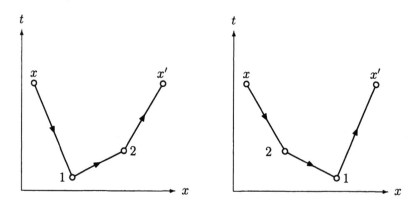

FIGURE 6.13: Second-order diagrams for pair creation.

6.9.4 Pair Annihilation

Finally, consider pair annihilation as shown in figure 6.14. In this case, we insert for $\Psi_i(y)$ a solution that reduces to $\Psi_i^{(+E)}(y)$ as $t \to -\infty$. This positive-energy solution represents an electron that propagates forward in time into the interaction volume, is scattered backwards in time, and emerges into a negative energy state. The nth order amplitude for electron scattering into a given final state $\psi_f^{(-E)}$, labeled by the physical quantum numbers $(\vec{p}_+, s_+)$, $\epsilon_f = -1$, is given by

$$S_{fi}^{(n)} = ie^n \int d^4y_1 \cdots \int d^4y_n \overline{\psi}_f^{(-E)}(y_n) \,A(y_n) S_F(y_n - y_{n-1})$$

$$\cdots A(y_1) \psi_i^{(+E)}(y_1). \qquad (6.129)$$

In hole theory, this is the nth order amplitude for a positive-energy electron to scatter into an electron state of negative three-momentum $-\vec{p}_+$ and spin $-s_+$.

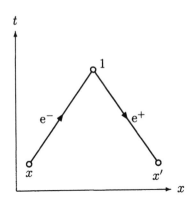

FIGURE 6.14: Pair annihilation.

6.10 Problems

1. Verify that the solution in equation 6.1 satisfies the wave equation asymptotically through terms of order $1/r$ in the region in which $V = 0$, for any form of the function $f(\theta, \phi)$.

2. Verify that equation 6.10 is a solution to equation 6.7.

3. Verify that equation 6.14 is a solution to equation 6.13.

4. Using the first Born approximation, calculate the cross section for electron scattering off a Coulomb potential due to an atomic nucleus.

5. [4] Show that the rate for an electron in the hydrogen-atom ground state to radiate and fall into empty negative-energy states (treated in Born approximation) in the energy interval $-mc^2$ to $-2mc^2$ is approximately $2\alpha^6 mc^2/\pi\hbar \approx 10^8$ s^{-1}.

6. [8] Show the following relationships for the nonrelativistic propagator.

   (a) If $t' > t_1 > t$,

   $$G^+(\vec{x}',t';\vec{x},t) = i \int d^3x_1 G^+(\vec{x}',t';\vec{x}_1,t_1)G^+(\vec{x}_1,t_1;\vec{x},t).$$

   (b) If $t' < t_1 < t$,

   $$G^-(\vec{x}',t';\vec{x},t) = -i \int d^3x_1 G^-(\vec{x}',t';\vec{x}_1,t_1)G^-(\vec{x}_1,t_1;\vec{x},t).$$

   (c) If $t > t_1$,

   $$\delta^3(\vec{x}-\vec{x}') = \int d^3x_1 G^+(\vec{x}',t';\vec{x}_1,t_1)G^-(\vec{x}_1,t_1;\vec{x},t).$$

   (d) If $t < t_1$,

   $$\delta^3(\vec{x}-\vec{x}') = \int d^3x_1 G^-(\vec{x}',t';\vec{x}_1,t_1)G^+(\vec{x}_1,t_1;\vec{x},t).$$

7. [8] Show

$$G^+(\vec{x}',t';\vec{x},t) = G^{-*}(\vec{x},t;\vec{x}',t').$$

8. Show

$$\theta(\tau) = -\frac{1}{2\pi i}\lim_{\epsilon\to 0}\int_{-\infty}^{\infty} d\omega \frac{e^{-i\omega\tau}}{\omega+i\epsilon}$$

by performing the ω-integral as a contour integral in the complex ω-plane.

Prove equation 6.57 by differentiating your result.

9. Show that using the integration contours in figure 6.8 leads to contributions from negative-energy waves propagating into the future or positive-energy waves propagating into the past.

10. Show that using the integration procedure in figure 6.9 is equivalent to using the integration contours in figure 6.7.

11. Show that $S_F(x', x)$ reduces to the free-particle retarded propagator of the Schrödinger equation in the nonrelativistic limit.

12. Verify equation 6.101 explicitly.

13. [4] Calculate $S_F(x)$ explicitly. How does it behave as $x \to \infty$, as $x \to 0$, and on the light cone?

14. [4] Suppose in our formalism we replace the vacuum by a Fermi gas with Fermi momentum k_F. How is the Feynman propagator modified? Compute the change in S_F in the low-density limit.

15. Show that the forms

$$S_F(x' - x) = -i \int \frac{d^3p}{(2\pi)^3} e^{i\vec{p}\cdot(\vec{x}'-\vec{x})} e^{-iE(t'-t)} \frac{+E\gamma_0 - \vec{p}\cdot\vec{\gamma} + m}{2E}$$

for $t' > t$ and

$$S_F(x' - x) = -i \int \frac{d^3p}{(2\pi)^3} e^{i\vec{p}\cdot(\vec{x}'-\vec{x})} e^{+iE(t'-t)} \frac{-E\gamma_0 - \vec{p}\cdot\vec{\gamma} + m}{2E}$$

for $t' < t$ can be combined into a single expression by using energy-projection operators.

16. Verify

$$\theta(t' - t)\psi^{(+)}(x') = i \int d^3x\, S_F(x' - x)\gamma_0\psi^{(+)}(x)$$

and

$$\theta(t - t')\psi^{(-)}(x') = -i \int d^3x\, S_F(x' - x)\gamma_0\psi^{(-)}(x),$$

and derive analogous relationships for the adjoint solutions $\overline{\psi}^{(+)}$ and $\overline{\psi}^{(-)}$.

17. Show that the S-matrix is symmetric under charge conjugation.

18. Show that the S-matrix is unitary.

Chapter 7

Photons

In this chapter, we examine the main properties of the photon that will be used in subsequent discussions. Starting from the classical Maxwell's equations and a discussion of gauge transformations, the wave function, polarizations, and propagator of the photon are developed. A complication in specifying the polarization states and propagator arises from the massless nature of the photon and gauge invariance.

7.1 Maxwell's Equations

Maxwell's equations of classical electrodynamics for the electric $\vec{E}(\vec{x}, t)$ and magnetic $\vec{B}(\vec{x}, t)$ fields in vacuum, but in the presence of a charge density $\rho(\vec{x}, t)$ and a current density $\vec{j}(\vec{x}, t)$, are

$$\vec{\nabla} \cdot \vec{E} = \rho, \tag{7.1}$$

$$\vec{\nabla} \times \vec{B} - \frac{1}{c}\frac{\partial \vec{E}}{\partial t} = \frac{\vec{j}}{c}, \tag{7.2}$$

$$\vec{\nabla} \cdot \vec{B} = 0, \tag{7.3}$$

$$\vec{\nabla} \times \vec{E} + \frac{1}{c}\frac{\partial \vec{B}}{\partial t} = 0, \tag{7.4}$$

where we are using Heaviside-Lorentz (rationalized Gaussian) units[1].

In three-dimensional tensor form, Maxwell's equations 7.1-7.4 can be written as

$$\frac{\partial E_k}{\partial x_k} = \frac{j_0}{c},$$

$$\varepsilon_{ijk}\frac{\partial B_k}{\partial x_j} - \frac{\partial E_i}{\partial x_0} = \frac{j_i}{c},$$

---

[1]The different systems of units for Maxwell's equations have already been reviewed in section 2.2.

$$\frac{\partial B_k}{\partial x_k} = 0,$$

$$\varepsilon_{ijk}\frac{\partial E_k}{\partial x_j} + \frac{\partial B_i}{\partial x_0} = 0, \tag{7.5}$$

where we have used $x^\mu = (ct, \vec{x})$, and introduced the four-current density $j^\mu = (c\rho, \vec{j})$. The above tensor forms suggest defining the second-rank electromagnetic contravariant field four-tensor,

$$F^{\mu\nu} = \begin{pmatrix} 0 & E_x & E_y & E_z \\ -E_x & 0 & B_z & -B_y \\ -E_y & -B_z & 0 & B_x \\ -E_z & B_y & -B_x & 0 \end{pmatrix}. \tag{7.6}$$

The components of this antisymmetric tensor can be written in the form $F^{0i} = E^i$ and $\frac{1}{2}\varepsilon^{ijk}F_{jk} = B^i$.

Using the field tensor, the inhomogeneous Maxwell's equations 7.1 and 7.2 can be written as

$$\partial_\nu F^{\mu\nu} = \frac{j^\mu}{c}, \tag{7.7}$$

and the homogenous equations 7.3 and 7.4 as

$$\partial^\lambda F^{\mu\nu} + \partial^\mu F^{\nu\lambda} + \partial^\nu F^{\lambda\mu} = 0. \tag{7.8}$$

In four-vector notation, current conservation

$$\partial_\nu j^\nu = 0 \tag{7.9}$$

follows automatically when contracting the antisymmetric field tensor $F^{\mu\nu}$ with the symmetric tensor operator $\partial^\mu \partial^\nu$.

Maxwell's equations in covariant form simplify further if we introduce potentials. We may introduce a vector potential $\vec{A}$, such that

$$\vec{B} = \vec{\nabla} \times \vec{A}. \tag{7.10}$$

Since the divergence of a curl vanishes, Maxwell's equation 7.3 is satisfied for all vector potentials satisfying equation 7.10. Similarly, the scalar potential ϕ is defined, such that

$$\vec{E} = -\frac{1}{c}\frac{\partial \vec{A}}{\partial t} - \vec{\nabla}\phi. \tag{7.11}$$

Using the definition of $\vec{A}$ in equation 7.10 and that the curl of a gradient vanishes, Maxwell's equation 7.4 is automatically satisfied for all potentials satisfying equations 7.10 and 7.11. In three-dimensional tensor form, the equations for the potentials are

$$B^i = \frac{\partial A^j}{\partial x_k} - \frac{\partial A^k}{\partial x_j} = F^{jk}, \tag{7.12}$$

$$E^i = \frac{\partial \phi}{\partial x_i} - \frac{1}{c}\frac{\partial A^i}{\partial t} = F^{oi}. \tag{7.13}$$

If we define the four-potential $A^\mu = (\phi, \vec{A})$, we can combine equations 7.12 and 7.13 to obtain

$$F^{\mu\nu} = \partial^\nu A^\mu - \partial^\mu A^\nu. \tag{7.14}$$

This form automatically satisfies the four-vector form of Maxwell's homogenous equation 7.8. In terms of the inhomogeneous Maxwell's equation 7.7, this form gives

$$\partial_\nu(\partial^\nu A^\mu - \partial^\mu A^\nu) = \frac{j^\mu}{c},$$

$$\Box A^\mu - \partial^\mu(\partial \cdot A) = \frac{j^\mu}{c}. \tag{7.15}$$

This result could have also been obtained by using the scalar and vector potentials along with Maxwell's two inhomogeneous equations in three-vector form. The form of the resulting two coupled inhomogeneous differential equations would not be as compact as equation 7.15, but the covariant form could be readily obtained by defining the four potential and four-current density (see problem 7.1).

7.2 Gauge Transformations

As we will see, there remains some freedom in the choices of A^μ. This freedom, called the "choice of gauge", can be utilized to impose relationships that the A^μ components must satisfy. We already introduced gauge transformations briefly in section 4.7.1.

The potentials ϕ and $\vec{A}$ are not uniquely determined. We may replace $\vec{A}$ by $\vec{A}' = \vec{A} + \vec{\nabla}\chi$, where χ is a scalar function – the gauge field. Then

$$\vec{B} = \vec{\nabla} \times \vec{A} \rightarrow \vec{B}' = \vec{\nabla} \times \vec{A}' = \vec{\nabla} \times \vec{A} = \vec{B}, \tag{7.16}$$

since the curl of a gradient vanishes. If in addition, ϕ is replaced by some ϕ', then

$$\vec{E} = -\vec{\nabla}\phi - \frac{1}{c}\frac{\partial \vec{A}}{\partial t} \rightarrow \vec{E}' = -\vec{\nabla}\phi' - \frac{1}{c}\frac{\partial \vec{A}'}{\partial t} = -\vec{\nabla}\phi' - \frac{1}{c}\frac{\partial \vec{A}}{\partial t} - \frac{1}{c}\vec{\nabla}\frac{\partial \chi}{\partial t}. \tag{7.17}$$

The electric field $\vec{E}$ will remain unchanged if

$$\phi' = \phi - \frac{1}{c}\frac{\partial \chi}{\partial t}. \tag{7.18}$$

In four-vector form these two transformations can be combined into a single gauge transformation

$$A^\mu \rightarrow A'^\mu - \partial^\mu \chi. \tag{7.19}$$

This means that there is a continuous family of four-vector potentials all of which give the same observable electromagnetic field. Since only the fields have direct physical interpretation – not the potentials – the theory is invariant under gauge transformations.

Gauge invariance is a very powerful concept that can lead to current conservation. The interaction of any electromagnetic current $J_\mu(x)$ with the vector potential $A_\mu(x)$ is given by

$$\int d^4x\, J_\mu(x) A^\mu(x). \tag{7.20}$$

The integral must be invariant under a gauge transformation (equation 7.19). Integrating the new term by parts implies

$$0 = \int d^4x\, J_\mu(x)\frac{\partial \chi(x)}{\partial x_\mu} = \int d^4x\, \frac{\partial J_\mu(x)}{\partial x_\mu}\chi(x). \tag{7.21}$$

Since $\chi(x)$ is an arbitrary function, this yields the condition of current conservation

$$\frac{\partial J_\mu(x)}{\partial x_\mu} = 0, \tag{7.22}$$

which can be written in momentum space as

$$k_\mu J^\mu(k) = 0, \tag{7.23}$$

where $J_\mu(x) = \int d^4x\, J_\mu(k) e^{-k\cdot x}$. We will see in section 7.4, this property is also shared by quantum-mechanical transition currents.

7.2.1 Lorentz Gauge

In the Lorentz gauge, we choose from among the family of vector potentials those satisfying the covariant equation

$$\partial_\nu A^\nu = \frac{1}{c}\frac{\partial \phi}{\partial t} + \vec{\nabla}\cdot\vec{A} = 0. \tag{7.24}$$

In this gauge, the inhomogeneous Maxwell's equation for A^μ reduces to

$$\Box A^{\mu} = \frac{j^{\mu}}{c}. \tag{7.25}$$

Notice that if some given A^{μ} does not satisfy the equation $\partial_{\nu} A^{\nu} = 0$, we can make a gauge transformation to a new $A'^{\mu} = A^{\mu} - \partial^{\mu}\chi$ such that $\partial_{\nu} A'^{\nu} = 0$ by finding a χ that solves the equation $\Box\chi = \partial \cdot A$. Solutions to this equation always exist.

Even after choosing a gauge, there is still some residual freedom in the choice of the potential A^{μ} in the Lorentz gauge. We can still make another gauge transformation

$$A'^{\mu} \rightarrow A''^{\mu} = A'^{\mu} - \partial^{\mu}\chi', \tag{7.26}$$

where χ' is any function that satisfies

$$\Box\chi' = 0. \tag{7.27}$$

This last equation ensures that the Lorentz condition is still satisfied.

The advantage of the Lorentz gauge is its Lorentz covariance. We will see, that in this gauge there are, in addition to the two transverse polarization states of the photon, longitudinal and scalar photon states.

7.2.2 Coulomb Gauge

The Coulomb Gauge is specified by

$$\vec{\nabla} \cdot \vec{A} = 0. \tag{7.28}$$

The advantage of the coulomb gauge is that it yields only two transverse-photon states, or after an appropriate transformation, two photons with helicity ± 1. The Coulomb gauge is also called the transverse gauge since the polarization vector is transverse to the direction of motion, or radiation gauge since the photon is real and represents radiation. By similar arguments as above, the required gauge fields needed to transform the potentials to those satisfying the Coulomb gauge are given by $\nabla^2\chi = -\vec{\nabla} \cdot \vec{A}$.

7.2.3 Other Gauges

Some other gauges include the time or temporal gauge, $A^0 = 0$, the axial gauge, $A^3 = 0$, and the unitary gauge, A^0 real. These gauges will not be used here, although $A^0 = 0$ will sometimes be imposed to satisfy the Lorentz gauge.

7.3   Polarization Vectors

Turning now from classical electrodynamics to quantum mechanics, a free photon can be represented by a wave function A^μ, which satisfies equation 7.15 in the absence of charges and currents. A^μ is a four-dimensional vector field and thus appears to have four degrees of freedom – labeled by λ. For now, we restrict ourselves to the Lorentz gauge, and thus the four-potential must satisfy equation 7.25 in the absence of charges and currents.

Consider a plane-wave solution to this equation:

$$A^\mu(x;k) = \frac{1}{\sqrt{2\omega V}} \left[\varepsilon^\mu(\vec{k},\lambda)e^{-ik\cdot x} + \varepsilon^{*\mu}(\vec{k},\lambda)e^{ik\cdot x} \right], \qquad (7.29)$$

where $k^\mu = (\omega, \vec{k})$ is the four-momentum and ε^μ is the unit polarization four-vector of the photon. The choice for ε^μ is arbitrary. They can be chosen differently for each $\vec{k}$. We will discuss the polarization and normalization presently.

On substituting this solution into equation 7.25 for the case of free electromagnetic fields, we find the condition

$$k^2 = 0, \quad \text{that is,} \quad m = 0. \qquad (7.30)$$

The photon is massless as required by special relativity.

In general, the polarization vectors have four components. The Lorentz condition $\partial_\mu A^\mu = 0$ gives

$$k \cdot \varepsilon = 0, \qquad (7.31)$$

and this reduces the number of independent components of ε^μ to three.

We may further restrict the gauge transformation, provided $\Box\chi = 0$ is satisfied. For a free field, one can choose a gauge in which $A^0 = 0$, whence the Lorentz condition reduces to $\vec{\nabla} \cdot \vec{A} = 0$. This is also a solution to Poisson's equation – which results from the absence of external sources – which vanishes at infinity. This gauge condition is not covariant and will be valid only in one particular reference frame. In this frame, $\varepsilon^0(\vec{k},\lambda) = 0$, and all the polarization vectors are purely space-like transverse vectors:

$$\varepsilon^\mu = (0, \vec{\varepsilon}(\vec{k},\lambda)) \qquad (7.32)$$

and

$$\vec{k} \cdot \vec{\varepsilon}(\vec{k},\lambda) = 0. \qquad (7.33)$$

If $\vec{\varepsilon}$ were along $\vec{k}$, $\vec{\varepsilon}$ would be associated with a helicity-zero photon. This state is missing because of the transversality condition. It can only be absent because the photon is massless. This additional gauge condition restricts the number of degrees of freedom of A^μ to two. Thus we have derived the well-known fact that photons can have only two polarization states, $\lambda = 1, 2$.

The polarization vectors are orthonormalized to unity:

$$\vec{\varepsilon}^*(\vec{k}, \lambda) \cdot \vec{\varepsilon}(\vec{k}, \lambda') = \delta_{\lambda\lambda'}, \qquad (7.34)$$

or

$$\varepsilon^*(\vec{k}, \lambda) \cdot \varepsilon(\vec{k}, \lambda') = -\delta_{\lambda\lambda'}. \qquad (7.35)$$

In this special reference frame, ε^μ is purely space-like. In an arbitrary reference frame ε^μ will remain space-like (see problem 3.4) and orthonormalized according to equation 7.35.

The polarization vector acts as the "spin part" of the wave function for the photon. The various cases which can occur with regards to the polarization of the photon are identical with the possible types of polarizations of the classical electromagnetic wave. Any polarization can be represented as a superposition of two mutually orthogonal polarizations chosen in some specified manner.

We see that there are only two independent polarization vectors and that they are both transverse to the three-momentum of the photon. As an example, for a photon traveling along the z-axis, we may take

$$\vec{\varepsilon}_1 = (1, 0, 0), \quad \vec{\varepsilon}_2 = (0, 1, 0), \qquad (7.36)$$

if $\vec{\varepsilon}(\vec{k}, \lambda)$ are real. These are referred to as linear polarization vectors. If $\vec{\varepsilon}(\vec{k}, \lambda)$ is complex[2], the linear combinations

$$\vec{\varepsilon}_R = -\frac{\vec{\varepsilon}_1 + i\vec{\varepsilon}_2}{\sqrt{2}} \quad +1 \text{ helicity}, \qquad (7.37)$$

$$\vec{\varepsilon}_L = +\frac{\vec{\varepsilon}_1 - i\vec{\varepsilon}_2}{\sqrt{2}} \quad -1 \text{ helicity}, \qquad (7.38)$$

are called circular polarization vectors.

We can continue to explore the consequences of the gauge freedom by noticing that, since $k^2 = 0$, any gauge field of the form

$$\chi = -ia(k)e^{\pm ik \cdot x}, \qquad (7.39)$$

with $a(k)$ an arbitrary real function of k only, will automatically satisfy $\Box\chi = 0$. The orthogonality condition continues to be satisfied. This, together with the solution for A^μ shows that the physics is unchanged by the replacement

$$\varepsilon^\mu \rightarrow \varepsilon'^\mu = \varepsilon^\mu + a(k)k^\mu. \qquad (7.40)$$

---

[2] When using the Feynman dagger notation in chapter 8 with polarization vectors we should be careful: $\bar{\slashed{\varepsilon}}^* = \varepsilon_\mu^* \gamma^\mu \neq (\slashed{\varepsilon})^*$.

In other words, two polarization vectors ε^μ and ε'^μ which differ by a multiple of k^μ describe the same photon. We may use this freedom to ensure that the time component of ε^μ vanishes: $\varepsilon^0 \equiv 0$. A free photon is thus described by its three-momentum $\vec{k}$ and a polarization three-vector $\vec{\varepsilon}$. Since $\vec{\varepsilon}$ transforms as a vector, we anticipate that it is associated with a particle of spin-1.

The intrinsic symmetry properties of a particle will manifest themselves in the rest frame of the particle. Symmetry with respect to all possible rotations about the center of the particle, i.e. with respect to the entire spherical symmetry group, must be considered when examining the intrinsic symmetry properties. The property which describes the symmetry of the particle with respect to this group is its spin s. The degree of degeneracy is the number of different wave functions which can be transformed into linear combinations of one another, and is given by $2s + 1$. In particular, a particle having a vector (three-component) wave function has spin-1.

A free particle with non-zero mass always has a rest frame in which to examine the symmetries. If the mass of a particle is zero, however, there is no rest frame, since it moves with the velocity of light in every reference frame. For such a particle, there is always a distinctive direction in space, the direction of the momentum vector $\vec{k}$ (the ζ-axis say). In such a case, there is clearly no symmetry with respect to the whole group of rotations in three dimensions, but only axial symmetry about the preferred ζ-axis.

When there is axial symmetry, only the helicity of the particle is conserved, i.e. the component of its angular momentum along the ζ-axis, which we denote by λ. If we also impose the condition of symmetry under reflections in planes passing through the ζ-axis, the states differing in sign of λ will be mutually degenerate, and when $\lambda \neq 0$ there is therefore twofold degeneracy. The states of a photon having a definite momentum in fact corresponds to one type of these doubly degenerate states. It is described by a "spin" wave function which is a vector $\vec{\varepsilon}$ in the plane orthogonal to the ζ-axis; the two components of this vector are transformed into combinations of each other by any rotation about the ζ-axis and any reflection in a plane passing through that axis.

The various cases of the polarization of the photon are in a certain relationship to the possible values of its helicity. The relationship connects the components of a vector wave function with those of the equivalent spinor of rank two. Vectors $\vec{\varepsilon}$ with only non-zero components $\varepsilon_1 + i\varepsilon_2$ or $\varepsilon_1 - i\varepsilon_2$ correspond to the components $\lambda = +1$ or -1, respectively. In other words, the values $\lambda = +1$ and -1 correspond to right-handed and left-handed circular polarization of the photon. Thus, the component of the photon angular momentum along the direction of its motion can have only the two values ± 1; the value zero is not possible.

The completeness relationship for polarization vectors in the transverse gauge is

$$\sum_{\lambda=1,2} \varepsilon_i(\vec{k}, \lambda)\varepsilon_j^*(\vec{k}, \lambda) = \delta_{ij} - \hat{k}_i\hat{k}_j. \tag{7.41}$$

This relationship can easily be shown to be correct by examining each case of i, j (see problem 7.2). The completeness relationship is clearly not Lorentz invariant. We shall return to this point in the next section.

The normalization constant of the plane wave for a photon is chosen such that the energy in the wave A^μ is just $\omega = k_0 = |\vec{k}|$, i.e. $E = \hbar\omega$ for a single photon. To verify this, we compute

$$E_{\text{photon}} = \frac{1}{2}\int d^3x \langle \vec{E}^2 + \vec{B}^2 \rangle. \tag{7.42}$$

This is for the Heaviside-Lorentz system of units[3]. Since $\phi = A^0 = 0$ in the Coulomb gauge,

$$\vec{E} = -\vec{\nabla}\phi - \frac{\partial \vec{A}}{\partial t} = 0 + i\sqrt{\frac{\omega}{2V}}\,\vec{\varepsilon}\,(e^{-ik\cdot x} - e^{ik\cdot x})$$

$$= \sqrt{\frac{2\omega}{V}}\vec{\varepsilon}\sin k\cdot x, \tag{7.43}$$

$$\vec{B} = \vec{\nabla}\times\vec{A} = i\sqrt{\frac{\omega}{2V}}\,\hat{k}\times\vec{\varepsilon}\,(e^{-ik\cdot x} - e^{ik\cdot x})$$

$$= \sqrt{\frac{2\omega}{V}}\,\hat{k}\times\vec{\varepsilon}\,\sin k\cdot x, \tag{7.44}$$

and

$$\begin{aligned}
(\hat{k}\times\vec{\varepsilon})\cdot(\hat{k}\times\vec{\varepsilon}) &= \epsilon_{ijk}\hat{k}_j\varepsilon_k\epsilon_{inm}\hat{k}_n\varepsilon_m \\
&= (\delta_{jn}\delta_{km} - \delta_{jm}\delta_{kn})\hat{k}_j\hat{k}_n\varepsilon_k\varepsilon_m \\
&= \hat{k}_j\hat{k}_j\varepsilon_k\varepsilon_k - \hat{k}_j\hat{k}_k\varepsilon_k\varepsilon_j \\
&= \hat{k}\cdot\hat{k}\vec{\varepsilon}\cdot\vec{\varepsilon} - \hat{k}\cdot\vec{\varepsilon}\hat{k}\cdot\vec{\varepsilon} \\
&= \vec{\varepsilon}\cdot\vec{\varepsilon} - (\hat{k}\cdot\vec{\varepsilon})^2 \\
&= -1.
\end{aligned} \tag{7.45}$$

Thus $\vec{E}^2 = \vec{B}^2$, and

$$E_{\text{photon}} = \frac{2\omega}{V}\int d^3x \sin^2(\omega t - \vec{k}\cdot\vec{x})$$

---

[3]In the Gaussian system of units, the constant in front of the integral in equation 7.42 would be $1/8\pi$.

$$= \frac{2\omega}{V} \int dx_1 dx_2 dx_3 \left(\frac{e^{i(\omega t - \vec{k} \cdot \vec{x})} - e^{-i(\omega t - \vec{k} \cdot \vec{x})}}{2i} \right)^2$$

$$= -\frac{\omega}{2V} \left((e^{2i\omega t} \int dx_1 e^{-2ik_1 x_1} \int dx_2 e^{-2ik_2 x_2} \int dx_3 e^{-2ik_3 x_3} \right.$$

$$+ e^{-2i\omega t} \int dx_1 e^{2ik_1 x_1} \int dx_2 e^{2ik_2 x_2} \int dx_3 e^{2ik_3 x_3}$$

$$\left. -2 \int dx_1 \int dx_2 \int dx_3 \right)$$

$$= -\frac{\omega}{2V} \left(\frac{e^{2i(\omega t - \vec{k} \cdot \vec{x})}}{8ik_1 k_2 k_3} - \frac{e^{-2i(\omega t - \vec{k} \cdot \vec{x})}}{8ik_1 k_2 k_3} - 2V \right)$$

$$= -\frac{\omega}{2V} \left(\frac{\sin 2(\omega t - \vec{k} \cdot \vec{x})}{4k_1 k_2 k_3} - 2V \right)$$

$$= \omega, \tag{7.46}$$

where we have taken the time average. By not quantizing the photon wave function, we are essentially treating the electromagnetic field semiclassically. We are treating $\vec{A}$ as an external field, though its magnitude is adjusted so that the total energy associated with the electromagnetic field corresponds to that carried by a single photon.

7.4 Photon Propagator

We now develop the propagator for the photon using Green-function techniques in a similar fashion to the Klein-Gordon and Dirac propagators. The propagator for a photon is not unique, on account of the gauge freedom in the choice of A^μ.

From equation 7.15, we see that the wave equation for a photon can be written in the form

$$(g^{\nu\lambda}\Box - \partial^\nu \partial^\lambda)A_\lambda = \frac{j^\nu}{c}. \tag{7.47}$$

In fact, a photon propagator cannot exist until we remove some of the gauge freedom of A_λ, i.e. the inverse of the "momentum space operator" $(g^{\nu\lambda}\Box - \partial^\nu \partial^\lambda)$ does not exist (see problem 7.3).

If we chose to work in the Lorentz class of gauges with $\partial_\lambda A^\lambda = 0$, the wave equation simplifies to

$$g^{\nu\lambda}\Box A_\lambda = \frac{j^\nu}{c}. \tag{7.48}$$

Using Green-function techniques, the wave equation for the Green function $D_{\mu\nu}(x' - x)$ is

$$g^{\nu\lambda} \Box D_{\mu\nu}(x' - x) = \delta_\mu{}^\lambda \delta(x' - x). \tag{7.49}$$

Because of the presence of $g^{\nu\lambda}$ in the wave equation, the Green function is a second-rank tensor. The Green-function equation is contracted over one of the indices only. This allows us to determine the four potential using

$$A_\mu(x) = \int d^4y\, D_{\mu\nu}(x - y) \frac{j^\nu(y)}{c}. \tag{7.50}$$

We could also add on a solution of the homogeneous equation $\Box A_\lambda = 0$. However, the solution which we wish for $A_\mu(x)$ is that which vanishes in the absence of a transition current density $j^\nu(y)$. This is a requirement we impose on all our Green-function solutions.

In momentum space, we have

$$g^{\nu\lambda}(-k^2) D_{\mu\nu}(k) = \delta_\mu{}^\lambda. \tag{7.51}$$

The most general form of $D_{\mu\nu}(k)$ satisfying Lorentz covariance is

$$D_{\mu\nu}(k) = D(k^2) g_{\mu\nu} + D^{(l)}(k^2) k_\mu k_\nu, \tag{7.52}$$

where $D(k^2)$ and $D^{(l)}(k^2)$ are scalar functions of k^2 only. We thus obtain

$$-k^2 \left[D(k^2) g^{\nu\lambda} g_{\mu\nu} + D^{(l)}(k^2) g^{\nu\lambda} k_\mu k_\nu \right] = \delta_\mu{}^\lambda,$$

$$D(k^2) \delta_\mu{}^\lambda + D^{(l)}(k^2) k_\mu k^\lambda = -\frac{1}{k^2} \delta_\mu{}^\lambda. \tag{7.53}$$

Therefore,

$$D(k^2) = \frac{-1}{k^2} \quad \text{and} \quad D^{(l)}(k^2) = 0. \tag{7.54}$$

The photon propagator in momentum space (in the Lorentz gauge) is thus

$$\boxed{D_F^{\mu\nu}(k^2) = \frac{-g^{\mu\nu}}{k^2 + i\epsilon}}. \tag{7.55}$$

The gauge condition $\partial_\lambda A^\lambda = 0$ was imposed covariantly, and the resulting covariant propagator is ideal for quantum-electrodynamic calculations. It is called the Feynman propagator. We have used Heaviside-Lorentz units[4]. The subscript F conventually is used to denote the Feynman propagator. In analogy with the Klein-Gordon and Dirac propagators, we have appended an

---

[4]If Gaussian units were chosen, the numerator in equation 7.55 would be multiplied by 4π.

infinitesimally small positive imaginary part to k^2; this is equivalent to adding a small negative imaginary mass. It is often convenient to discard the tensor indices, and write $D_F^{\mu\nu}(k^2) = g^{\mu\nu}D_F(k)$.

The condition $\partial_\lambda A^\lambda = 0$ does not fully define the propagator. We are at liberty to rewrite the wave equation as

$$\left[g^{\nu\lambda}\Box - \left(1 - \frac{1}{\zeta}\right)\partial^\nu\partial^\lambda\right]A_\lambda = \frac{j^\nu}{c}, \tag{7.56}$$

where ζ is a real parameter. For $\zeta \neq 0$, the system is not really a Maxwell system. However, if we confine ourselves to currents that are conserved, the Lorentz condition will be satisfied. Therefore, in the Lorentz gauge everything is equivalent to classical electrodynamics. Equation 7.56 can be viewed as an extension of the Maxwell system beyond the "current-conservation shell".

Performing a similar calculation to that performed previously, we have

$$\left[g^{\nu\lambda}\Box - \left(1 - \frac{1}{\zeta}\right)\partial^\nu\partial^\lambda\right]D_{\mu\nu}(x - y) = \delta_\mu{}^\lambda\delta(x - y),$$

$$\left[g^{\nu\lambda}(-k^2) + \left(1 - \frac{1}{\zeta}\right)k^\nu k^\lambda\right]\left[g_{\mu\nu}D(k^2) + k_\mu k_\nu D^{(l)}(k^2)\right] = \delta_\mu{}^\lambda$$

$$-D(k^2)k^2\delta_\mu{}^\lambda + \left[D(k^2)\left(1 - \frac{1}{\zeta}\right) - \frac{1}{\zeta}k^2 D^{(l)}(k^2)\right]k_\mu k^\lambda = \delta_\mu{}^\lambda, \tag{7.57}$$

$$D(k^2) = -\frac{1}{k^2} \quad \text{and} \quad D^{(l)}(k^2) = \frac{1 - \zeta}{k^4}. \tag{7.58}$$

In this case, the propagator is

$$D_{\mu\nu}(k) = \frac{1}{k^2}\left[-g_{\mu\nu} + (1 - \zeta)\frac{k_\mu k_\nu}{k^2}\right]. \tag{7.59}$$

$\zeta = 1$ is referred to as the Feynman gauge[5]. $\zeta = 0$ is referred to as the Landau gauge. In this form, the propagator obeys the condition $k^\mu D_{\mu\nu}(k) = 0$. This gauge appears to be more complicated but the extra term in the propagator vanishes in quantum-electrodynamic calculations, in which the virtual photon is coupled to conserved currents, which satisfy $k_\mu j^\mu = k_\nu j^\nu = 0$.

When the photon propagator appears in physical quantities, such as scattering amplitudes, it is multiplied by the transition currents of two Dirac particles, as shown in figure 7.1. That is, it appears in combinations of the form $(j^\mu)_{21}D_{\mu\nu}(j^\nu)_{43}$, where $(j^\mu)_{21} = \bar{\psi}_2\gamma^\mu\psi_1$ and $(j^\nu)_{43} = \bar{\psi}_4\gamma^\nu\psi_3$. From current conservation $(\partial \cdot j = 0)$, the transversality conditions $k_\mu(j^\mu)_{21} = 0$ and $k_\nu(j^\nu)_{43} = 0$, where $k = p_2 - p_1 = p_4 - p_3$, will be satisfied. We therefore see that terms in the photon propagator involving k_μ and k_ν will not effect physical quantities.

---

[5]It is really the Lorentz gauge.

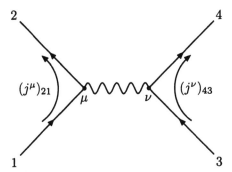

FIGURE 7.1: Virtual photon exchange between two charged currents.

All physical results will be unchanged by the substitution

$$D_{\mu\nu} \rightarrow D_{\mu\nu} + \chi_\mu k_\nu + \chi_\nu k_\mu, \qquad (7.60)$$

where χ_μ and χ_ν are any functions of $\vec{k}$ and ω. This arbitrariness in the choice of $D_{\mu\nu}$ corresponds to the arbitrariness in the choice of the gauge for A^μ. The arbitrariness of the gauge can violate the Lorentz invariant form of $D_{\mu\nu}$ if the quantities χ_μ and χ_ν are not four-vectors. Even if we consider only Lorentz invariant forms of the propagator, the choice of the function $D^{(l)}(k^2)$ is entirely arbitrary; it does not effect any physical results, and can be chosen in any convenient manner.

The determination of the propagator function amounts to that of a single gauge-invariant function $D(k^2)$. If we take the z-axis in the direction of $\vec{k}$, the transformation given by equation 7.60 will not effect the transverse components of $D_{\mu\nu}$. We see that $D_{11} = D_{22} = -D(k^2)$, and it is therefore sufficient to calculate a single transverse component, D_{11} say, using any gauge for the potentials.

In order to obtain the propagator for the Coulomb gauge, we put

$$\chi_0 = \frac{k_0}{2k^2\vec{k}^2} \quad \text{and} \quad \chi_i = -\frac{k_i}{2k^2\vec{k}^2} \qquad (7.61)$$

into equation 7.60 to get

$$D_{ij} = -\frac{g_{ij}}{k^2} - \frac{k_i k_j}{k^2\vec{k}^2} = \frac{1}{k^2}\left(\delta_{ij} - \frac{k_i k_j}{\vec{k}^2}\right). \qquad (7.62)$$

The remaining components are

$$D_{0i} = \frac{k_0}{2k^2\vec{k}^2}k_i - \frac{k_i}{2k^2\vec{k}^2}k_0 = 0 \qquad (7.63)$$

and

$$D_{00} = -\frac{g_{00}}{k^2} + \frac{k_0 k_0}{k^2 \vec{k}^2} = -\frac{1}{k^2 \vec{k}^2}(\vec{k}^2 - k_0^2) = \frac{1}{\vec{k}^2}. \qquad (7.64)$$

We notice that D_{00} is the Fourier component of the Coulomb potential, and we will elaborate on this point soon.

The general form of the propagator of a virtual particle is

$$\frac{\sum_{\text{spins}}}{p^2 - m^2}. \qquad (7.65)$$

The spin sum is the completeness relationship for the spin wave functions of the corresponding particle. It includes all possible spin states of the propagating particle. For the Klein-Gordon propagator $\sum_{\text{spins}} = 1$ and for the Dirac propagator $\sum_{\text{spins}} = \sum_s u \bar{u}$. In the case of the photon, $m = 0$ and

$$\boxed{\sum_{r=0}^{3} \varsigma_r \varepsilon^{\mu *}(\vec{k}, r) \varepsilon^{\nu}(\vec{k}, r) = -g^{\mu\nu}}, \qquad (7.66)$$

where $\varsigma_0 = -1$ and $\varsigma_1 = \varsigma_2 = \varsigma_3 = 1$.

For a real photon, the completeness relationship for the polarization vectors was given by equation 7.41. On the other hand, we have associated with the virtual photon the covariant propagator $-g_{\mu\nu}/k^2$, where $-g_{\mu\nu}$ implies we are summing over four polarization states. In addition to summing the two transverse components, we now have the longitudinal component $\varepsilon_\mu(\vec{k}, 3)$ and the scalar component $\varepsilon_\mu(\vec{k}, 0)$. The longitudinal and scalar components are important in the completeness relationship but they do not correspond to physical photons. For real photons, the longitudinal and scalar polarization components cancel (see problem 7.4). For a virtual photon the longitudinal and scalar components cannot be neglected. These terms become the instantaneous Coulomb interaction between the charges of the two particles, as we will now show and as was hinted at in equation 7.64.

Consider a more general analysis of the photon propagator. If we do not pick the z-axis as the direction of motion, we can write the four orthonormal polarization vectors as

$$\varepsilon^\mu(\vec{k}, 0) = n^\mu \equiv (1, 0, 0, 0), \qquad (7.67)$$
$$\varepsilon^\mu(\vec{k}, r) = (0, \vec{\varepsilon}(\vec{k}, r)) \quad r = 1, 2, 3 \qquad (7.68)$$

with

$$\vec{\varepsilon}(\vec{k}, r) = \vec{\varepsilon}(\vec{k}, \lambda) \quad r = \lambda = 1, 2, \qquad (7.69)$$
$$\vec{\varepsilon}(\vec{k}, 3) = \frac{\vec{k}}{|\vec{k}|}. \qquad (7.70)$$

The longitudinal polarization vector can also be expressed in the form

$$\varepsilon^\mu(\vec{k}, 3) = \frac{k^\mu - (k \cdot n)n^\mu}{\sqrt{(k \cdot n)^2 - k^2}}. \tag{7.71}$$

By inserting our specific representation for the polarization vectors into equation 7.66 and using equation 7.65, we obtain the general photon propagator in momentum space

$$D^{\mu\nu}(k) = \frac{1}{k^2 + i\epsilon}\left[\sum_{\lambda=1,2}\varepsilon^{*\mu}(\vec{k},\lambda)\varepsilon^\nu(\vec{k},\lambda)\right.$$
$$\left.+\frac{(k^\mu - (k\cdot n)n^\mu)(k^\nu - (k\cdot n)n^\nu)}{(k\cdot n)^2 - k^2} - n^\mu n^\nu\right]. \tag{7.72}$$

The first term on the right-hand side represents the exchange of transverse photons:

$$D^{\mu\nu}_{\text{trans}}(k) = \frac{1}{k^2 + i\epsilon}\sum_{\lambda=1,2}\varepsilon^{*\mu}(\vec{k},\lambda)\varepsilon^\nu(\vec{k},\lambda). \tag{7.73}$$

We separate the remainder of the expression in equation 7.72, i.e. the second and third terms, into two parts:

$$\begin{aligned}D^{\mu\nu}_{\text{coul}}(k) &= \frac{1}{k^2 + i\epsilon}\left[\frac{(k\cdot n)^2 n^\mu n^\nu}{(k\cdot n)^2 - k^2} - n^\mu n^\nu\right]\\ &= \frac{k^2}{k^2 + i\epsilon}\frac{n^\mu n^\nu}{(k\cdot n)^2 - k^2}\\ &= \frac{n^\mu n^\nu}{\vec{k}^2}\end{aligned} \tag{7.74}$$

and

$$D^{\mu\nu}_{\text{red}}(k) = \frac{1}{k^2 + i\epsilon}\left[\frac{k^\mu k^\nu - (k\cdot n)(k^\mu n^\nu + n^\mu k^\nu)}{(k\cdot n)^2 - k^2}\right]. \tag{7.75}$$

In coordinate space, the Coulomb piece is

$$\begin{aligned}D^{\mu\nu}_{\text{coul}}(x) &= n^\mu n^\nu \int \frac{d^4k}{(2\pi)^4}e^{-ik\cdot x}\frac{1}{|\vec{k}|^2}\\ &= g^{\mu 0}g^{\nu 0}\int \frac{d^3k}{(2\pi)^3}\frac{e^{ik\cdot x}}{|\vec{k}|^2}\int \frac{dk^0}{2\pi}e^{-ik^0 x^0}\\ &= g^{\mu 0}g^{\nu 0}\frac{1}{|\vec{x}|}\delta(x^0).\end{aligned} \tag{7.76}$$

This part of the propagator represents the instantaneous Coulomb interaction. The longitudinal and scalar photons thus yield the instantaneous Coulomb interaction between charged particles. In the Coulomb gauge only transverse photons occurred. The Coulomb interaction now no longer occurs explicitly in the theory, but is contained as the exchange of scalar and longitudinal photons in the propagator.

The remaining term $D_{\text{red}}^{\mu\nu}$ makes no physical contribution, and is redundant since it comprises terms proportional to k^μ and k^ν, which do not contribute in matrix-element calculations.

7.5 Problems

1. Obtain equation 7.15 by using the scalar and vector potentials along with Maxwell's two inhomogeneous equations in three-vector form.

2. Show that equation 7.41 is correct.

3. Verify that the inverse of the "momentum space operator" in the equation

$$(g^{\mu\nu}\Box - \partial^\mu\partial^\nu)A_\nu = j^\mu$$

   does not exist.

4. For real photons, show that the longitudinal and scalar polarization components cancel.

Chapter 8

Quantum Electrodynamic Processes

During the period from 1929 to 1936, a number of cross sections were calculated to lowest order in powers of α for various electrodynamic processes, such as those shown in figures 8.1 and 8.2. These lowest-order calculations gave finite results in reasonable agreement with the experimental data. The successful agreement with measurements demonstrated the calculational power and practicality of the theory of quantum electrodynamics.

We will now develop the practical abilities to calculate lowest-order quantum electrodynamical processes (figures 8.1 and 8.2). That is, we will apply the propagator formalism to problems involving electrons, positrons, and photons. As we go, we will derive general rules for the calculation of transition probabilities and cross sections: the Feynman rules.

We continue the pedagogical approach of this book. Some of the calculations performed in this chapter take about ten pages to complete. Some textbooks do these calculations in a single page. While this may show off the calculational power of quantum electrodynamics, and the author, our intent is to explain each step in detail so that they can be adapted or modified by the reader in more complicated calculations.

8.1 Lifetimes and Cross Sections

In this section, we present an overview of the general approach taken in the following calculations. Only an outline is given here. The details will be worked out as needed.

The S-matrix permits the calculation of two types of observable quantities: lifetimes and cross sections. Both can be calculated from the transition probability per unit space-time volume. If there were no interactions between the particles, the state of the system would be unchanged, which corresponds to a unit S-matrix δ_{fi} (absence of scattering). It is convenient to separate out this unit matrix, and writing the scattering matrix in the form

$$
\begin{aligned}
S_{fi} &= \delta_{fi} + R_{fi} \\
&= \delta_{fi} + i(2\pi)^4 \delta^4(p_f - p_i) T_{fi},
\end{aligned}
\tag{8.1}
$$

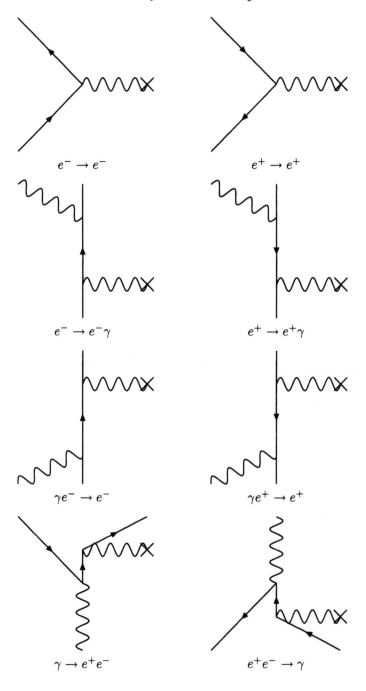

FIGURE 8.1:   Lowest order $1 \to 1$, $1 \to 2$, and $2 \to 1$ processes involving electrons, positrons, and photons. Time runs upward in the diagrams.

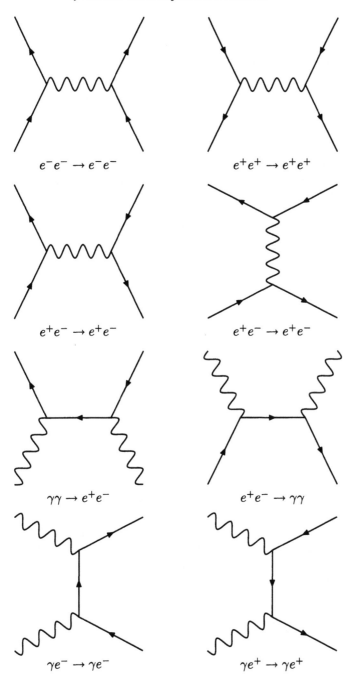

FIGURE 8.2: Lowest order $2 \to 2$ processes involving electrons, positrons, and photons. Time runs upward in the diagrams.

where T_{fi} is another matrix – the transition matrix[1]. In the second term, we have written separately the four-dimensional Dirac delta function, which we will see comes from the integral

$$\frac{1}{(2\pi)^4} \int e^{i(p_f - p_i) \cdot x} d^4 x = \delta^4(p_f - p_i). \tag{8.2}$$

The product of exponentials in this expression comes from the product of the initial and final particle wave functions. The Dirac delta-function expresses the law of conservation of four-momentum – p_f and p_i being the sums of the four-momenta of all the particles in the final and initial states – the other factor of $i(2\pi)^4$ is included for subsequent convenience.

If one of the colliding particles is sufficiently heavy, and its state is unaltered by the collision, it acts only as a fixed source of a constant field in which the other particle is scattered. Since the energy – though not the momentum – of the system is conserved in a constant field, we will write the S-matrix elements in the form

$$S_{fi} = \delta_{fi} + i(2\pi)\delta(E_f - E_i)T_{fi}. \tag{8.3}$$

The structure of the scattering amplitude T_{fi} in equations 8.1 and 8.3 is of the form

$$T_{fi} = u_1^* u_2^* \cdots Q \cdots u_2 u_1, \tag{8.4}$$

where on the left-hand side we have the amplitude of wave functions of the final particles, and on the right-hand side those of the initial particles; Q is some matrix relating the indices of the wave amplitude components of all the particles.

The matrix elements of the R-matrix in equation 8.1, $\langle f|R|i \rangle$, are the transition probability amplitudes for transitions to take place over all space and all time from the infinite past to the infinite future. The corresponding probability, $|\langle f|R|i \rangle|^2$, is not a meaningful quantity – and not at all a probability – since observations are carried out over finite times, and only the transition probability per unit time is essentially measurable. Indeed, $|\langle f|R|i \rangle|^2$ is infinite and simply expresses the fact that, during an infinite time, a nonzero incident flux of particles will cause an infinite number of repetitions of the elementary process under consideration.

The interesting quantity is the transition probability per unit time, or for convenience in our covariant formalism, the transition probability Γ per unit space-time volume. The latter can be obtained as a limit from a finite space-time volume:

---

[1] Actually, the matrix elements T_{fi} which remain after separation of the delta function are called the scattering amplitudes.

$$\Gamma = \lim_{\substack{V \to \infty \\ T \to \infty}} \frac{|\langle f|R|i\rangle|^2_{VT}}{VT}. \tag{8.5}$$

Here, $|\langle f|R|i\rangle|^2_{VT}$ is $|\langle f|R|i\rangle|^2$ calculated for a finite space-time volume VT.

We remember that the expression for $\langle f|R|i\rangle$ results from an integration over infinite space-time,

$$\langle f|R|i\rangle = i\langle f|T|i\rangle \int e^{i(p_f - p_i)\cdot x} d^4 x. \tag{8.6}$$

Therefore,

$$\langle f|R|i\rangle_{VT} = i\langle f|T|i\rangle \int_{VT} e^{i(p_f - p_i)\cdot x} d^4 x. \tag{8.7}$$

To obtain a probability from an amplitude, the moduli $|S_{fi}|$ are squared. The square of the delta function appears, and is to be interpreted as follows. If another such integral is calculated with $p_f = p_i$ – since one delta function is already present – and if the integral is taken over some large finite volume V and time interval T, the result is $VT/(2\pi)^4$.

The evaluation of Γ involves the limit

$$\lim_{\substack{V \to \infty \\ T \to \infty}} \frac{1}{VT} \left| \int_{VT} e^{i(p_f - p_i)\cdot x} d^4 x \right|^2 = (2\pi)^4 \delta^4(p_f - p_i). \tag{8.8}$$

We find for the transition probability per unit space-time volume,

$$\Gamma = (2\pi)^4 \delta^4(p_f - p_i)|\langle f|T|i\rangle|^2, \tag{8.9}$$

where Γ depends on the details of the observation – the experiment. For example, if the spin of the outgoing electron is not observed, a summation over the final spin states must be carried out. Or, if the initial spin state is unknown, a suitable average must be found.

Let S indicate a generalized summation symbol representing integrations over momenta, and summations over electron spins and photon polarizations, depending on the type of process. The average transition probability per unit space-time volume is

$$\overline{\Gamma} = (2\pi)^4 S_f \overline{S}_i \delta^4(p_f - p_i)|\langle f|T|i\rangle|^2, \tag{8.10}$$

where the bar over S_i indicates the average process. To calculate the cross section, we take an incoherent average in the sense that we average the cross sections rather than the amplitudes.

In our expression, it is not necessary to restrict the considerations to a single initial electron, positron, or photon. Let us then consider one single initial system of particles. The transition probability per unit time into all other possible states will be

$$\Gamma_{\text{tot}} = V\Gamma, \tag{8.11}$$

which is independent of V.

The most important cases are those where the initial state comprises only one or two particles: decay and scattering, respectively. We can form simplified expressions for special cases like a two-body decay or a $2 \to 2$ scattering process; the 2 on either side of the arrow represents 2 particles. For a decay, the characteristic time is given by the inverse of Γ_{tot},

$$\tau = \frac{1}{\Gamma_{\text{tot}}}, \tag{8.12}$$

which is called the lifetime of the system.

For the case of two initial systems of particles, the ratio of the transition probability per unit volume Γ and the flux density of the initial state I is called the cross section

$$\bar{\sigma} = \frac{\overline{\Gamma}}{I}. \tag{8.13}$$

The flux density can be written in such a way that it is valid in any reference frame:

$$I = \frac{1}{V}\frac{1}{2E_1 2E_2}F, \tag{8.14}$$

where

$$F = 4\sqrt{(p_1 \cdot p_2)^2 - m_1^2 m_2^2}. \tag{8.15}$$

The flux factor F will be first encountered in equation 8.104 and is calculated for each case in appendix A. The cross sections can be written as

$$\bar{\sigma} = 4(2\pi)^4 \frac{VE_1 E_2}{|F|} S_f \overline{S}_i \delta^4(p_f - p_i)|\langle f|T|i\rangle|^2. \tag{8.16}$$

According to the summation in S_f there exists various partial cross sections $\bar{\sigma}_f$. If the momentum vectors of the final state fall within a certain differential range, we call the right-hand side of equation 8.16 the differential cross section.

Generally, one wants to express a cross section in units of cm^2, and a lifetime in units of seconds. We use $\hbar = c = 1$, and will express everything else in MeV. The cross section will have the dimensions of $(\text{MeV})^{-2}$, the decay rate is of dimension MeV (and lifetime $(\text{MeV})^{-1}$). To go to cm^2 the cross section must be multiplied by $(\hbar c)^2$. To go from MeV to s^{-1} the decay rate must be divided by $\hbar$ and the lifetime is thus $\hbar/\Gamma$.

8.2 Coulomb Scattering of Electrons

As our first practical calculation, we consider the relativistic treatment of Rutherford scattering of an electron from a fixed Coulomb potential. Rutherford derived this historic formula using classical concepts. The formula is of great importance since it was used to interpret Rutherford's famous experiments involving the collisions of alpha particles with heavier nuclei.

The most important approximation to the elastic scattering of electrons by atoms consists in neglecting the electron cloud surrounding the nucleus. The atom is then considered as an infinitely heavy positive point charge of magnitude Ze. The system under consideration is thereby simplified to the motion of an electron in a Coulomb field (see figure 8.3).

FIGURE 8.3: Scattering of an electron from a fixed Coulomb potential.

In dealing with the expansion of the S-matrix, we have a double expansion, one in e which describes the radiation field of the electron, and one in Ze which corresponds to a power series expansion in the external field. We shall refer to Coulomb scattering as the scattering of an electron in a Coulomb field to all orders in the external field, but to lowest order in the radiation field of the electron. The effects of virtual (or real) photons emitted and reabsorbed by the scattered electron in this process can be found in Jauch & Rohrlich [15].

In terms of diagrams, the three lowest orders in the Coulomb field can be depicted as in figure 8.4. The static external field is represented using a wavy line with a cross. In this section, we evaluate the diagram in figure 8.4a.

The S-matrix element for a scattered electron is

$$S_{fi} = ie \int d^4x \overline{\psi}_f(x) \, \mathcal{A}(x) \Psi_i(x), \tag{8.17}$$

where the final state f differs from the initial state i, and the electron charge is $-e = -|e| < 0$.

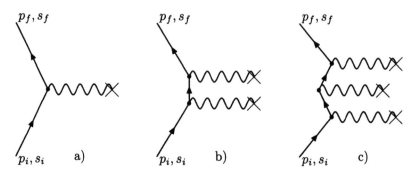

FIGURE 8.4: Coulomb scattering: a) first order, b) second order, and c) third order.

In lowest order, $\Psi_i(x)$ is approximated by the incoming wave $\psi_i(x)$ of an electron. In this approximation, the wavefunctions reduce to the plane waves

$$\psi_i(x) = \sqrt{\frac{m}{E_i V}} u(p_i, s_i) e^{-ip_i \cdot x} \qquad (8.18)$$

and

$$\overline{\psi}_f(x) = \sqrt{\frac{m}{E_f V}} \overline{u}(p_f, s_f) e^{ip_f \cdot x}, \qquad (8.19)$$

where we have normalized to unit probability in a box of volume V, and are using natural units ($\hbar = c = 1$). We can set $V = 1$. The advantage of using V will be apparent when calculating physical rates, lifetimes, or cross sections. The normalization volume V cancels out of the quantity of physical interest; any unphysical quantity will vanish or become infinitely large as $V \to \infty$. If we had employed the continuum normalization with periodic boundary conditions, then $V \to (2\pi)^3$, and typical scattering amplitudes would contain as many as 12 powers of 2π. Using the box normalization goes a long way towards keeping the factors of 2π straight. The further modification of m/E keeps the box normalization covariant and is useful for relativistic particles (see problem 8.2).

The Coulomb potential from a positive point charge Ze is

$$A_0(x) = \frac{Ze}{4\pi|\vec{x}|} \quad \text{and} \quad \vec{A}(x) = 0. \qquad (8.20)$$

We are working in the Heaviside-Lorentz (rationalized Gaussian) system of units[2]. For a nucleus, $Ze > 0$ and the validity of the perturbation expansion is valid provided Z is not too large.

---

[2]If we were working in the Gaussian system of units, the Coulomb potential would be $A_0(x) = Ze/|\vec{x}|$.

Inserting the plane waves and Coulomb potential into the S-matrix element, we have

$$
\begin{aligned}
S_{fi} &= \frac{iZe^2}{4\pi} \frac{1}{V} \frac{m}{\sqrt{E_f E_i}} \bar{u}(p_f, s_f)\gamma^0 u(p_i, s_i) \int d^4x \frac{e^{i(p_f - p_i)\cdot x}}{|\vec{x}|} \\
&= \frac{iZ(4\pi\alpha)}{4\pi V} \frac{m}{\sqrt{E_f E_i}} \bar{u}(p_f, s_f)\gamma^0 u(p_i, s_i)[2\pi\delta(E_f - E_i)] \\
&\quad \cdot \int d^3x \frac{e^{-i(\vec{p}_f - \vec{p}_i)\cdot \vec{x}}}{|\vec{x}|}.
\end{aligned}
\tag{8.21}
$$

We have expressed the result in terms of the dimensionless fine-structure constant $\alpha \approx 1/137$. The delta function comes from the integral over the time coordinate x_0, and expresses energy conservation between the initial and final states in a static potential. Energy conservation will always occur for time independent potentials.

The space integral is the Fourier transform of the Coulomb potential

$$
\int d^3x \frac{e^{-i\vec{q}\cdot\vec{x}}}{|\vec{x}|} = \frac{4\pi}{|\vec{q}|^2},
\tag{8.22}
$$

where $\vec{q} = \vec{p}_f - \vec{p}_i$ is defined in figure 8.5. Therefore,

$$
S_{fi} = \frac{2i(2\pi)Z\alpha}{V} \frac{m}{\sqrt{E_f E_i}} \frac{\bar{u}(p_f, s_f)\gamma^0 u(p_i, s_i)}{|\vec{q}|^2} [2\pi\delta(E_f - E_i)].
\tag{8.23}
$$

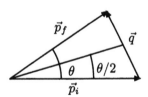

FIGURE 8.5: Definition of the three-momentum transfer.

The number of final states in the momentum interval $\vec{p}_f$ to $\vec{p}_f + d\vec{p}_f$ is $V d^3 p_f/(2\pi)^3$. To see this, we notice that standing waves in a cubic box of volume $V = L^3$ require

$$
\begin{aligned}
k_x L &= 2\pi n_x, \\
k_y L &= 2\pi n_y, \\
k_z L &= 2\pi n_z,
\end{aligned}
\tag{8.24}
$$

with integer numbers n_x, n_y, n_z. For large L, the discrete set of $\vec{k}$-values approaches a continuum of values. The number of states is (setting $\hbar = 1$)

$$
\begin{aligned}
dN &= dn_x dn_y dn_y \\
&= \frac{1}{(2\pi)^3} L^3 dk_x dk_y dk_y \\
&= \frac{V}{(2\pi)^3} d^3k.
\end{aligned}
\tag{8.25}
$$

The transition probability per particle into these states is

$$
|S_{fi}|^2 V \frac{d^3p_f}{(2\pi)^3} = \frac{4(Z\alpha)^2 m^2}{2\pi E_i V} \frac{|\bar{u}(p_f, s_f)\gamma^0 u(p_i, s_i)|^2}{|\vec{q}|^4} \frac{d^3p_f}{E_f} [2\pi\delta(E_f - E_i)]^2.
\tag{8.26}
$$

The square of the delta function is mathematically not well defined. It is a divergent quantity and has to be specified by a limiting procedure. We can reason it to be

$$
[2\pi\delta(E_f - E_i)]^2 = \lim_{T \to \infty} \int_{-T/2}^{T/2} dt e^{i(E_f - E_i)t} [2\pi\delta(E_f - E_i)].
\tag{8.27}
$$

If $E_f = E_i$ as required by the other remaining delta function, then

$$
2\pi\delta(E_f - E_i) = \lim_{T \to \infty} \int_{-T/2}^{T/2} dt e^{i(E_f - E_i)t}.
\tag{8.28}
$$

For $E_f = E_i$,

$$
2\pi\delta(0) = \lim_{T \to \infty} \int_{-T/2}^{T/2} dt = \lim_{T \to \infty} T
\tag{8.29}
$$

and

$$
[2\pi\delta(E_f - E_i)]^2 = 2\pi T\delta(E_f - E_i).
\tag{8.30}
$$

Another way of viewing this is that the finite time interval smears the delta function:

$$
\begin{aligned}
\lim_{T \to \infty} \frac{1}{T} \left| \int_{-T/2}^{T/2} dt e^{i(E_f - E_i)t} \right|^2 &= \lim_{T \to \infty} \frac{1}{T} \left| \frac{e^{i(E_f - E_i)t}}{i(E_f - E_i)} \Big|_{-T/2}^{T/2} \right|^2 \\
&= \lim_{T \to \infty} \frac{1}{T} \left[\frac{\sin[(E_f - E_i)T/2]}{(E_f - E_i)/2} \right]^2 \\
&= \frac{2\pi\delta(E_f - E_i)}{T}.
\end{aligned}
\tag{8.31}
$$

Using wave packets in the S-matrix element (equation 8.17) instead of plane waves would remove the need to square the delta function.

If we consider transitions in a time T and divide out the time, we obtain the rate, which is the number R of transitions per unit time into the momentum interval $d^3 p_f$:

$$R = |S_{fi}|^2 \frac{V}{T} \frac{d^3 p_f}{(2\pi)^3} = \frac{4(Z\alpha)^2 m^2}{E_i V} \frac{|\bar{u}(p_f, s_f)\gamma^0 u(p_i, s_i)|^2}{|\vec{q}|^4} \frac{d^3 p_f}{E_f} \delta(E_f - E_i).$$
(8.32)

A cross section is defined as the transition rate R divided by the flux of incident particles $J^\mu_{inc} = \bar{\psi}_i(x)\gamma^\mu \psi_i(x)$, where μ denotes the vector component along the incident velocity $\vec{v}_i = \vec{p}_i/E_i$, and again we are using $c = 1$. With the normalization we have adapted, the flux is

$$|J^\mu_{inc}| = \frac{m}{E_i V} \bar{u}_i \gamma^\mu u_i = \frac{m}{E_i V} \frac{p_i^\mu}{m} = \frac{p_i^\mu}{V E_i},$$
(8.33)

where we have used the Gordon decomposition (equation 5.290)

$$\bar{u}(p)\gamma^\mu u(q) = \frac{1}{2m} \bar{u}(p) \left[(p+q)^\mu + i\sigma^{\mu\nu}(p-q)_\nu \right] u(q).$$
(8.34)

Therefore the flux of incident electrons is

$$|J_{inc}| = \frac{|\vec{p}_i|}{V E_i} = \frac{\beta}{V}.$$
(8.35)

Thus the differential cross section $d\sigma$ per unit solid angle $d\Omega$ is

$$\frac{d\sigma}{d\Omega} = \frac{R}{|J_{inc}|} = \int \frac{4(Z\alpha)^2 m^2}{\beta E_i} \frac{|\bar{u}(p_f, s_f)\gamma^0 u(p_i, s_i)|^2}{|\vec{q}|^4} \frac{p_f^2 dp_f}{E_f} \delta(E_f - E_i),$$
(8.36)

where we have used spherical coordinates $d^3 p_f = p_f^2 dp_f d\Omega$, and the solid angle is defined in figure 8.6.

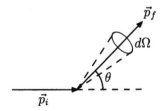

FIGURE 8.6: Definition of the solid angle for electron scattering.

Since $p_f dp_f = E_f dE_f$, we obtain

$$\frac{d\sigma}{d\Omega} = 4(Z\alpha)^2 m^2 \int \frac{|\overline{u}(p_f, s_f)\gamma^0 u(p_i, s_i)|^2}{|\vec{q}|^4} dE_f \frac{|\vec{p_f}|}{|\vec{p_i}|}\delta(E_f - E_i)$$

$$= \frac{4(Z\alpha)^2 m^2}{|\vec{q}|^4}|\overline{u}(p_f, s_f)\gamma^0 u(p_i, s_i)|^2, \qquad (8.37)$$

where it is understood that $E_i = E_f$ and $|\vec{p_i}| = |\vec{p_f}|$ in the above expression. In the nonrelativistic limit,

$$|\overline{u}(p_f, s_f)\gamma^0 u(p_i, s_i)|^2 \rightarrow \left|(1\ 0)\begin{pmatrix} 1 & 0 \\ 0 & 1 \end{pmatrix}\begin{pmatrix} 1 \\ 0 \end{pmatrix}\right|^2 = 1. \qquad (8.38)$$

Calculating $|\vec{q}|^4$ and using $E = p^2/2m$, equation 8.37 can be shown to reduce to the Rutherford scattering formula (see problem 8.3).

The resulting differential cross section in equation 8.37 can in principle be applied to calculate the scattering of an electron with initial spin s_i to the final spin s_f (see problem 8.4). In general, one does not know the initial spin of the electron. If the probability of either spin state is equal, we sum over both polarization directions with weight $1/2$, or we average over initial-state polarizations. This is an incoherent average in the sense that we average the cross sections rather than the amplitudes. Furthermore, most experiments do not measure the final polarization of the electron, and we simply detect all of them. We thus sum over all final-state polarizations.

The unpolarized scattering cross section is

$$\frac{d\bar{\sigma}}{d\Omega} = \frac{4(Z\alpha)^2 m^2}{2|\vec{q}|^4} \sum_{\pm s_f, \pm s_i} |\overline{u}(p_f, s_f)\gamma^0 u(p_i, s_i)|^2, \qquad (8.39)$$

where the bar over σ denotes average. By unpolarized we mean that no information about the electron spin is recorded by the experiment.

The double sum over spins can be written in component form

$$\sum_{\pm s_f, \pm s_i} \overline{u}_\alpha(p_f, s_f)\gamma^0_{\alpha\beta} u_\beta(p_i, s_i) u^\dagger_\lambda(p_i, s_i)(\gamma^0_{\lambda\delta})^\dagger (\gamma^0_{\delta\sigma})^\dagger u_\sigma(p_f, s_f)$$

$$= \sum_{\pm s_f, \pm s_i} \overline{u}_\alpha(p_f, s_f)\gamma^0_{\alpha\beta} u_\beta(p_i, s_i)\overline{u}_\delta(p_i, s_i)\gamma^0_{\delta\sigma} u_\sigma(p_f, s_f)$$

$$= \sum_{\pm s_f} \overline{u}_\alpha(p_f, s_f)\gamma^0_{\alpha\beta}\left(\sum_{\pm s_i} u_\beta(p_i, s_i)\overline{u}_\delta(p_i, s_i)\right)\gamma^0_{\delta\sigma} u_\sigma(p_f, s_f). \qquad (8.40)$$

The completeness relationship (equation 5.204) for Dirac spinors can be written as

$$u_\alpha(p, s)\bar{u}_\beta(p, s) + u_\alpha(p, -s)\bar{u}_\beta(p, -s)$$
$$-v_\alpha(p, s)\bar{v}_\beta(p, s) - v_\alpha(p, -s)\bar{v}_\beta(p, -s) = \delta_{\alpha\beta}. \tag{8.41}$$

If we operate on equation 8.41 with a positive energy projection operator $\Lambda_+(p)$, we have

$$u_\alpha(p, s)\bar{u}_\beta(p, s) + u_\alpha(p, -s)\bar{u}_\beta(p, -s) = [\Lambda_+(p)]_{\alpha\beta}$$
$$\sum_{\pm s} u_\alpha(p, s)\bar{u}_\beta(p, s) = [\Lambda_+(p)]_{\alpha\beta}. \tag{8.42}$$

The spin sum (equation 8.40) thus becomes

$$\sum_{\pm s_f} \bar{u}_\alpha(p_f, s_f) \left(\gamma^0 \frac{\not{p}_i + m}{2m} \gamma^0 \right)_{\alpha\beta} u_\beta(p_f, s_f)$$
$$= \left(\gamma^0 \frac{\not{p}_i + m}{2m} \gamma^0 \right)_{\alpha\beta} \left(\frac{\not{p}_f + m}{2m} \right)_{\beta\alpha}$$
$$= \left(\gamma^0 \frac{\not{p}_i + m}{2m} \gamma^0 \frac{\not{p}_f + m}{2m} \right)_{\alpha\alpha}, \tag{8.43}$$

which is a trace. Therefore, the differential cross section is

$$\frac{d\bar{\sigma}}{d\Omega} = \frac{2(Z\alpha)^2 m^2}{|\vec{q}|^4} \text{Tr} \left[\gamma_0 \frac{\not{p}_i + m}{2m} \gamma_0 \frac{\not{p}_f + m}{2m} \right]. \tag{8.44}$$

Using trace theorems, which we shall develop in the next section, we have

$$\text{Tr} \left[\gamma_0 \frac{\not{p}_i + m}{2m} \gamma_0 \frac{\not{p}_f + m}{2m} \right] = \frac{1}{4m^2} \left(\text{Tr}[\gamma_0 \not{p}_i \gamma_0 \not{p}_f] + m^2 \text{Tr}[(\gamma_0)^2]) \right)$$
$$= \frac{1}{m^2} \left(p_i^0 p_f^0 + p_f^0 p_i^0 - p_i \cdot p_f + m^2 \right)$$
$$= \frac{1}{m^2} \left(2E_i E_f - p_i \cdot p_f + m^2 \right) \tag{8.45}$$

and

$$\frac{d\bar{\sigma}}{d\Omega} = \frac{2(Z\alpha)^2}{|\vec{q}|^4} \left(2E_i E_f - p_i \cdot p_f + m^2 \right). \tag{8.46}$$

The differential cross section can be written in terms of the scattering energy E and scattering angle θ. The scattering angle is the angle between the incoming electron three-momentum and the outgoing electron three-momentum (figure 8.6). Since $E_i = E_f \equiv E$ as required by our energy-conserving delta function, $|\vec{p}_i| = |\vec{p}_f| \equiv |\vec{p}|$, but $\vec{p}_i \neq \vec{p}_f$. The kinematical relationships become

$$
\begin{aligned}
p_i \cdot p_f &= E^2 - \vec{p}^2 \cos\theta \\
&= m^2 + \vec{p}^2(1 - \cos\theta) \\
&= m^2 + 2\beta^2 E^2 \sin^2 \frac{\theta}{2}
\end{aligned} \tag{8.47}
$$

and

$$
\begin{aligned}
|\vec{q}|^2 &= |\vec{p}_f - \vec{p}_i|^2 \\
&= 2\vec{p}^2 - 2\vec{p}_f \cdot \vec{p}_i \\
&= 2\vec{p}^2(1 - \cos\theta) \\
&= 4\beta^2 E^2 \sin^2 \frac{\theta}{2}.
\end{aligned} \tag{8.48}
$$

Reintroducing the correct powers of c, we find

$$
\boxed{\frac{d\bar{\sigma}}{d\Omega} = \frac{(Z\alpha)^2}{4(\gamma\beta^2)^2(mc^2)^2 \sin^4(\theta/2)} \left(1 - \beta^2 \sin^2 \frac{\theta}{2}\right).} \tag{8.49}
$$

This formula is known as the Mott cross section and was first calculated to order α^3 by Mott[3].

In the nonrelativistic limit $\beta \to 0$ and the second term of the Mott cross section vanishes. This is in agreement with our previous result shown in equation 8.38. One can also show (see problem 8.5) that the equivalent cross section using spinless Klein-Gordon theory is

$$
\frac{d\sigma}{d\Omega} = \frac{(Z\alpha)^2}{4(\gamma\beta^2)^2(mc^2)^2 \sin^4(\theta/2)}. \tag{8.50}
$$

Thus we see in three different ways that $-\beta^2 \sin^2(\theta/2)$ is a spin correction to the Rutherford formula. The physical origin of the spin correction is due to the fact that the Dirac electron has a magnetic moment interacting with the magnetic field of the scattering center (viewed from rest frame of electron). For small velocities this effect is negligible.

The remaining term in the cross section (equation 8.49) reduces in the nonrelativistic limit $\gamma \to 1$ to

---

[3]N.F. Mott, "The Scattering of Fast Electrons by Atomic Nuclei", Proc. Roy. Soc. **124** (1929) 425-442.

$$\frac{d\bar{\sigma}}{d\Omega} \approx \frac{(Z\alpha)^2}{4v^4 m^2 \sin^4(\theta/2)}$$

$$= \frac{(Z\alpha)^2}{16E^2 \sin^4(\theta/2)}, \qquad (8.51)$$

where we have used $E = p^2/2m$. Thus the Mott cross section reduces to the Rutherford formula in the nonrelativistic limit. Apart from the spin-dependent term, the relativistic result (equation 8.49) differs from the Rutherford formula by a kinematic factor $1/\gamma^2 = 1 - \beta^2$.

In the extreme-relativistic limit $\beta \to 1$ and

$$\frac{d\bar{\sigma}}{d\Omega} \approx \frac{(Z\alpha)^2 \cos^2(\theta/2)}{4E^2 \sin^4(\theta/2)}, \qquad (8.52)$$

which differs from spinless case by factor of $\cos^2(\theta/2)$. The physical origin of this result can be argued in terms of chirality (see Holstein [14]).

In all cases the cross section diverges for small momentum transfers (small θ). This is due to the infinite range of the Coulomb potential, and that it decreases only slowly with distance. Hence electrons passing even at a great distance from the atom experience small angle scattering. For the same reason the total cross section is infinite.

The Rutherford cross section is known to be an exact formula in the non-relativistic limit. We have derived the Rutherford cross section exactly using only a first approximation to the scattering matrix without including the contributions from other diagrams. It has been shown that the second-order process in figure 8.4b leads to a divergent integral. This is connected with the infinite range of the Coulomb field. In the nonrelativistic limit, a far-distant particle cannot be described by a plane wave, as was assumed. We must represent the incoming state using a "Coulomb wave", i.e. a distorted plane wave, to be able to physically interpret the higher-order processes.

8.3 Trace Theorems

We must now digress and establish useful properties of traces of products of Dirac gamma matrices. These properties will allow us to calculate cross sections without ever looking directly at a Dirac matrix or spinor. They are derived from the commutation algebra of the γ's (equation 5.105) and the cyclic property of the trace, and are valid independently of the choice of representation for the γ's. The order of the theorems below is of no particular significance, and the numbering of the theorems is arbitrary.

Theorem 1

The trace of an odd number of γ matrices is zero.
Proof: For n odd,

$$
\begin{aligned}
\text{Tr}[\not{a}_1 \cdots \not{a}_n] &= \text{Tr}[\not{a}_1 \cdots \not{a}_n \gamma_5 \gamma_5] \\
&= \text{Tr}[\gamma_5 \not{a}_1 \cdots \not{a}_n \gamma_5] \\
&= (-1)^n \text{Tr}[\not{a}_1 \cdots \not{a}_n \gamma_5 \gamma_5] \\
&= 0 \quad \text{for } n \text{ odd.}
\end{aligned}
\tag{8.53}
$$

Theorem 2

$$
\text{Tr}[I] = 4.
\tag{8.54}
$$

$$
\begin{aligned}
\text{Tr}[\not{a}\,\not{b}] = \text{Tr}[\not{b}\,\not{a}] &= \frac{1}{2}\text{Tr}[\not{a}\,\not{b} + \not{b}\,\not{a}] \\
&= \frac{1}{2}\text{Tr}[a_\mu b_\nu (\gamma^\mu \gamma^\nu + \gamma^\nu \gamma^\mu)] \\
&= \frac{1}{2}\text{Tr}[a_\mu b_\nu (2g^{\mu\nu})] \\
&= \text{Tr}[a \cdot b] \\
&= a \cdot b\,\text{Tr}[I] \\
&= 4a \cdot b.
\end{aligned}
\tag{8.55}
$$

This also shows

$$
\begin{aligned}
\text{Tr}[a_\mu \gamma^\mu b_\nu \gamma^\nu] &= 4a_\mu b^\mu, \\
a_\mu b_\nu \text{Tr}[\gamma^\mu \gamma^\nu] &= 4a_\mu b_\nu g^{\mu\nu}, \\
\Rightarrow \quad \text{Tr}[\gamma^\mu \gamma^\nu] &= 4g^{\mu\nu}.
\end{aligned}
\tag{8.56}
$$

Theorem 3

$$
\begin{aligned}
\text{Tr}[\not{a}_1 \cdots \not{a}_n] = a_1 \cdot a_2 \text{Tr}[\not{a}_3 \cdots \not{a}_n] &- a_1 \cdot a_3 \text{Tr}[\not{a}_2 \not{a}_4 \cdots \not{a}_n] + \cdots \\
&+ a_1 \cdot a_n \text{Tr}[\not{a}_2 \cdots \not{a}_{n-1}].
\end{aligned}
\tag{8.57}
$$

In particular,

$$
\text{Tr}[\not{a}_1 \not{a}_2 \not{a}_3 \not{a}_4] = 4(a_1 \cdot a_2 a_3 \cdot a_4 + a_1 \cdot a_4 a_2 \cdot a_3 - a_1 \cdot a_3 a_2 \cdot a_4).
\tag{8.58}
$$

Proof: Using $\rlap{/}{a}_1 \rlap{/}{a}_2 = -\rlap{/}{a}_2 \rlap{/}{a}_1 + 2a_1 \cdot a_2$,

$$\mathrm{Tr}[\rlap{/}{a}_1 \rlap{/}{a}_2 \cdots \rlap{/}{a}_n] = 2a_1 \cdot a_2 \mathrm{Tr}[\rlap{/}{a}_3 \cdots \rlap{/}{a}_n] - \mathrm{Tr}[\rlap{/}{a}_2 \rlap{/}{a}_1 \rlap{/}{a}_3 \cdots \rlap{/}{a}_n]$$
$$= 2a_1 \cdot a_2 \mathrm{Tr}[\rlap{/}{a}_3 \cdots \rlap{/}{a}_n] - \cdots$$
$$+ 2a_1 \cdot a_n \mathrm{Tr}[\rlap{/}{a}_2 \cdots \rlap{/}{a}_{n-1}] - \mathrm{Tr}[\rlap{/}{a}_2 \cdots \rlap{/}{a}_n \rlap{/}{a}_1]. \quad (8.59)$$

Using the cyclic property of the trace proves the theorem. This also shows

$$b_\mu d_\nu \mathrm{Tr}[\rlap{/}{a}\gamma^\mu \rlap{/}{c}\gamma^\nu] = 4(a^\mu b_\mu c^\nu d_\nu + a^\nu d_\nu b_\mu c^\mu - a \cdot c b_\mu d^\mu)$$
$$= b_\mu d_\nu 4(a^\mu c^\nu + a^\nu c^\mu - a \cdot c g^{\mu\nu}),$$
$$\Rightarrow \quad \mathrm{Tr}[\rlap{/}{a}\gamma^\mu \rlap{/}{c}\gamma^\nu] = 4(a^\mu c^\nu + a^\nu b^\mu - a \cdot c g^{\mu\nu}). \quad (8.60)$$

Theorem 4

$$\mathrm{Tr}[\gamma_5] = \mathrm{Tr}[i\gamma^0\gamma^1\gamma^2\gamma^3]$$
$$= -\mathrm{Tr}[i\gamma^1\gamma^0\gamma^2\gamma^3]$$
$$= \mathrm{Tr}[i\gamma^1\gamma^2\gamma^0\gamma^3]$$
$$= -\mathrm{Tr}[i\gamma^1\gamma^2\gamma^3\gamma^0]$$
$$= -\mathrm{Tr}[\gamma_5]$$
$$= 0. \quad (8.61)$$

This leads to

$$\mathrm{Tr}[\gamma_5 \, \rlap{/}{a} \, \rlap{/}{b}] = 0 \quad (8.62)$$

(see problem 8.7), and

$$\mathrm{Tr}[\gamma_5 \, \rlap{/}{a} \, \rlap{/}{b} \, \rlap{/}{c} \, \rlap{/}{d}] = 4i\epsilon_{\alpha\beta\gamma\delta} a^\alpha b^\beta c^\gamma d^\delta, \quad (8.63)$$

where the totally antisymmetric tensor $\epsilon_{\alpha\beta\gamma\delta}$ is $+1$ for $(\alpha, \beta, \gamma, \delta)$, an even permutation of $(0, 1, 2, 3)$ and is -1 for an odd permutation, and is 0 if two indices are the same.

Proof: For a non-vanishing contribution, all components of a, b, c, d must be different and the total contribution is the sum of the various combinations of components multiplied by the sign of the permutation. To fix the overall sign take an example case:

$$\mathrm{Tr}[\gamma_5\gamma_0\gamma_1\gamma_2\gamma_3 a^0 b^1 c^2 d^3] = i\epsilon_{0123} a^0 b^1 c^2 d^3 \mathrm{Tr}[\gamma_5^2]$$
$$= 4i\epsilon_{0123} a^0 b^1 c^2 d^3. \quad (8.64)$$

Theorem 5

$$\gamma_\mu \gamma^\mu = 4I. \tag{8.65}$$

$$
\begin{aligned}
\gamma_\mu \,\rlap{/}{a}\gamma^\mu &= a_\nu \gamma_\mu \gamma^\nu \gamma^\mu \\
&= -a_\nu \gamma_\mu \gamma^\mu \gamma^\nu + 2a_\nu \gamma^\mu g^{\nu\mu} \\
&= -4a_\nu \gamma^\nu + 2\,\rlap{/}{a} \\
&= -4\,\rlap{/}{a} + 2\,\rlap{/}{a} \\
&= -2\,\rlap{/}{a}.
\end{aligned} \tag{8.66}
$$

$$
\begin{aligned}
\gamma_\mu \,\rlap{/}{a}\,\rlap{/}{b}\gamma^\mu &= \gamma_\mu \,\rlap{/}{a}b_\nu \gamma^\nu \gamma^\mu \\
&= -\gamma_\mu \,\rlap{/}{a}b_\nu \gamma^\mu \gamma^\nu + 2\gamma_\mu \,\rlap{/}{a}b_\nu g^{\nu\mu} \\
&= 2\,\rlap{/}{a}\,\rlap{/}{b} + 2\,\rlap{/}{b}\,\rlap{/}{a} \\
&= 2a_\mu b_\nu (\gamma^\mu \gamma^\nu + \gamma^\nu \gamma^\mu) \\
&= 4a_\mu b_\nu g^{\mu\nu} \\
&= 4a \cdot b.
\end{aligned} \tag{8.67}
$$

This leads to

$$\gamma_\mu \,\rlap{/}{a}\,\rlap{/}{b}\,\rlap{/}{c}\gamma^\mu = -2\,\rlap{/}{c}\,\rlap{/}{b}\,\rlap{/}{a}, \tag{8.68}$$

$$\gamma_\mu \,\rlap{/}{a}\,\rlap{/}{b}\,\rlap{/}{c}\,\rlap{/}{d}\gamma^\mu = 2[\rlap{/}{d}\,\rlap{/}{a}\,\rlap{/}{b}\,\rlap{/}{c} + \rlap{/}{c}\,\rlap{/}{b}\,\rlap{/}{a}\,\rlap{/}{d}]. \tag{8.69}$$

Theorem 6

$$\mathrm{Tr}[\rlap{/}{a}_1 \,\rlap{/}{a}_2 \cdots \rlap{/}{a}_{2n}] = \mathrm{Tr}[\rlap{/}{a}_{2n} \cdots \rlap{/}{a}_2 \,\rlap{/}{a}_1]. \tag{8.70}$$

Proof: From the charge conjugation discussion, recall that there exists a matrix C such that $C\gamma_\mu C^{-1} = -\gamma_\mu^T$. Then

$$
\begin{aligned}
\mathrm{Tr}[\rlap{/}{a}_1 \,\rlap{/}{a}_2 \cdots \rlap{/}{a}_{2n}] &= \mathrm{Tr}[C\,\rlap{/}{a}_1 C^{-1} C\,\rlap{/}{a}_2 C^{-1} \cdots C\,\rlap{/}{a}_{2n} C^{-1}] \\
&= (-1)^{2n}\mathrm{Tr}[\rlap{/}{a}_1^T \,\rlap{/}{a}_2^T \cdots \rlap{/}{a}_{2n}^T] \\
&= \mathrm{Tr}[\rlap{/}{a}_{2n} \cdots \rlap{/}{a}_2 \,\rlap{/}{a}_1]^T \\
&= \mathrm{Tr}[\rlap{/}{a}_{2n} \cdots \rlap{/}{a}_2 \,\rlap{/}{a}_1].
\end{aligned} \tag{8.71}
$$

A general form which we shall often encounter, where Γ is some combination of Dirac matrices, is

$$\begin{aligned}
|\bar{u}(f)\Gamma u(i)|^2 &= |u^\dagger(f)\gamma^0\Gamma u(i)|^2 \\
&= [u^\dagger(f)\gamma^0\Gamma u(i)][u^\dagger(i)\Gamma^\dagger(\gamma^0)^\dagger u(f)] \\
&= [\bar{u}(f)\Gamma u(i)][\bar{u}(i)\gamma^0\Gamma^\dagger\gamma^0 u(f)] \\
&= [\bar{u}(f)\Gamma u(i)][\bar{u}(i)\bar{\Gamma}u(f)],
\end{aligned} \tag{8.72}$$

where $\bar{\Gamma} = \gamma^0\Gamma^\dagger\gamma^0$. Therefore,

$$\begin{aligned}
\overline{\gamma^\mu} &= \gamma^0(\gamma^\mu)^\dagger\gamma^0 = \gamma^\mu, \\
\overline{i\gamma^5} &= \gamma^0(i\gamma^5)^\dagger\gamma^0 = i\gamma^5, \\
\overline{\gamma^\mu\gamma^5} &= \gamma^0(\gamma^5)^\dagger(\gamma^\mu)^\dagger\gamma^0 = \gamma^\mu\gamma^5, \\
\overline{\slashed{a}\,\slashed{b}\,\slashed{c}\cdots\slashed{p}} &= \gamma^0\,\slashed{p}^\dagger\cdots\slashed{c}^\dagger\,\slashed{b}^\dagger\,\slashed{a}^\dagger\gamma^0 = \slashed{p}\cdots\slashed{c}\,\slashed{b}\,\slashed{a}.
\end{aligned} \tag{8.73}$$

When averaging over initial-spin states and summing over final-spin states, we often encounter the form

$$\frac{1}{2}\sum_{r=1}^{2}\sum_{s=1}^{2}|\bar{u}_f^r(p_f)\Gamma u_i^s(p_i)|^2$$

$$\begin{aligned}
&= \frac{1}{2}\sum_{r=1}^{2}\sum_{s=1}^{2}\sum_{\alpha,\beta,\gamma,\delta}\bar{u}_f^r(p_f)_\beta\Gamma_{\beta\alpha}u_i^s(p_i)_\alpha\bar{u}_i^s(p_i)_\delta\bar{\Gamma}_{\delta\gamma}u_f^r(p_f)_\gamma \\
&= \frac{1}{2}\sum_{\alpha,\beta,\gamma,\delta}\Gamma_{\beta\alpha}\left(\frac{\slashed{p}_i + m}{2m}\right)_{\alpha\delta}\bar{\Gamma}_{\delta\gamma}\left(\frac{\slashed{p}_f + m}{2m}\right)_{\gamma\beta} \\
&= \frac{1}{8m^2}\text{Tr}[\Gamma(\slashed{p}_i + m)\bar{\Gamma}(\slashed{p}_f + m)].
\end{aligned} \tag{8.74}$$

The other "trick" we will use is to introduce the unit four-vector $a_\mu = (1,0,0,0)$ to put γ^0 into the slash notation: $a\gamma^0 = \slashed{a}$. For example,

$$\text{Tr}[\gamma^0\,\slashed{b}] = \text{Tr}[\slashed{a}\,\slashed{b}] = 4a\cdot b = 4b^0. \tag{8.75}$$

8.4 Coulomb Scattering of Positrons

To help elucidate the similarities and differences between electron and positron scattering, we perform one calculation with the electron replaced by a positron. Consider the scattering of a positron in the Coulomb field of a nucleus of charge Ze. For the Coulomb scattering of an electron, the S-matrix was proportional to e^2. We therefore expect the form of the cross section for positron scattering to be similar to the Coulomb scattering of an electron. In fact, we expect the result to be identical to lowest order in e^2. It would be identical to all orders in e if the Coulomb field was due to a negatively charged nucleus of charge $-Ze$. Figure 8.7 shows the diagram for positron scattering in a coulomb field. Throughout this book we will draw positrons as negative-energy electrons with the arrows indicating motion backwards in time (cf. section 6.9.2).

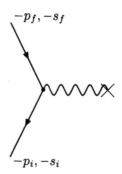

FIGURE 8.7: Scattering of a positron from a fixed Coulomb potential.

The S-matrix element for the process in figure 8.7 is

$$S_{fi} = -ie \int d^4x \overline{\psi}_f(x) \, A(x) \Psi_i^{(-)}(x). \qquad (8.76)$$

Here the incoming state is in the future and is to be interpreted as a negative-energy electron of four-momentum $-p_f$ and spin $-s_f$ running backwards in time. Using plane waves in lowest order, the incident wave function is

$$\psi_i(x) = \sqrt{\frac{m}{E_f V}} v(p_f, s_f) e^{+ip_f \cdot x}. \qquad (8.77)$$

Similarly, the outgoing state is the negative-energy electron running backwards into the past. Its wave function is

$$\psi_f(x) = \sqrt{\frac{m}{E_i V}} v(p_i, s_i) e^{+ip_i \cdot x}, \qquad (8.78)$$

which represents the incident positron with momentum $\vec{p}_i$ and spin $\vec{s}_i$ before the scattering. The S-matrix element becomes (cf. equation 8.21)

$$S_{fi} = -\frac{iZe^2}{4\pi} \frac{1}{V} \frac{m}{\sqrt{E_f E_i}} \overline{v}(p_i, s_i) \gamma^0 v(p_f, s_f) \int d^4x \frac{e^{i(p_f - p_i) \cdot x}}{|\vec{x}|}. \qquad (8.79)$$

By the same calculation as in equation 8.39, we find the differential cross section is

$$\frac{d\bar{\sigma}}{d\Omega} = \frac{2(Z\alpha)^2 m^2}{|\vec{q}|^4} \sum_{\pm s_f, s_i} |\bar{v}(p_i, s_i)\gamma^0 v(p_f, s_f)|^2. \tag{8.80}$$

Again the spin sum may be reduced to a trace, using the completeness relationship for positron spinors

$$\sum_{\pm s_i} v_\alpha(p_i, s_i)\bar{v}_\beta(p_i, s_i) = -\left(\frac{-\not{p}_i + m}{2m}\right)_{\alpha\beta}. \tag{8.81}$$

The first minus sign comes from the normalization of the negative-energy spinors and the relative minus sign between the two terms comes from the negative-energy projection operator.

The differential cross section now becomes

$$\frac{d\bar{\sigma}}{d\Omega} = \frac{(Z\alpha)^2}{2|\vec{q}|^4} \text{Tr}\left[\gamma^0(\not{p}_i - m)\gamma^0(\not{p}_f - m)\right]. \tag{8.82}$$

This is the same as the result of equation 8.44 for the electron with mass m replaced by $-m$. Since our answer for electron scattering was even in m, this shows that the positron scattering cross section is equal to the electron scattering cross section to lowest order in α.

We could have anticipated this result from charge-conjugation invariance of the S-matrix. We could equally well write

$$S_{fi} = ie \int d^4x (\bar{\psi}_C)_i \not{A} (\psi_C)_f, \tag{8.83}$$

where we have written the S-matrix element for electron scattering with positron wave functions and an electromagnetic potential unchanged from electron scattering (cf. equation 8.76). Using the definition of ψ_C, we write

$$S_{fi} = ie \int d^4x \left(-\psi_i^T C^{-1}\right) \not{A} \left(C\bar{\psi}_f^T\right)$$

$$= -ie \int d^4x \psi_i^T \left(C^{-1} \not{A} C\right) \bar{\psi}_f^T. \tag{8.84}$$

Using equation 5.257 and the fact that A^μ is real, we write $C^{-1} \not{A} C = -\not{A}^T$, and thus

$$S_{fi} = ie \int d^4x \psi_i^T \not{A}^T \bar{\psi}_f^T$$

$$= ie \int d^4x (\bar{\psi}_f \not{A} \psi_i)^T$$

$$= ie \int d^4x \bar{\psi}_f \not{A} \psi_i, \tag{8.85}$$

which is the same as equation 8.17. The last step follows from the fact that $\overline{\psi}_f \, A\!\!\!/ \, \psi_i$ is just an ordinary number. In this picture, the positron runs forward in time and $\psi_{Cf}(x) = C\gamma^0\psi_f^*$ is the wave function of the initial positron.

In section 5.16, we saw that for each solution of the electron in the potential A_μ there is a corresponding solution of the positron in the potential $-A_\mu$, that is, the scattering of an electron from the potential $+Ze/4\pi r$ is the same as that of a positron from potential $-Ze/4\pi r$. However, since the calculated cross section depends only on α^2, the sign of A_μ does not matter. This is not true for the α^3 correction which comes from the product of the first- and second-order scattering amplitudes, which have opposite signs for electrons and positrons.

Figure 8.8 compares electron with positron Coulomb scattering to second order. The first-order amplitudes for electron and positron scattering (figures 8.8a and 8.8c) differ only in the sign of Ze^2, while the amplitudes are identical at second order in $(Ze^2)^2$. When squaring the amplitudes to calculate the measurable cross section, the results to first order will be equal. The second-order contributions will also be equal (figures 8.8b and 8.8d), but the cross term resulting from the combination of first order and second-order amplitudes will differ in sign at order $(Ze^2)^3$. Therefore electron and positron Coulomb scattering will differ in second order due to the interference between the first- and second-order amplitudes[4].

We should also notice that the positron cross section is obtained from that of the electron cross section by replacing $u(p_i)$ with $\overline{v}(-p_f)$ and $\overline{u}(p_f)$ with $v(-p_i)$; this is a general feature of the relativistic theory, and is one of the substitution rules described in the next section.

8.5   Crossing Symmetry and Substitution Rules

One of the most powerful consequences of the structure of the S-matrix is crossing symmetry. Consider figure 8.9. If two processes S' and S differ only in one external particle, such that this particle is an outgoing photon, electron, or positron in S' and an ingoing photon, positron, or electron in S, respectively, then the S-matrix elements associated with S' and S are related as shown in table 8.1. In the case of circular polarization, right-circular $\leftrightarrow$ left-circular polarization. The double arrow indicates that the substitution necessary to obtain S from S' is reversible, so that one can also obtain S' from S.

---

[4]W.A. McKinley & H. Feshbach, "The Coulomb Scattering of Relativistic Electrons by Nuclei", Phys. Rev. **74** (1948) 1759-1763.

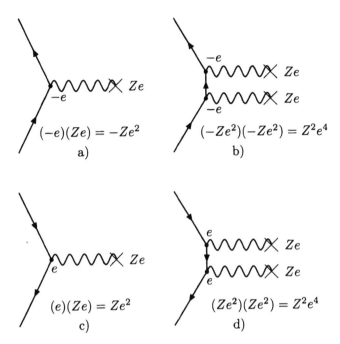

FIGURE 8.8: Coulomb scattering: a) electron first order, b) electron second order, c) positron first order, and d) positron second order.

These substitution rules can also be applied to the square of the matrix elements $\overline{\mathcal{M}}$ after the spin summations have been carried out. In this case, we have transformed to momentum space and projection operators take the place of the wave functions. The substitutions can then be carried out directly in the projection operators as shown in table 8.2. This simply means that there is also an overall sign change of the trace in addition to the momentum substitution.

Repeated application of the substitution rules allow one to calculate all permutations of a process involving the same number and types of particles from a single calculation. Some of the resulting processes many not be possible

TABLE 8.1: Substitution rules for photons, electrons, and positrons.

Process S'		Process S		Kinematics		Spin	
photon	k' out	photon	k in	$k' \leftrightarrow -k$		$\varepsilon^{*\prime} \leftrightarrow \varepsilon$	
electron	p' out	positron	q in	$p' \leftrightarrow -q$		$\overline{u}(p') \leftrightarrow v(q)$	
positron	q' out	electron	p in	$q' \leftrightarrow -p$		$\overline{v}(q') \leftrightarrow u(p)$	

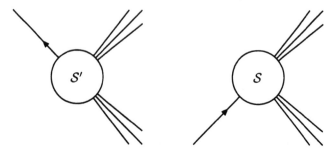

FIGURE 8.9: Two arbitrary processes related by crossing symmetry.

TABLE 8.2: Substitution rules for spin sums.

$\overline{\mathcal{M}}$		Projection Operator	
electron	$p' \leftrightarrow q$	$\Lambda_-(p') \quad \leftrightarrow$	$-\Lambda_+(q) = -\Lambda_-(-q)$
positron	$q' \leftrightarrow p$	$\Lambda_+(q) \quad \leftrightarrow$	$-\Lambda_-(p) = -\Lambda_+(-p)$

because they violate the conservation of energy and momentum.

An example of an application of the substitution rules is the calculation of electron-photon scattering, electron-positron annihilation, and electron-positron production, as shown in figure 8.10. After any one of these processes is calculated, the others follow by the substitution rules (see problem 8.9). However, the original matrix elements – or traces – must be known in complete generality, without restriction to a special coordinate system, in order to make the procedure work.

8.6 Electron Scattering from a Dirac Proton

The treatment of electron scattering from a fixed potential in section 8.2 was not relativistic covariant; energy was conserved but momentum was not. This was because the scattering center was assumed to be fixed. In order to deal with a more realistic situation, we consider the scattering of an electron from a freely movable nucleus, in particular, a proton. Now all the recoil effects are present.

As a first approximation, we treat the proton as a structureless – point-like – Dirac particle. By Dirac particle we mean a particle having a gyromagnetic ratio $g = 2$, which obeys the Dirac equation.

With two particles – the electron and proton – there is no external classical electromagnetic potential in the calculation. Since the electron and proton are charged, they act as sources of electromagnetic fields. We can picture the process as one in which each particle scatters off the "virtual" field produced

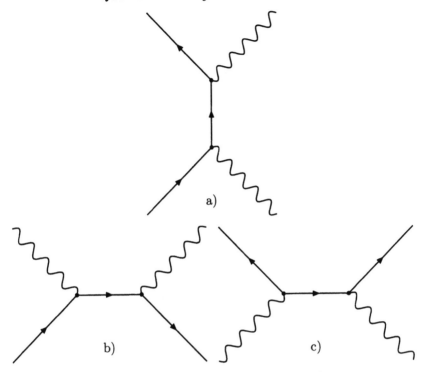

FIGURE 8.10: Processes related by the crossing symmetry: a) electron-photon scattering, b) electron-positron annihilation, and c) electron-positron productions.

by the other. We will consider the electron scattering off the electromagnetic field of the proton; although we should expect the physical results to be equivalent to proton scattering off the electromagnetic field of the electron.

If we know the current of the proton $J_\nu(y)$, we can calculate the electromagnetic potential it generates using Maxwell's equations and the Green-function techniques of section 7.4:

$$A^\mu(x) = \int d^4y D_F^{\mu\nu}(x-y)J_\nu(y). \qquad (8.86)$$

This equation is the same as equation 7.50, and needs the Feynman propagator in configuration space for electromagnetic radiation:

$$D_F^{\mu\nu}(x-y) = g^{\mu\nu}D_F(x-y) = \int \frac{d^4q}{(2\pi)^4} e^{-iq\cdot(x-y)}\left(\frac{-g^{\mu\nu}}{q^2+i\epsilon}\right). \qquad (8.87)$$

To lowest order in α, we can consider the radiation as a photon exchanged between the electron and proton. The interaction photon is said to be virtual, or "off-mass-shell"; its effective mass squared is q^2. This photon is said to be virtual since for real photons, the energy-momentum vector q_μ satisfies $q^2 = 0$. Hence the propagator becomes infinite and does not exist for real photons; it has its biggest contribution close to mass shell. Likewise for the Klein-Gordon and Dirac equations, the propagator becomes infinite when the particle is on its mass shell: $p^2 = m^2$.

Using plane waves for the electron, the S-matrix element is

$$
\begin{aligned}
S_{fi} &= ie \int d^4x \overline{\psi}_f(x)\, \slashed{A}(x)\Psi_i^{(+)}(x) \\
&= ie \int d^4x \overline{\psi}_f(x)\gamma_\mu \int d^4y\, g^{\mu\nu} D_F(x-y) J_\nu(y)\psi_i(x) \\
&= -\int d^4x d^4y \left[-e\overline{\psi}_f(x)\gamma_\mu \psi_i(x) \right] \left[iD_F(x-y) \right] J^\mu(y).
\end{aligned} \tag{8.88}
$$

In this calculation, we group the factors in mathematical expressions in what might appear a strange fashion. The reason is to show the correspondence between factors in the scattering amplitude and pieces of the diagrams for the process. We recognize $-e\overline{\psi}_f(x)\gamma_\mu \psi_i(x)$ as the current of the electron. It is a matrix element of the current operator between initial and final electron states, and is usually referred to as a transition current.

Now we must face the problem of what to choose for the proton current $J^\mu(y)$. Since we have identified the electron transition current, we may want to choose a similar form for the proton current. This is particularly justified since we could have equally well taken the approach at the beginning of the problem of considering proton scattering off the electromagnetic field produced by the electron. It is thus reasonable to try

$$
J^\mu(y) = e_P \overline{\psi}_f^P(y)\gamma^\mu \psi_i^P(y), \tag{8.89}
$$

where $e_P = +e$ is the proton electric charge. $\psi_i^P(y)$ and $\overline{\psi}_f^P(y)$ represent the initial and final state plane-wave solutions for a free Dirac proton. Using plane-wave solutions for the proton gives

$$
J^\mu(y) = \sqrt{\frac{M^2}{E_f' E_i'}}\, \frac{e}{V}\, e^{i(P_f - P_i)\cdot y}\overline{u}(P_f, S_f)\gamma^\mu u(P_i, S_i), \tag{8.90}
$$

where P_i, S_i, E_i' and P_f, S_f, E_f' are the four-momentum, spin, and energy of the initial and final state protons. The mass of the proton is M.

We now write for the S-matrix element

$$
S_{fi} = -\int d^4x d^4y \left[\sqrt{\frac{m}{E_i V}} \sqrt{\frac{m}{E_f V}}\, e^{i(p_f - p_i)\cdot x}\overline{u}(p_f.s_f)(ie\gamma_\mu)u(p_i, s_i) \right]
$$

$$\cdot \int \frac{d^4q}{(2\pi)^4} e^{-iq\cdot(x-y)} \left(\frac{-ig^{\mu\nu}}{q^2+i\epsilon}\right)$$

$$\cdot \left[\sqrt{\frac{M}{E_i'V}}\sqrt{\frac{M}{E_f'V}} e^{i(P_f-P_i)\cdot y}\overline{u}(P_f,S_f)(-ie\gamma_\nu)u(P_i,S_i)\right]$$

$$= -\int \frac{d^4x\,d^4y\,d^4q}{(2\pi)^4}\sqrt{\frac{m^2}{E_iE_fV^2}}\sqrt{\frac{M^2}{E_i'E_f'V^2}} e^{i(p_f-p_i-q)\cdot x}e^{i(P_f-P_i+q)\cdot y}$$

$$\cdot [\overline{u}(p_f,s_f)(i\gamma_\mu)u(p_i,s_i)]\frac{-ig^{\mu\nu}}{q^2+i\epsilon}[\overline{u}(P_f,S_f)(-ie\gamma_\nu)u(P_i,S_i)]. \quad (8.91)$$

The x- and y-integrations can be performed using the definitions of the Dirac delta function:

$$S_{fi} = -\int d^4q\sqrt{\frac{m^2}{E_fE_iV^2}}\sqrt{\frac{M^2}{E_f'E_i'}}V^2(2\pi)^4\delta^4(p_f-p_i-q)\delta^4(P_f-P_i+q)$$

$$\cdot [\overline{u}(p_f,s_f)(ie\gamma_\mu)u(p_i,s_i)]\frac{-g^{\mu\nu}}{q^2+i\epsilon}[\overline{u}(P_f,S_f)(-ie\gamma_\nu)u(P_i,S_i)]$$

$$= i(2\pi)^4\delta^4(P_f-P_i+p_f-p_i)\sqrt{\frac{m^2}{E_fE_iV^2}}\sqrt{\frac{M^2}{E_f'E_i'V^2}}\mathcal{M}_{fi}, \quad (8.92)$$

where

$$i\mathcal{M}_{fi} = [\overline{u}(p_f,s_f)(ie\gamma_\mu)u(p_i,s_i)]\frac{-ig^{\mu\nu}}{(p_f-p_i)^2+i\epsilon}[\overline{u}(P_f,S_f)(-ie\gamma_\nu)u(P_i,S_i)]. \quad (8.93)$$

The amplitude $i\mathcal{M}_{fi}$ is the Lorentz invariant matrix element for the process under consideration. It is usually simply called the invariant amplitude. The choice of this name is quite natural since the matrix element consists of scalar products of four-vectors, which are Lorentz invariant. It is understood that the momenta in $i\mathcal{M}_{fi}$ are restricted by the four-momentum conserving delta function $\delta^4(P_f-P_i+p_f-p_i)$. Equations 8.92 and 8.93 give the electron-proton scattering amplitude to lowest order in e. Higher order interaction effects which distort the plane waves that were inserted in the currents have been ignored.

Comparing this result with the amplitude for Coulomb scattering of electrons, equation 8.23, shows two differences:

$$\frac{Z\gamma^0}{|\vec{q}|^2} \to \gamma_\mu\left(\frac{-1}{q^2+i\epsilon}\right)\sqrt{\frac{M^2}{E_f'E_i'}}\overline{u}(P_f,S_f)\gamma^\mu u(P_i,S_i) \quad (8.94)$$

and

$$V \rightarrow (2\pi)^3 \delta^3 (\vec{P}_f - \vec{P}_i + \vec{p}_f - \vec{p}_i). \qquad (8.95)$$

The latter difference guarantees momentum conservation, which was a problem in Coulomb scattering, which we are trying to circumvent in this calculation.

The invariant matrix element (equation 8.93) shows a clear symmetry between the electron and proton variables, giving us faith in our choice of proton current. In addition, we notice that the two factors $\bar{u}(f)\gamma^\mu u(i)$ are the momentum-space versions of the x-dependent current matrix elements. Thus the amplitude $i\mathcal{M}_{fi}$ has the form of two currents connected together by a photon propagator.

Based on the arguments at the end of section 7.4, it can be shown (see problem 8.10) that the covariant amplitude includes contributions from the exchange of transversely polarized photons and from the familiar Coulomb potential, showing that the result is truly a relativistic extension of static Coulomb scattering.

The expression for the S-matrix element in momentum space may be represented by a "Feynman diagram" as shown in figure 8.11. S_{fi} always contains a four-dimensional delta function expressing overall energy-momentum conservation. In addition, for each line and intersection of the diagram there corresponds a unique factor in the invariant matrix element. A solid line with an arrow pointing toward positive time represents the electron and a double line the proton. The double line is to represent structure, which a real proton has. The wavy line represents the influence of the electromagnetic interaction, which is expressed in the matrix element by the reciprocal of the square of the momentum transfer, or the inverse d'Alembert operator in momentum space – the propagator. We refer to this line as representing a virtual photon exchanging four-momentum $q = p_f - p_i = P_i - P_f$ between the electron and proton. The amplitude for the virtual photon to propagate between the two currents is $-i(q^2 + i\epsilon)^{-1}$. At the points – or vertices – on which the photon lands there are operators $-ie\gamma^\mu$ sandwiched between spinors $\sqrt{m/E}\,u(p,s)$ representing the real incident and outgoing free particles. To get the spinor factor in expressions such as these, the rule is to start at the ingoing fermion line u and follow the line in the direction of the arrow through until the end, inserting vertices and propagators in the right order, until you reach the outgoing state $\bar{u}$. For the factors of i, there is a uniform rule: $-i$ for each vertex and i for each internal line in the diagram.

It should be made clear that the internal particles in a Feynman diagram are not on their mass shell: $q^2 \neq m^2$. The energy component q_0 and the momentum components $\vec{q}$ of the four-momentum q^μ of the internal line are independent variables.

The exchange photon is not to be thought of as either emitted by the electron and absorbed by the proton, or the other way around; rather, it includes both processes. This must be the case since we do not have any measuring

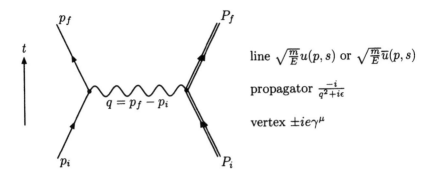

line $\sqrt{\frac{m}{E}}u(p,s)$ or $\sqrt{\frac{m}{E}}\bar{u}(p,s)$

propagator $\frac{-i}{q^2+i\epsilon}$

vertex $\pm ie\gamma^\mu$

FIGURE 8.11: Feynman diagram for electron-proton scattering.

instrument inside the interaction region to tell us about the sequence of what goes on inside; it is a non-physical question.

We now form a transition rate per unit volume w_{fi} by dividing $|S_{fi}|^2$ by the time interval of observation T and by the spatial volume V of the interaction region.

$$w_{fi} = \frac{|S_{fi}|^2}{VT} = (2\pi)^4\delta^4(P_f+p_f-P_i-p_i)\frac{1}{V^4}\frac{m^2}{E_fE_i}\frac{M^2}{E'_fE'_i}|\mathcal{M}_{fi}|^2, \quad (8.96)$$

where

$$\mathcal{M}_{fi} = [\bar{u}(p_f,s_f)\gamma_\mu u(p_i,s_i)]\frac{e^2}{q^2+i\epsilon}[\bar{u}(P_f,S_f)\gamma^\mu u(P_i,S_i)]. \quad (8.97)$$

We have used, in analogy to equations 8.27-8.31, the square of the delta function

$$[(2\pi)^4\delta^4(P_f+p_f-P_i-p_i)]^2 = (2\pi)^4\delta^4(0)(2\pi)^4\delta^4(P_f+p_f-P_i-p_i),$$
$$\rightarrow VT(2\pi)^4\delta^4(P_f+p_f-P_i-p_i), \quad (8.98)$$

since the four-dimensional delta function is just the product of four one-dimensional delta functions.

We divide the transition rate per unit volume by the flux of incident particles $|J_{\text{inc}}|$ and by the number of target particles per unit volume, which is just $1/V$, since the normalization of the wave functions was performed in such a way that there is just one particle in the normalized volume V.

To get a physical cross section, we must integrate over a given group of final states of the electron and proton corresponding to the laboratory conditions for observing the process. The number of final states in the momentum interval $d^3p_f d^3P_f$ is

$$V \frac{d^3 p_f}{(2\pi)^3} V \frac{d^3 P_f}{(2\pi)^3}, \tag{8.99}$$

and thus the six-fold differential cross section for transitions to the final states is

$$
\begin{aligned}
d\sigma &= \int V^2 \frac{d^3 p_f}{(2\pi)^3} \frac{d^3 P_f}{(2\pi)^3} \frac{V}{|J_{\text{inc}}|} w_{fi} \\
&= \int \frac{m}{E_f} \frac{d^3 p_f}{(2\pi)^3} \frac{M}{E_f'} \frac{d^3 P_f}{(2\pi)^3} \frac{mM}{E_i E_i'} \frac{(2\pi)^4 \delta^4(P_f + p_f - P_i - p_i)}{|J_{\text{inc}}|V} |\mathcal{M}_{fi}|^2.
\end{aligned}
\tag{8.100}
$$

The physics lies in $|\mathcal{M}_{fi}|^2$, the square of the invariant amplitude. There is a factor m/E for each external fermion line, that is, for each Dirac particle incident upon or emerging from the interaction. The phase-space factor for each final-state particle is $d^3 p_f/(2\pi)^3$. Thus each final-state particle gives rise to the factor $\frac{m}{E} \frac{d^3 p}{(2\pi)^3}$. We have deliberately kept these factors together. We now show that this factor is Lorentz invariant by working some of the results of section 4.4 backwards. The following combination forms a Lorentz-invariant volume element in momentum space.

$$
\begin{aligned}
\frac{d^3 p}{2E} &= \int_0^\infty dp_0 \frac{\delta(p_0 - E)}{2p_0} d^3 p + \int_0^\infty dp_0 \frac{\delta(p_0 + E)}{2p_0} d^3 p \\
&= \int_0^\infty dp_0 \delta(p_0^2 - E^2) d^3 p \\
&= \int_0^\infty dp_0 \delta(p^2 - m^2) d^3 p,
\end{aligned}
\tag{8.101}
$$

$$\boxed{\frac{d^3 p}{2E} = \int_{-\infty}^\infty d^4 p \, \delta(p^2 - m^2) \theta(p_0)} , \tag{8.102}$$

which is invariant provided p^μ is time-like, as is the case here.

In the factor $1/(V|J_{\text{inc}}|)$, $|J_{\text{inc}}|$ is the flux. For collinear beams, it is the number of particles per unit area which run by each other per unit time,

$$|J_{\text{inc}}| = \frac{|\vec{v}_i - \vec{V}_i|}{V}, \tag{8.103}$$

which is the particle density times the relative velocity.

We have required that the velocity vectors are collinear. When $V|J_{\text{inc}}|$ is combined with the normalization factors for two incident particles, it forms a Lorentz-invariant expression

$$\frac{mM}{E_i E_i' |\vec{v}_i - \vec{V}_i|} = \frac{mM}{|\vec{p}_i|E_i' + |\vec{P}_i|E_i} = \frac{mM}{\sqrt{(p_i \cdot P_i)^2 - m^2 M^2}}. \tag{8.104}$$

The last expression can be seen by working backwards:

$$\begin{aligned}
(p_i \cdot P_i)^2 - m^2 M^2 &= (E_i E_i' + |\vec{p}_i||\vec{P}_i|)^2 - (E_i^2 - |\vec{p}_i|^2)(E_i'^2 - |\vec{P}_i|^2) \\
&= 2E_i E_i' |\vec{p}_i||\vec{P}_i| + E_i^2 |\vec{P}_i|^2 + E_i'^2 |\vec{p}_i|^2 \\
&= (|\vec{p}_i|E_i' + |\vec{P}_i|E_i)^2. \tag{8.105}
\end{aligned}$$

In the case of collinear collisions, both results are identical. Consequently the naïve flux factor in the cross section can be replaced by the Lorentz-invariant flux factor so that the total cross section is Lorentz invariant. This shows that the total cross section is invariant under Lorentz transformations along the direction of motion of the incident beams. We write the invariant form

$$\begin{aligned}
d\sigma &= \frac{mM}{\sqrt{(p_i \cdot P_i)^2 - m^2 M^2}} |\mathcal{M}_{fi}|^2 \\
&\cdot (2\pi)^4 \delta^4(P_f - P_i + p_f - p_i) \frac{m d^3 p_f}{E_f (2\pi)^3} \frac{M d^3 P_f}{E_f' (2\pi)^3}. \tag{8.106}
\end{aligned}$$

This is a general form for a $2 \to 2$ process

We see that the cross section consists of three invariant quantities: the Lorentz-invariant flux factor equation 8.104, the Lorentz-invariant matrix element, and the Lorentz-invariant phase-space factor:

$$\text{dLips}(s; p_f, P_f) = (2\pi)^4 \delta^4(p_f + P_f - p_i - P_i) \frac{m}{(2\pi)^3} \frac{d^3 p_f}{2E_f} \frac{M}{(2\pi)^3} \frac{d^3 P_f}{2E_f'}, \tag{8.107}$$

where $s = (p_i + P_i)^2$ is the s Mandelstam variable. Equation 8.107 is referred to as the two-particle Lorentz-invariant phase space factor, and will be elaborated on further in appendix B.

Like in the case of Coulomb scattering, we will calculate the unpolarized cross section by averaging over initial and summing over final spin states. This gives four sums over spin and two factors of $1/2$. We have (see problem 8.11)

$$\begin{aligned}
|\overline{\mathcal{M}}_{fi}|^2 &= \frac{1}{4} \sum_{s_f, s_i, S_f, S_i} \left| \bar{u}(p_f, s_f) \gamma^\mu u(p_i, s_i) \frac{e^2}{q^2 + i\epsilon} \bar{u}(P_f, S_f) \gamma_\mu u(P_i, S_i) \right|^2 \\
&= \left(\frac{e^2}{q^2}\right)^2 \frac{1}{4} \text{Tr} \left[\frac{(\not{p}_f + m)}{2m} \gamma^\mu \frac{(\not{p}_i + m)}{2m} \gamma^\nu \right] \\
&\cdot \text{Tr} \left[\frac{(\not{P}_f + M)}{2M} \gamma_\mu \frac{(\not{P}_i + M)}{2M} \gamma_\nu \right], \tag{8.108}
\end{aligned}$$

where the bar over $\mathcal{M}_{fi}$ denotes spin average.

Note that squaring the amplitude, which contained the scalar product of two Lorentz four-vectors $\bar{u}(f)\gamma^\mu u(i)$, has led to the contraction of two tensors, i.e. a double sum. One often abbreviates this as

$$|\overline{\mathcal{M}}_{fi}|^2 = \left(\frac{e^2}{q^2}\right)^2 L^{\mu\nu} H_{\mu\nu}, \qquad (8.109)$$

were $L^{\mu\nu}$ is the leptonic, i.e. electron in this case, tensor and $H_{\mu\nu}$ the hadronic, i.e. proton in this case, tensor:

$$
\begin{aligned}
L^{\mu\nu} &= \frac{1}{2}\sum_{s_i,s_f} \bar{u}(p_f,s_f)\gamma^\mu u(p_i,s_i)\bar{u}(p_i,s_i)\gamma^\nu u(p_f,s_f) \\
&= \frac{1}{2}\text{Tr}\left[\frac{\slashed{p}_f + m}{2m}\gamma^\mu \frac{\slashed{p}_i + m}{2m}\gamma^\nu\right].
\end{aligned} \qquad (8.110)
$$

And similarly,

$$H_{\mu\nu} = \frac{1}{2}\text{Tr}\left[\frac{\slashed{P}_f + M}{2M}\gamma_\mu \frac{\slashed{P}_i + M}{2M}\gamma_\nu\right]. \qquad (8.111)$$

This factorization remains meaningful as long as a single virtual photon is exchanged in the scattering process, even if the transition currents are more complicated than those here (see problem 8.12).

Now consider the evaluation of the traces. The first trace is

$$
\begin{aligned}
\text{Tr}\left[\frac{\slashed{p}_f + m}{2m}\gamma^\mu \frac{\slashed{p}_i + m}{2m}\gamma^\nu\right] &= \frac{1}{4m^2}\text{Tr}[\slashed{p}_f\gamma^\mu \slashed{p}_i\gamma^\nu + m^2\gamma^\mu\gamma^\nu] \\
&= \frac{1}{4m^2}[4p_f^\mu p_i^\nu + 4p_f^\nu p_i^\mu - 4p_i\cdot p_f g^{\mu\nu} + 4m^2 g^{\mu\nu}] \\
&= \frac{1}{m^2}[p_f^\mu p_i^\nu + p_i^\mu p_f^\nu - g^{\mu\nu}(p_f\cdot p_i - m^2)]. \quad (8.112)
\end{aligned}
$$

The second trace has the same form:

$$\text{Tr}\left[\frac{\slashed{P}_f + M}{2M}\gamma_\mu \frac{\slashed{P}_i + M}{2M}\gamma_\nu\right] = \frac{1}{M^2}[P_{f\,\mu}P_{i\nu} + P_{i\mu}P_{f\,\nu} - g_{\mu\nu}(P_f\cdot P_i - M^2)]. \tag{8.113}$$

Both are second-rank tensors consisting of products of four-vectors.

The square of the invariant amplitude now becomes, by contracting the two tensors,

$$|\overline{\mathcal{M}}_{fi}|^2 = \frac{e^4}{4m^2 M^2 q^4}\left[p_f^\mu p_i^\nu + p_i^\mu p_f^\nu - g^{\mu\nu}(p_f\cdot p_i - m^2)\right]$$

$$\cdot \left[P_{f\,\mu} P_{i\nu} + P_{i\mu} P_{f\,\nu} - g_{\mu\nu}(P_f \cdot P_i - M^2) \right]$$

$$= \frac{e^4}{2m^2 M^2 q^4} \left[(p_i \cdot P_i)(p_f \cdot P_f) + (p_i \cdot P_f)(p_f \cdot P_i) \right.$$
$$- p_i \cdot p_f (P_i \cdot P_f - M^2) - P_i \cdot P_f (p_i \cdot p_f - m^2)$$
$$\left. + 2(p_i \cdot p_f - m^2)(P_i \cdot P_f - M^2) \right]$$

$$= \frac{e^4}{2m^2 M^2 q^4} \left[(P_f \cdot p_f)(P_i \cdot p_i) + (P_f \cdot p_i)(P_i \cdot p_f) \right.$$
$$\left. - m^2 (P_f \cdot P_i) - M^2 (p_f \cdot p_i) + 2m^2 M^2 \right]. \tag{8.114}$$

To evaluate the scattering cross section further, the frame of reference has to be specified. Usually calculations take their simplest form in the center-of-mass reference system (see problem 8.13). However, electron-proton scattering experiments are traditionally performed using a beam of electrons and a fixed target of atoms, corresponding to protons at rest in the laboratory frame. Therefore we evaluate the differential cross section in the laboratory frame in which the initial proton is at rest (figure 8.12):

$$p_f = (E', \vec{p}'), \tag{8.115}$$
$$p_i = (E, \vec{p}), \tag{8.116}$$
$$P_i = (M, 0), \tag{8.117}$$
$$P_f = (E'_f, \vec{P}'_f). \tag{8.118}$$

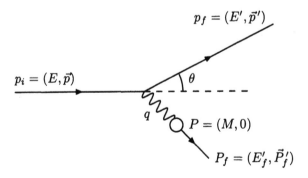

FIGURE 8.12: Electron-proton scattering in the rest frame of the proton.

In the rest frame of the proton, the square of the invariant matrix element becomes

$$
\begin{aligned}
|\mathcal{M}_{fi}|^2 &= \frac{e^4}{2m^2 M^2 q^4} [ME p_f \cdot (P_i + p_i - p_f) + M E' p_i \cdot (P_i + p_i - p_f) \\
&\quad - p_i \cdot p_f M^2 - m^2 M (M + E - E') + 2m^2 M^2] \\
&= \frac{e^4}{2m^2 M^2 q^4} [ME(ME' + p_f \cdot p_i - m^2) + M E'(ME + m^2 - p_i \cdot p_f) \\
&\quad - p_i \cdot p_f M^2 - m^2 M^2 - m^2 ME + m^2 ME' + 2m^2 M^2] \\
&= \frac{e^4}{2m^2 M^2 q^4} \{ 2M^2 EE' + 2m^2 M (E' - E) \\
&\quad - p_i \cdot p_f [M^2 + M(E' - E)] + m^2 M^2 \}.
\end{aligned}
\tag{8.119}
$$

We calculate the differential cross section for electron scattering into a given solid-angle element $d\Omega$ centered around the scattering angle θ, as shown in figure 8.6. We thus integrate the differential cross section over all momentum variables except for the direction of $\vec{p}_f$. The volume element in spherical coordinates is written as

$$
d^3 p' = |\vec{p}'|^2 d|\vec{p}'| d\Omega = |\vec{p}'| E' dE' d\Omega,
\tag{8.120}
$$

where $|\vec{p}'| d|\vec{p}'| = E' dE'$ has been used. For the final-state proton, we use equation 8.102:

$$
\frac{d^3 P_f}{2E'_f} = \int_{-\infty}^{\infty} d^3 P_f \delta(P_f^2 - M^2) \theta(P_f^0).
\tag{8.121}
$$

The invariant flux factor reduces to

$$
\frac{mM}{\sqrt{(p_i \cdot P_i) - m^2 M^2}} = \frac{mM}{\sqrt{(EM)^2 - m^2 M^2}} = \frac{m}{|\vec{p}|}.
\tag{8.122}
$$

Combining the results for the flux factor and final-state phase-space integrals, the differential cross section in the rest frame of the proton becomes

$$
\begin{aligned}
d\sigma &= \frac{m}{|\vec{p}|} |\mathcal{M}_{fi}|^2 (2\pi)^4 \delta^4 (P_f - P_i + p_f - p_i) \\
&\quad \cdot \frac{m}{(2\pi)^3} |\vec{p}'| dE' d\Omega \frac{2M}{(2\pi)^3} d^4 P_f \delta(P_f^2 - M^2) \theta(P_f^0).
\end{aligned}
\tag{8.123}
$$

Integrating over dE' and $d^3 P_f$, the differential cross section becomes

$$
\begin{aligned}
\frac{d\bar{\sigma}}{d\Omega} &= \frac{2m^2 M}{|\vec{p}|} \int \frac{|\vec{p}'| dE'}{(2\pi)^2} |\mathcal{M}_{fi}|^2 d^4 P_f \delta^4 (P_f^2 - M^2) \theta(P_f^0) \delta^4 (P_f + p' - P_i - p) \\
&= \frac{m^2 M}{2\pi^2 |\vec{p}|} \int |\vec{p}'| dE' |\overline{\mathcal{M}}_{fi}|^2 \delta[(P_i - p' + p)^2 - M^2] \theta(M - E' + E)
\end{aligned}
$$

$$= \frac{m^2 M}{2\pi^2 |\vec{p}|} \int_m^{M+E} |\vec{p}'| dE' |\overline{\mathcal{M}}_{fi}|^2 \delta[(P_i - p' + p)^2 - M^2].$$

$$(8.124)$$

The lower bound on the integration over E' is m because the rest mass is the lowest energy of the electron. The upper bound comes from the requirement of the step function: $M - E' + E > 0 \Rightarrow E' < M + E$.

Expressing the argument of the delta function in equation 8.124 in terms of the kinematic variables in the rest frame of the proton, we have

$$
\begin{aligned}
f(E') &\equiv (P_i - p' + p)^2 - M^2 \\
&= P_i^2 + p'^2 + p^2 - 2P_i \cdot p' + 2P_i \cdot p - 2p' \cdot p - M^2 \\
&= M^2 + m^2 + m^2 - 2ME' + 2ME - 2EE' + 2|\vec{p}'||\vec{p}|\cos\theta - M^2 \\
&= 2m^2 + 2M(E - E') - 2EE' + 2|\vec{p}||\vec{p}'|\cos\theta.
\end{aligned}
$$

$$(8.125)$$

Using the identity for a delta function of an arbitrary function (equation 2.32), we calculate

$$\frac{df(E')}{dE'} = -2\left[M + E - |\vec{p}|(E'/|\vec{p}'|)\cos\theta\right], \qquad (8.126)$$

where we have used $|\vec{p}'|d|\vec{p}'| = E'dE'$. With this expression, the differential cross section becomes

$$\frac{d\bar{\sigma}}{d\Omega} = \frac{m^2 M}{4\pi^2} \frac{|\vec{p}'|}{|\vec{p}|} \frac{|\overline{\mathcal{M}}_{fi}|^2}{M + E - |\vec{p}|(E'/|\vec{p}'|)\cos\theta}. \qquad (8.127)$$

This differential cross section is a function of E, $|\vec{p}|$, E', $|\vec{p}'|$, and θ. However, $|\vec{p}|$ and $|\vec{p}'|$ are fixed by $|\vec{p}|^2 = E^2 - m^2$ and $|\vec{p}'|^2 = E'^2 - m^2$. In addition, E, E', and θ are constrained by the energy-conserving delta function. The argument of the delta function is

$$m^2 + M(E - E') - EE' + |\vec{p}||\vec{p}'|\cos\theta = 0, \qquad (8.128)$$

which is a quadratic equation in E'. Thus the differential cross section is a function of the two variables E and θ. That is, for a given initial electron energy E and a measured scattering angle θ, we can solve the three constraint equations to determine all the kinematic variables in the invariant matrix element 8.114, and hence the cross section 8.127.

The differential cross section result in equation 8.127 is general. However, it is difficult to interpret given its complexity, and this is best left for numerical calculations. In order to understand the scattering formula (equation 8.127), we will investigate its low-energy and high-energy limits.

A low-energy limit is obtained when the energy of the electron is much less than the proton rest mass, $E/M \ll 1$. We anticipate that in this limit we will reproduce the Mott cross section. Since $E/M \ll 1$ implies $m/M \ll 1$

and $|\vec{p}|/M \ll 1$, equation 8.128 leads to $M(E - E') \approx 0$, or $E \approx E'$, which is complete elastic scattering.

In this limit, the differential cross section equation 8.127 becomes

$$\frac{d\bar{\sigma}}{d\Omega} = \frac{m^2}{4\pi^2}|\overline{\mathcal{M}}_{fi}|^2 \quad \text{for} \quad \frac{E}{M} \ll 1, \tag{8.129}$$

where

$$|\overline{\mathcal{M}}_{fi}|^2 = \frac{8\pi^2\alpha^2}{m^2q^4}(2E^2 - p_f \cdot p_i + m^2) \quad \text{for} \quad \frac{E}{M} \ll 1, \tag{8.130}$$

or

$$\frac{d\bar{\sigma}}{d\Omega} = \frac{2\alpha^2}{|\vec{q}|^4}(2E^2 - p_f \cdot p_i + m^2) \quad \text{for} \quad \frac{E}{M} \ll 1, \tag{8.131}$$

where we realize that the momentum transfer q has no time component, i.e. $q^2 = -\vec{q}^2$. The result is exactly the Mott cross section in equation 8.46 for $Z = 1$. In this limit the proton does not recoil.

When the proton recoil becomes important, the electron may be treated as extremely relativistic and the electron rest mass is negligible with respect to the electron energy, $m/E \ll 1$. In this limit, $|\vec{p}| = \sqrt{E^2 - m^2} \approx E$ and $|\vec{p}'| = \sqrt{E'^2 - m^2} \approx E'$. The differential cross section equation 8.127 becomes

$$\frac{d\bar{\sigma}}{d\Omega} = \frac{m^2 M}{4\pi^2} \frac{E'}{E} \frac{|\overline{\mathcal{M}}_{fi}|^2}{M + E - E\cos\theta}$$
$$= \frac{m^2}{4\pi^2} \frac{E'/E}{1 + (2E/M)\sin^2(\theta/2)}|\overline{\mathcal{M}}_{fi}|^2 \quad \text{for} \quad \frac{m}{E} \ll 1. \tag{8.132}$$

To evaluate the square of the invariant matrix element, we express the scalar product $p_i \cdot p_f$ in terms of the square of the momentum transfer through

$$q^2 = (p_f - p_i)^2 = p_f^2 p_i^2 - 2p_f \cdot p_i = 2(m^2 - p_i \cdot p_f). \tag{8.133}$$

We obtain

$$|\overline{\mathcal{M}}_{fi}|^2 = \frac{e^4}{2m^2M^2q^4}\left\{ 2M^2EE' + 2m^2M(E' - E) \right.$$
$$\left. + \left(\frac{q^2}{2} - m^2\right)[M^2 + M(E' - E)] + m^2M^2 \right\}$$
$$= \frac{16\pi^2\alpha^2 EE'}{m^2q^4}\left[1 + \frac{q^2}{4EE'}\left(1 + \frac{E' - E}{M}\right) + \frac{m^2}{2EE'}\frac{(E' - E)}{M} \right]. \tag{8.134}$$

This expression for the square of the invariant matrix element is still exact and no approximations have been made.

In the extreme-relativistic limit, the square of the momentum transfer can be related to the scattering angle:

$$q^2 = -2EE'(1 - \cos\theta) = -4EE'\sin^2\frac{\theta}{2}. \tag{8.135}$$

In the limit $m/E \to 0$, conservation of energy (equation 8.128) gives

$$
\begin{aligned}
M(E - E') - EE' + EE'\cos\theta &\approx 0, \\
M(E - E') &= EE'(1 - \cos\theta), \\
\frac{E - E'}{M} &= \frac{EE'}{M^2}(1 - \cos\theta) \\
&= \frac{2EE'}{M^2}\sin^2\frac{\theta}{2} \\
&= -\frac{q^2}{2M^2}.
\end{aligned} \tag{8.136}
$$

Thus equation 8.134 becomes

$$
\begin{aligned}
|\mathcal{M}_{fi}|^2 &= \frac{16\pi^2\alpha^2 EE'}{m^2 16E^2 E'^2 \sin^4(\theta/2)}\left[1 + \frac{q^2}{4EE'}\left(1 + \frac{q^2}{2M^2}\right)\right] \\
&= \frac{\pi^2\alpha^2}{m^2 EE' \sin^4(\theta/2)}\left[1 - \sin^2\frac{\theta}{2}\left(1 + \frac{q^2}{2M^2}\right)\right] \\
&= \frac{\pi^2\alpha^2}{m^2 EE' \sin^4(\theta/2)}\left(\cos^2\frac{\theta}{2} - \frac{q^2}{2M^2}\sin^2\frac{\theta}{2}\right) \quad \text{for} \quad \frac{m}{E} \ll 1.
\end{aligned} \tag{8.137}
$$

The differential cross section thus becomes

$$\boxed{\frac{d\bar{\sigma}}{d\Omega} = \frac{\alpha^2}{4E^2 \sin^4\frac{\theta}{2}}\frac{\cos^2\frac{\theta}{2} - \frac{q^2}{2M^2}\sin^2\frac{\theta}{2}}{1 + \frac{2E}{M}\sin^2\frac{\theta}{2}} \quad \text{for} \quad \frac{m}{E} \ll 1} . \tag{8.138}$$

The total cross section is of limited interest, so we do not integrate the differential cross section here.

In the limit of $E \ll M$, but still $E \gg m$, we have

$$\frac{d\bar{\sigma}}{d\Omega} = \frac{\alpha^2}{4E^2 \sin^4\frac{\theta}{2}}\cos^2\frac{\theta}{2}, \tag{8.139}$$

which is equal to the extreme-relativistic limit of the Mott cross section (equation 8.52). The result in equation 8.138 differs from the Mott cross section in the limit as $\beta \to 1$ in two important ways. First, the denominator of the second factor in equation 8.138,

$$\frac{E'}{E} = \frac{1}{1 + \frac{2E}{M}\sin^2\frac{\theta}{2}}, \qquad (8.140)$$

arises from the recoil of the target proton. Second, the q^2-dependent second term in the numerator of the second factor in equation 8.138 originates from the fact that the target is a spin-1/2 particle. This term is absent in the calculation of the collision of electrons with spin-0 particles (see problem 8.14). We can now understand the $\sin^{-4}(\theta/2)$ factor in the Rutherford formula, Mott cross section, and equation 8.138 in terms of the exchange of a massless quantum, via the propagator factor $(1/q^2)^2$.

The derivation in this section has treated the proton as a heavy "electron" of mass M. The description is incomplete since it fails to take into account the proton structure and proton anomalous magnetic moment. Our result would however apply with great accuracy to the scattering of electrons off muons, which are structureless Dirac particles. A complete description of the proton leads to modifications, which are important at high energies exceeding several hundred MeV. This is because the de Broglie wavelength of the electron, $\lambda = \hbar c/E \sim 10^{-13}$ cm, at these energies is so small that the structure of the proton becomes detectable. In a complete treatment at very high energies, the formula has to be modified by introducing electric and magnetic form factors, which represent the internal structure of the proton. The result yields the so-called Rosenbluth formula (see problem 8.15).

A powerful technique for exploring the internal structure of a target is to bombard it with a beam of high-energy electrons and to observe the angular distribution and energy of the scattered electrons. Such experiments have repeatedly led to major advances in our understanding of the structure of matter.

8.7   Bremsstrahlung

When an electron scatters off an electromagnetic field, it can emit real photons. This process is called bremsstrahlung because it involves an acceleration – in German "bremson" – of the electron. More accurately, a deceleration of the electron occurs. We will see that the emission of a single photon is a well-defined process only within certain kinematical limits. The simultaneous emission of very soft, or low-energy, photons can never be excluded from consideration if they are too soft to be observed within the accuracy of the experimental energy resolution of the incident and outgoing electron energy measurements. In fact, bremsstrahlung radiation is always present, even in so-called elastic scattering. It is thus impossible to make a clean physical distinction between bremsstrahlung and radiationless scattering when the emitted photon is very soft. In the following calculation, we shall restrict ourselves to the emission of one not-too-soft photon.

Consider the emission of radiation of a charged particle (electron) in the presence of an external field. The four-vector potential of a photon with momentum $k^\mu = (\omega, \vec{k})$ and polarization $\varepsilon^\mu(\vec{k}, \lambda)$ is written in the Heaviside-Lorentz system of units as the plane wave

$$A^\mu(x; k, \lambda) = \frac{\varepsilon^\mu(\vec{k}, \lambda)}{\sqrt{2\omega V}}(e^{-ik\cdot x} + e^{ik\cdot x}), \qquad (8.141)$$

where we take the polarization vectors to be real (cf. equation 7.29). In working with photon plane waves, we add the two exponential solutions together since a photon is its own antiparticle. One of the terms represents photon emission and the other photon absorption. This two-termed plane-wave solution will give rise to terms in the S-matrix element which are other than those we are interested in. Rather than restrict the two terms in the photon plane-wave solution at the beginning, we keep them both to help elucidate the meaning of the other terms in the S-matrix element. For this same reason, we do not put arrows on external real photons in Feynman diagrams.

To illustrate the concepts of bremsstrahlung, we return to the static approximation and replace the proton by an external Coulomb field. Bremsstrahlung in electron-proton scattering will be calculated in problem 8.16. We will calculate the S-matrix element to lowest non-vanishing order in the electric charge e. There can be no first-order emission of radiation by a free electron in the absence of an external field (figure 8.13). This is kinematically forbidden, since it is impossible to conserve energy and momentum simultaneously. For a real photon $k^2 = 0$, while conservation of four-momentum gives $k^2 = (p_i - p_f)^2 < 0$, which is a contradiction.

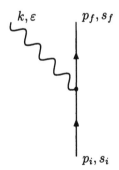

FIGURE 8.13:   First-order bremsstrahlung. This process is forbidden because energy and momentum cannot be conserved simultaneously.

The Feynman diagrams for the lowest-order bremsstrahlung process are shown in figure 8.14. There are two diagrams since we can not tell if electron scattering occurs before or after photon emission. The separation of the matrix element into terms corresponding to the individual diagrams, although extremely useful, has in general no physical meaning. Only the sum of both diagrams is observable. The process shown in figure 8.14 is second order with one vertex for the interaction of the electron with the Coulomb field and one for the emission of the bremsstrahlung photon. This is our first example with a real photon and the Dirac propagator, represented by the internal electron line in figure 8.14.

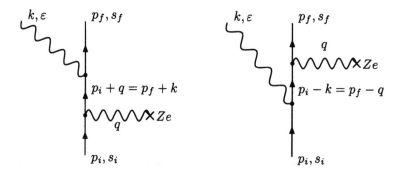

FIGURE 8.14:   Feynman diagrams for bremsstrahlung in a Coulomb field.

The calculations we have carried out thus far are valid only to lowest non-vanishing order in e^2. To obtain the next higher-order corrections in e we must return to equations 6.112 and 6.124, and consider the amplitude for second-order interactions between the electron and an external electromagnetic potential. The S-matrix element is given by

$$S_{fi}^{(2)} = -ie^2 \int d^4x d^4y \overline{\psi}_f(x) \, A\!\!\!/(x)) S_F(x-y) \, A\!\!\!/(y) \psi_i(y). \qquad (8.142)$$

For bremsstrahlung production in a Coulomb field, the lowest-order nonvanishing terms in the S-matrix element becomes

$$S_{fi} = \int d^4x d^4y \overline{\psi}_f(x)[(ie \, A\!\!\!/(x;k))(iS_F(x-y))(ie\gamma^0)A_0^{\text{coul}}(y)$$
$$+ (ie\gamma^0)A_0^{\text{coul}}(x)(iS_F(x-y))(ie \, A\!\!\!/(y;k))]\psi_i(y), \qquad (8.143)$$

where the two terms correspond to the two time orderings of the vertices (figure 8.14). We have again included a factor of i with the propagator and a factor of $ie\gamma^\mu$ at each vertex involving an electron, where $e = |e| > 0$. $A_0^{\text{coul}} = \frac{Ze}{4\pi|\vec{x}|}$ is as before, with $Ze > 0$.

It is convenient to transform to momentum space by Fourier expanding all factors and carrying out the coordinate integrations.

$$S_{fi} = \int d^4x d^4y \sqrt{\frac{m}{E_f V}} \bar{u}(p_f, s_f) e^{ip_f \cdot x}$$

$$\cdot \left[\frac{ie \not{\epsilon}}{\sqrt{2\omega V}} (e^{-ik\cdot x} + e^{ik\cdot x}) \int \frac{d^4q}{(2\pi)^4} e^{-iq\cdot(x-y)} \frac{i}{\not{q} - m} (ie\gamma_0) \left(\frac{Ze}{4\pi|\vec{y}|} \right) \right.$$

$$\left. + (ie\gamma_0) \left(\frac{Ze}{4\pi|\vec{x}|} \right) \int \frac{d^4q}{(2\pi)^4} e^{ieq\cdot(x-y)} \frac{i}{\not{q} - m} \frac{ie \not{\epsilon}}{\sqrt{2\omega V}} (e^{-ik\cdot y} + e^{ik\cdot y}) \right]$$

$$\cdot \sqrt{\frac{m}{E_i V}} u(p_i, s_i) e^{-ip_i \cdot y}$$

$$= \int \frac{d^4x d^4y d^4q}{(2\pi)^4 V^{3/2}} \sqrt{\frac{m^2}{2\omega E_i E_f}} \bar{u}(p_f, s_f)$$

$$\cdot \left[(ie \not{\epsilon}) \frac{i}{\not{q} - m} (ie\gamma_0) \frac{Ze}{4\pi|\vec{y}|} e^{ix\cdot(p_f \mp k - q)} e^{iy\cdot(q - p_i)} \right.$$

$$\left. + (ie\gamma_0) \frac{Ze}{4\pi|\vec{x}|} \frac{i}{\not{q} - m} (ie \not{\epsilon}) e^{ix\cdot(p_f - q)} e^{iy\cdot(q \mp k - p_i)} \right] u(p_i, s_i), \quad (8.144)$$

where the use of $\mp$ here is unconventional. Rather than it representing either $-$ or $+$, here it represents the sum of two terms: one term with $+$, plus an identical term with $-$. Continuing,

$$S_{fi} = \frac{Ze}{4\pi} \int \frac{d^4q}{V^{3/2}} \sqrt{\frac{m^2}{2\omega E_i E_f}} \bar{u}(p_f, s_f)$$

$$\cdot \left[(ie \not{\epsilon}) \frac{i}{\not{q} - m} (ie\gamma_0) \int \frac{d^4y}{|\vec{y}|} \delta^4(p_f \mp k - q) e^{iy\cdot(q - p_i)} \right.$$

$$\left. + (ie\gamma_0) \int \frac{d^4x}{|\vec{x}|} \frac{i}{\not{q} - m} (ie \not{\epsilon}) e^{ix\cdot(p_f - q)} \delta^4(q \mp k - p_i) \right] u(p_i, s_i)$$

$$= \frac{Ze}{4\pi} \frac{1}{V^{3/2}} \sqrt{\frac{m^2}{2\omega E_i E_f}} \bar{u}(p_f, s_f)$$

$$\cdot \left[(ie \not{\epsilon}) \frac{i}{\not{p}_f \mp \not{k} - m} (ie\gamma_0) \int \frac{d^4y}{|\vec{y}|} e^{iy\cdot(p_f \mp k - p_i)} \right.$$

$$\left. + (ie\gamma_0) \int \frac{d^4x}{|\vec{x}|} \frac{i}{\not{p}_i \pm \not{k} - m} (ie \not{\epsilon}) e^{ix\cdot(p_f - p_i \mp k)} \right] u(p_i, s_i)$$

$$= \frac{Ze}{V^{3/2}} 2\pi \delta(E_f + \omega - E_i) \frac{1}{\sqrt{2\omega}} \sqrt{\frac{m^2}{E_f E_i}} \frac{1}{|\vec{q}|^2} \bar{u}(p_f, s_f)$$

$$\cdot \left[(ie\ \rlap{/}{\epsilon})\frac{i}{\rlap{/}{p}_f + \rlap{/}{k} - m}(ie\gamma_0) + (ie\gamma_0)\frac{i}{\rlap{/}{p}_i - \rlap{/}{k} - m}(ie\ \rlap{/}{\epsilon}) \right] u(p_i, s_i),$$

$$(8.145)$$

where we have taken the $E_f + \omega - E_i$ solution only. The other solution comes from the first term of the photon wave function, $e^{-ik\cdot x}$, which gives rise to the energy delta function $\delta(E_i + \omega - E_f)$. This term describes absorption of energy (a photon) in the scattering process (figure 8.15), and is not the process of interest here, i.e. the bremsstrahlung process in which the incident electron gives up energy to the radiation field and emerges with $E_f = E_i - \omega < E_i$.

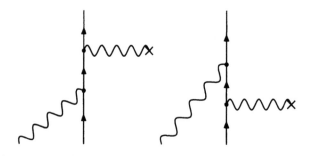

FIGURE 8.15: Feynman diagrams for photon absorption by an electron in a Coulomb field.

In equation 8.145, $\vec{q} = \vec{p}_f + \vec{k} - \vec{p}_i$ is the three-momentum transfer to the nucleus. There is no energy transfer to the nucleus since it was assumed to be infinitely heavy.

We notice the factor $ie\rlap{/}{\epsilon}$ appears at the vertex, where a free photon of polarization ϵ^μ is emitted, and $1/\sqrt{2\omega V}$ appears as the normalization factor for a photon wave function. We add these to our Feynman rules (see appendix C).

The invariant matrix element in equation 8.145 can be identified by writing the S-matrix element as

$$S_{fi} = i2\pi\delta(E_f + \omega - E_i)\frac{1}{\sqrt{2\omega V}}\sqrt{\frac{m^2}{E_f E_i V^2}}\frac{Ze}{|\vec{q}|^2}\varepsilon \cdot \mathcal{M}_{fi}(k), \qquad (8.146)$$

where

$$\varepsilon_\mu \mathcal{M}_{fi}^\mu(k)$$

$$= -e^2 \bar{u}(p_f, s_f)\left[\rlap{/}{\epsilon}\frac{1}{\rlap{/}{p}_f + \rlap{/}{k} - m}\gamma_0 + \gamma_0\frac{1}{\rlap{/}{p}_i - \rlap{/}{k} - m}\rlap{/}{\epsilon} \right] u(p_i, s_i)$$

$$= -e^2 \bar{u}(p_f, s_f) \left[\not{\varepsilon} \frac{\not{p}_f + \not{k} + m}{(p_f + k)^2 - m^2} \gamma_0 + \gamma_0 \frac{\not{p}_i - \not{k} + m}{(p_i - k)^2 - m^2} \not{\varepsilon} \right] u(p_i, s_i)$$

$$= -e^2 \bar{u}(p_f, s_f) \left[\not{\varepsilon} \frac{\not{p}_f + \not{k} + m}{2k \cdot p_f} \gamma_0 + \gamma_0 \frac{\not{p}_i - \not{k} + m}{-2k \cdot p_i} \not{\varepsilon} \right] u(p_i, s_i), \quad (8.147)$$

where we have used $p_i^2 = p_f^2 = m^2$ and $k^2 = 0$ in the last step. The amplitude $i\varepsilon \cdot \mathcal{M}_{fi}(k)$ is the Lorentz-invariant matrix element for the bremsstrahlung process. Besides depending on the initial- and final-state electrons, it now also depends on the photon polarization ε and four-momentum k.

The general result of this calculation is known as the Bethe-Heitler formula[5] (see problem 8.17). To illustrate the concepts, we limit the derivation presented here to the emission of a very soft photon. In the limit as $k \to 0$, the invariant matrix element becomes

$$i\varepsilon \cdot \mathcal{M}_{fi}(k) \approx -ie^2 \bar{u}(p_f, s_f) \left[\not{\varepsilon} \frac{\not{p}_f + m}{2k \cdot p_f} \gamma_0 + \gamma_0 \frac{\not{p}_i + m}{-2k \cdot p_i} \not{\varepsilon} \right] u(p_i, s_i). \quad (8.148)$$

The order of the $\not{p}$ and $\not{\varepsilon}$ matrices can been reversed by using the anticommutation relationship for the gamma matrices, $\not{\varepsilon} \not{p} + \not{p} \not{\varepsilon} = 2\varepsilon \cdot p$:

$$i\varepsilon \cdot \mathcal{M}_{fi}(k)$$

$$= -ie^2 \bar{u}(p_f, s_f) \left[\frac{2\varepsilon \cdot p_f - \not{p}_f \not{\varepsilon} + m}{2k \cdot p_f} \gamma_0 + \gamma_0 \frac{2\varepsilon \cdot p_i - \not{\varepsilon} \not{p}_i + m}{-2k \cdot p_i} \right] u(p_i, s_i)$$

$$= -ie^2 \bar{u}(p_f, s_f) \left[\frac{2\varepsilon \cdot p_f - (\not{p}_f - m) \not{\varepsilon}}{2k \cdot p_f} \gamma_0 + \gamma_0 \frac{2\varepsilon \cdot p_i - \not{\varepsilon}(\not{p}_i - m)}{-2k \cdot p_i} \right] u(p_i, s_i).$$

$$(8.149)$$

The reason why we reversed the order of $\not{p}$ and $\not{\varepsilon}$ is so that we can use the Dirac equation $(\not{p}_i - m)u(p_i, s_i) = 0$ and $\bar{u}(p_f, s_f)(\not{p}_f - m) = 0$ to simplify the matrix element to

$$i\varepsilon \cdot \mathcal{M}_{fi}(k) = -i4\pi\alpha \bar{u}(p_f, s_f)\gamma_0 u(p_i, s_i) \left(\frac{\varepsilon \cdot p_f}{k \cdot p_f} - \frac{\varepsilon \cdot p_i}{k \cdot p_i} \right). \quad (8.150)$$

Proceeding to the cross section, we square S_{fi}, divide by the flux $|\vec{v}_i|/V = |\vec{p}_i|/(E_i V)$ and by $2\pi\delta(0)$, formerly T, to form a rate, and integrate over the final states $(V^2 d^3 k \, d^3 p_f)/(2\pi)^6$ in the observed interval of phase space. We obtain

---

[5] H. Bethe & W. Heitler, "On the Stopping of Fast Particles and on the Creation of Positron Electrons", Proc. Roy. Soc. **146** (1934) 83-112.

$$
\begin{aligned}
d\sigma &= \frac{VE_i}{|\vec{p}_i|} \frac{|S_{fi}|^2}{2\pi\delta(0)} \frac{V d^3 k}{(2\pi)^3} \frac{V d^3 p_f}{(2\pi)^3} \\
&= \frac{E_i}{|\vec{p}_i|} \frac{[2\pi\delta(E_f + \omega - E_i)]^2}{2\pi\delta(0)} \frac{1}{2\omega} \frac{m^2}{E_f E_i} \frac{(Ze)^2}{|\vec{q}|^4} |\varepsilon \cdot \mathcal{M}_{fi}(k)|^2 \frac{d^3 k\, d^3 p_f}{(2\pi)^6} \\
&= \frac{Z^2(4\pi\alpha)m^2}{2\omega|\vec{p}_i|E_f} \frac{|\varepsilon \cdot \mathcal{M}_{fi}(k)|^2}{|\vec{q}|^4} \delta(E_f + \omega - E_i) \frac{d^3 k\, d^3 p_f}{(2\pi)^5} \\
&= \frac{4Z^2\alpha^3 m^2}{\omega|\vec{p}_i|E_f} \frac{|\bar{u}(p_f, s_f)\gamma_0 u(p_i, s_i)|^2}{|\vec{q}|^4} \\
&\quad \cdot \left(\frac{\varepsilon \cdot p_f}{k \cdot p_f} - \frac{\varepsilon \cdot p_i}{k \cdot p_i}\right)^2 \delta(E_f + \omega - E_i) \frac{d^3 k\, d^3 p_f}{(2\pi)^2}. \quad (8.151)
\end{aligned}
$$

We use the energy-conserving delta function to integrate over the final-state electron energy using $d^3 p_f = |\vec{p}_f|^2 dp_f d\Omega_f = |\vec{p}_f| E_f dE_f d\Omega_f$, such that

$$
\begin{aligned}
d\sigma &= \frac{4Z^2\alpha^3 m^2}{|\vec{q}|^4} |\bar{u}(p_f, s_f)\gamma_0 u(p_i, s_i)|^2 \\
&\quad \cdot \left(\frac{\varepsilon \cdot p_f}{k \cdot p_f} - \frac{\varepsilon \cdot p_i}{k \cdot p_i}\right)^2 \delta(E_f + \omega - E_i) \frac{|\vec{p}_f|}{|\vec{p}_i|} \frac{dE_f d\Omega_f d^3 k}{\omega(2\pi)^2}. \quad (8.152)
\end{aligned}
$$

Since $k \to 0$, we can take $|\vec{p}_f|/|\vec{p}_i| \to 1$, and using

$$
\int_m^\infty dE_f \delta(E_f + \omega - E_i) = \int_{-\infty}^\infty dE_f \delta(E_f + \omega - E_i)\theta(E_f - m)
$$
$$
= \theta(E_i - \omega - m), \quad (8.153)
$$

we have

$$
\begin{aligned}
d\sigma &= \frac{4Z^2\alpha^3 m^2}{|\vec{q}|^4} |\bar{u}(p_f, s_f)\gamma_0 u(p_i, s_i)|^2 \\
&\quad \cdot \left(\frac{\varepsilon \cdot p_f}{k \cdot p_f} - \frac{\varepsilon \cdot p_i}{k \cdot p_i}\right)^2 \theta(E_i - \omega - m) \frac{d\Omega_f d^3 k}{\omega(2\pi)^2}. \quad (8.154)
\end{aligned}
$$

We identify a set of factors with the elastic-scattering cross section from equation 8.37:

$$
\left(\frac{d\sigma}{d\Omega_f}\right)_{\text{elastic}} = \frac{4(Z\alpha)^2 m^2}{|\vec{q}|^4} |\bar{u}(p_f, s_f)\gamma_0 u(p_i, s_i)|^2. \quad (8.155)
$$

Thus

$$\frac{d\sigma}{d\Omega_f d\Omega_k d\omega} = \left(\frac{d\sigma}{d\Omega_f}\right)_{\text{elastic}} \alpha \left(\frac{\varepsilon \cdot p_f}{k \cdot p_f} - \frac{\varepsilon \cdot p_i}{k \cdot p_i}\right)^2 \frac{\omega}{(2\pi)^2} \theta(E_i - \omega - m),$$

(8.156)

where we have used $d^3 k = \omega^2 d\omega d\Omega_k$. This is the cross section for the electron to be observed in a solid angle $d\Omega_f$, and for a soft photon of polarization ε to emerge with momentum $\vec{k}$ in the interval $d\Omega_k d\omega$. It is natural that the bremsstrahlung cross section for soft photon emission is proportional to the elastic-scattering cross section for the electron at the same energy and scattering angle, since the amount of energy and momentum carried off by the photon is very small.

If the cross section for unpolarized electrons is to be calculated, one has to sum over the final spin states and average over the initial spin states of the electrons. Owing to the factorization property of equation 8.156 this is easily achieved. One simply replaces the elastic cross section by the unpolarized expression, equation 8.37, since the remaining factors in equation 8.156 do not depend on the electron spin. Factorization also shows that the bremsstrahlung cross section for soft photon emission from a positron is identical to that of an electron to lowest nonvanishing order in α.

The factorization of equation 8.156 is more general than one might expect. It has been shown, in the limit as $k \to 0$, that the amplitude for any process leading to photon emission can be factorized according to

$$\lim_{k \to 0} M(k) = \sqrt{\alpha} \left(\frac{\varepsilon \cdot p_f}{k \cdot p_f} - \frac{\varepsilon \cdot p_i}{k \cdot p_i}\right) M_0,$$

(8.157)

where M_0 is the amplitude for the same process without photon emission. This result is true for any kind of process, irrespective of the spin or internal structure of the charged particle.

We notice that the photon energy spectrum in equation 8.156 behaves as $d\omega/\omega$ and therefore the probability to emit a zero-energy photon is infinite. This is called the infrared catastrophe. For a consistent comparison with experiment, we must include both elastic and inelastic cross sections calculated to the same order in the electric charge e. Since the bremsstrahlung contribution is of order e^2 higher than elastic scattering, we must also include so-called radiative corrections to $(d\sigma/d\Omega)_{\text{elastic}}$ to the same order in e. These correspond to second-order scattering of the electron in a Coulomb field. We must also take into account the interaction of the electron with itself via the radiation field as shown in figure 8.16. The amplitudes coming from these processes contain a divergent term which precisely cancels the divergence in equation 8.156 at $k = 0$.

In section 7.2, we saw how gauge invariance of the electromagnetic field puts a condition on the electromagnetic current in momentum space: $k_\mu J^\mu(k) = 0$. This property of a conserved current in momentum space is shared also by

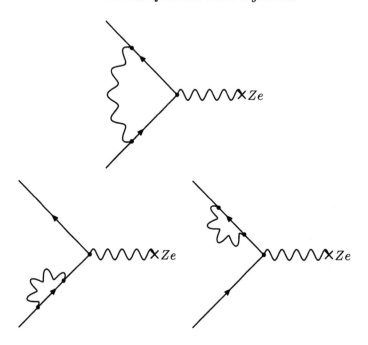

FIGURE 8.16:   Radiative corrections to Coulomb scattering (self interactions).

quantum-mechanical transitions currents. Thus we can expect that the matrix element $\mathcal{M}_\mu(k)$ satisfies

$$k_\mu \mathcal{M}^\mu_{fi}(k) = 0, \tag{8.158}$$

since $\mathcal{M}_{fi}$ is the transition current for bremsstrahlung, up to a numerical factor. Using the matrix element in equation 8.150, this condition is easily shown to be true (see problem 8.18).

In most cases of soft-photon bremsstrahlung, we will not observe the final photon polarization; we thus sum over them. The quantity of interest is

$$\overline{|\varepsilon \cdot \mathcal{M}|^2} = \sum_{\lambda=1,2} |\varepsilon_\mu(\vec{k}, \lambda)\mathcal{M}^\mu(k)|^2$$

$$= \sum_{\lambda=1,2} \varepsilon_\mu(\vec{k}, \lambda)\varepsilon_\nu^*(\vec{k}, \lambda)\mathcal{M}^\mu(k)\mathcal{M}^{*\nu}(k). \tag{8.159}$$

Since this is a scalar, we can evaluate it in an arbitrary Lorentz frame. We orient the axis such that $k^\mu = \omega(1, 0, 0, 1)$, where $\omega = |\vec{k}|$, since $k^2 = 0$. We choose $A^0(x) = 0$. In the time gauge, the polarizations are transverse to the direction of motion, and the two independent transverse polarizations may be taken as

$$\varepsilon(\vec{k}, 1) = (0, 1, 0, 0), \tag{8.160}$$

$$\varepsilon(\vec{k}, 2) = (0, 0, 1, 0). \tag{8.161}$$

Therefore $\varepsilon(\vec{k}, 1) \cdot k = 0$ and $\varepsilon(\vec{k}, 2) \cdot k = 0$, $\varepsilon(\vec{k}, 1) \cdot \varepsilon(\vec{k}, 2) = 0$ and $\varepsilon(\vec{k}, 1) \cdot \varepsilon(\vec{k}, 1) = \varepsilon(\vec{k}, 2) \cdot \varepsilon(\vec{k}, 2) = -1$.

Summing over polarizations, we have

$$\overline{|\varepsilon \cdot \mathcal{M}|^2} = \mathcal{M}^1 \mathcal{M}^{*1} + \mathcal{M}^2 \mathcal{M}^{*2}. \tag{8.162}$$

Invoking our condition of current conservation (equation 8.158), we have

$$k \cdot \mathcal{M} = \omega(\mathcal{M}^0 - \mathcal{M}^3) = 0, \tag{8.163}$$

which implies $\mathcal{M}^0 = \mathcal{M}^3$. We transform equation 8.162 into a four dimensional scalar product by adding a vanishing contribution

$$\overline{|\varepsilon \cdot \mathcal{M}|^2} = \mathcal{M}^1 \mathcal{M}^{*1} + \mathcal{M}^2 \mathcal{M}^{*2} + \mathcal{M}^3 \mathcal{M}^{*3} - \mathcal{M}^0 \mathcal{M}^{*0} = -\mathcal{M}_\mu \mathcal{M}^{*\mu}. \tag{8.164}$$

Since this result is covariant, we compare it with equation 8.159 to obtain

$$\sum_{\lambda=1,2} \varepsilon_\mu(\vec{k}, \lambda) \varepsilon_\nu^*(\vec{k}, \lambda) = -g_{\mu\nu} + \text{gauge terms}. \tag{8.165}$$

The additional gauge terms need not be specified in detail. They are proportional to k_μ and k_ν, and thus do not contribute to any observable quantity, since our result will be multiplied with conserved currents which satisfy $k \cdot J = 0$. Nevertheless these terms have to be present since a complete basis in four-dimension space of Lorentz vectors has to contain four elements. The contribution of longitudinal $\varepsilon_\mu(\vec{k}, 3)$ and scalar $\varepsilon_\mu(\vec{k}, 0)$ photons to the completeness relationship makes their appearance on the right-hand side of our result. However, they do not correspond to physical photons.

We now apply the completeness relationship for photon polarizations to the bremsstrahlung cross section. The sum over polarizations is

$$\sum_{\lambda=1,2} \left(\frac{\varepsilon \cdot p_f}{k \cdot p_f} - \frac{\varepsilon \cdot p_i}{k \cdot p_i} \right)^2 = - \left(\frac{p_f}{k \cdot p_f} - \frac{p_i}{k \cdot p_i} \right)^2$$

$$= - \frac{p_f^2}{(k \cdot p_f)^2} - \frac{p_i^2}{(k \cdot p_i)^2} + \frac{2 p_f \cdot p_i}{(k \cdot p_f)(k \cdot p_i)}, \tag{8.166}$$

and thus the differential cross section given by equation 8.156 becomes

$$\frac{d\bar{\sigma}}{d\Omega_f} = \left(\frac{d\sigma}{d\Omega_f}\right)_{\text{elastic}} \frac{\alpha}{(2\pi)^2} \int \omega d\omega \int d\Omega_k \theta(E_i - \omega - m)$$
$$\cdot \left[\frac{2p_f \cdot p_i}{(k \cdot p_f)(k \cdot p_i)} - \frac{m^2}{(k \cdot p_f)^2} - \frac{m^2}{(k \cdot p_i)^2}\right]. \tag{8.167}$$

Using $\vec{\beta}_f = \vec{p}_f/E_f$ and $\vec{\beta}_i = \vec{p}_i/E_i$ to write

$$k \cdot p_f = \omega E_f - \vec{k} \cdot \vec{p}_f = \omega(E_f - \hat{k} \cdot \vec{p}_f) = \omega E_f(1 - \hat{k} \cdot \vec{\beta}_f), \tag{8.168}$$
$$k \cdot p_i = \omega E_i - \vec{k} \cdot \vec{p}_i = \omega(E_i - \hat{k} \cdot \vec{p}_i) = \omega E_i(1 - \hat{k} \cdot \vec{\beta}_i), \tag{8.169}$$
$$p_f \cdot p_i = E_f E_i - \vec{p}_f \cdot \vec{p}_i = E_f E_i(1 - \vec{\beta}_f \cdot \vec{\beta}_i), \tag{8.170}$$

we have

$$\frac{d\bar{\sigma}}{d\Omega_f} = \left(\frac{d\sigma}{d\Omega_f}\right)_{\text{elastic}} \frac{\alpha}{\pi} \int \frac{d\omega}{\omega} \int \frac{d\Omega_k}{4\pi} \left[\frac{2(1 - \vec{\beta}_f \cdot \vec{\beta}_i)}{(1 - \hat{k} \cdot \vec{\beta}_f)(1 - \hat{k} \cdot \vec{\beta}_i)}\right.$$
$$\left. - \frac{m^2}{E_f^2(1 - \hat{k} \cdot \vec{\beta}_f)^2} - \frac{m^2}{E_i^2(1 - \hat{k} \cdot \vec{\beta}_i)^2}\right]. \tag{8.171}$$

Integrating over all photon emission angles and energies in the interval $0 < \omega_{\min} \leq \omega \leq \omega_{\max} \ll E_i - m$ gives

$$\frac{d\bar{\sigma}}{d\Omega_f} = \left(\frac{d\sigma}{d\Omega_f}\right)_{\text{elastic}} \frac{\alpha}{\pi} \int_{\min}^{\max} \frac{d\omega}{\omega} \int \frac{d\Omega_k}{4\pi} \left[\frac{2(1 - \vec{\beta}_f \cdot \vec{\beta}_i)}{(1 - \hat{k} \cdot \vec{\beta}_f)(1 - \hat{k} \cdot \vec{\beta}_i)}\right.$$
$$\left. - \frac{m^2}{E_f^2(1 - \hat{k} \cdot \vec{\beta}_f)^2} - \frac{m^2}{E_i^2(1 - \hat{k} \cdot \vec{\beta}_i)^2}\right]$$
$$= \left(\frac{d\sigma}{d\Omega_f}\right)_{\text{elastic}} \frac{\alpha}{\pi} \ln \frac{\omega_{\max}}{\omega_{\min}} \int \frac{d\Omega_k}{4\pi} \left[\frac{2(1 - \vec{\beta}_f \cdot \vec{\beta}_i)}{(1 - \hat{k} \cdot \vec{\beta}_f)(1 - \hat{k} \cdot \vec{\beta}_i)}\right.$$
$$\left. - \frac{m^2}{E_f^2(1 - \hat{k} \cdot \vec{\beta}_f)^2} - \frac{m^2}{E_i^2(1 - \hat{k} \cdot \vec{\beta}_i)^2}\right]. \tag{8.172}$$

If the emitted bremsstrahlung photon is very soft, the initial and final energies of the electron are almost the same $|\vec{\beta}_i| = |\vec{\beta}_f| = \beta$, and we get for numerator of the first term

$$2(1 - \vec{\beta}_f \cdot \vec{\beta}_i) = 2(1 - \beta^2 \cos\Theta), \tag{8.173}$$

where Θ is the scattering angle of the electron.

To integrate the last two terms in equation 8.172, we use

$$\int \frac{d\Omega_f}{4\pi} \frac{m^2}{E^2(1 - \vec{\beta} \cdot \hat{k})^2} = \frac{m^2}{E^2} \int_{-1}^{1} \frac{d\cos\theta}{2(1 - \beta\cos\theta)^2}$$

$$= \left(\frac{m}{E}\right)^2 \int_{-1}^{1} \frac{dz}{2(1 - \beta z)^2}$$

$$= \left(\frac{m}{E}\right)^2 \left(-\frac{1}{\beta}\right) \int_{1+\beta}^{1-\beta} \frac{dx}{2x^2}$$

$$= -\left(\frac{m}{E}\right)^2 \left(\frac{1}{2\beta}\right) \left[-\frac{1}{x}\right]_{1+\beta}^{1-\beta}$$

$$= \left(\frac{m}{E}\right)^2 \frac{1}{(1-\beta)^2}$$

$$= 1, \tag{8.174}$$

where we have used $(m/E)^2 = 1/\gamma^2 = 1 - \beta^2$. The differential cross section now becomes

$$\frac{d\bar{\sigma}}{d\Omega_f} = \left(\frac{d\sigma}{d\Omega_f}\right)_{\text{elastic}} \frac{2\alpha}{\pi} \ln\frac{\omega_{\text{max}}}{\omega_{\text{min}}} \left[(1 - \beta^2\cos\Theta)I - 1\right], \tag{8.175}$$

where

$$I = \int \frac{d\Omega_k}{4\pi} \frac{1}{(1 - \hat{k} \cdot \vec{\beta}_f)(1 - \hat{k} \cdot \vec{\beta}_i)}. \tag{8.176}$$

To evaluate this integral we use

$$\frac{1}{ab} = \int_0^1 \frac{dx}{[ax + b(1 - x)]^2} \tag{8.177}$$

to obtain

$$I = \int_0^1 dx \int \frac{d\Omega_f}{4\pi} \frac{1}{[(1 - \hat{k} \cdot \vec{\beta}_f)x + (1 - \hat{k} \cdot \vec{\beta}_i)(1 - x)]^2}$$

$$= \int_0^1 dx \int \frac{d\Omega_f}{4\pi} \frac{1}{\{1 - \hat{k} \cdot [\vec{\beta}_f x + \vec{\beta}_i(1 - x)]\}^2}$$

$$= \int_0^1 dx \int_{-1}^{1} \frac{d\cos\vartheta}{2} \frac{1}{[1 - |\vec{\beta}_f x + \vec{\beta}_i(1 - x)|\cos\vartheta]^2}$$

$$= \int_0^1 dx \frac{1}{2}\left(\frac{-1}{\zeta}\right) \int_{1+\zeta}^{1-\zeta} \frac{dz}{z^2}$$

$$= \int_0^1 dx \frac{1}{1 - \zeta^2}$$

$$= \int_0^1 dx \frac{1}{1 - |\vec{\beta}_f x + \vec{\beta}_i(1 - x)|^2}$$

$$= \int_0^1 dx \frac{1}{1 - \beta^2 + 4\beta^2 x(1 - x)\sin^2(\Theta/2)}, \tag{8.178}$$

where we have temporarily defined $\zeta = |\vec{\beta}_f x + \vec{\beta}_i(1 - x)|$.

The integral in equation 8.177 is a simple example of a set of integrals referred to as Feynman integrals. Using these integrals is a useful trick, particularly if one is calculating higher-order processes. The method transforms the denominator factors into a single quadratic polynomial in $\cos\theta$, raised to a power. We can then shift $\cos\theta$ by a constant to complete the square in the polynomial, and evaluate the remaining spherically-symmetric integral. The price is the introduction of an auxiliary parameter x to be integrated over.

The integral in equation 8.178 can be solved in closed form using

$$\int \frac{dx}{a + bx + cx^2} = \frac{1}{\sqrt{-d}} \ln \frac{2cx + b - \sqrt{-d}}{2cx + b + \sqrt{-d}} \quad \text{if} \quad d < 0, \tag{8.179}$$

where $d = 4ac - b^2$. Including the limits of integration, we have

$$\int_0^1 \frac{dx}{a + bx + cx^2} = \frac{1}{\sqrt{-d}} \ln \frac{(2c + b - \sqrt{-d})(b + \sqrt{-d})}{(2c + b + \sqrt{-d})(b - \sqrt{-d})}. \tag{8.180}$$

For the case of $c = -b$,

$$\int_0^1 \frac{dx}{1 + bx(1 - x)} = \frac{2}{\sqrt{-d}} \ln \frac{\sqrt{-d} + b}{\sqrt{-d} - b}, \tag{8.181}$$

where $-d = b(b + 4a)$. For our case $a = 1 - \beta^2$ and $b = 4\beta^2 \sin^2(\Theta/2)$, giving

$$\sqrt{-d} = 2\beta \sin(\Theta/2)\sqrt{4\beta^2 \sin^2(\Theta/2) + 4(1 - \beta^2)}$$

$$= 4\beta \sin(\Theta/2)\sqrt{1 - \beta^2 \cos^2(\Theta/2)}. \tag{8.182}$$

Therefore the integral in equation 8.178 becomes

$$I = \frac{1}{2\beta \sin(\Theta/2)\sqrt{1 - \beta^2 \cos^2(\Theta/2)}} \ln \frac{\sqrt{1 - \beta^2 \cos^2(\Theta/2)} + \beta \sin(\Theta/2)}{\sqrt{1 - \beta^2 \cos^2(\Theta/2)} - \beta \sin(\Theta/2)}. \tag{8.183}$$

The integral in equation 8.183 gives the general cross section for soft photon emission. We now simplify the cross section expression by taking the nonrelativistic and extreme-relativistic limits.

In the nonrelativistic limit ($\beta \ll 1$), the factors in equation 8.183 can be Taylor expanded as power series in β, such that

$$I = \frac{1 + 1/2\beta^2 \cos^2(\Theta/2) - \mathcal{O}(\beta^4)}{2\beta \sin(\Theta/2)}$$
$$\cdot \{\ln[1 + \beta \sin(\Theta/2) - 1/2\beta^2 \cos^2(\Theta/2) + \mathcal{O}(\beta^4)]$$
$$- \ln[1 - \beta \sin(\Theta/2) - 1/2\beta^2 \cos^2(\Theta/2) + \mathcal{O}(\beta^4)]\}$$
$$= \frac{1 + 1/2\beta^2 \cos^2(\Theta/2) - \mathcal{O}(\beta^4)}{2\beta \sin(\Theta/2)}$$
$$\cdot \{[\beta \sin(\Theta/2) - 1/2\beta^2 \cos^2(\Theta/2)]$$
$$- 1/2[\beta \sin(\Theta/2) - 1/2\beta^2 \cos^2(\Theta/2)]^2$$
$$+ 1/3[\beta \sin(\Theta/2) - 1/2\beta^2 \cos^2(\Theta/2)]^3$$
$$- [-\beta \sin(\Theta/2) - 1/2\beta^2 \cos^2(\Theta/2)]$$
$$+ 1/2[-\beta \sin(\Theta/2) - 1/2\beta^2 \cos^2(\Theta/2)]^2$$
$$- 1/3[-\beta \sin(\Theta/2) - 1/2\beta^2 \cos^2(\Theta/2)]^3 + \mathcal{O}(\beta^4)\}$$
$$= \frac{1 + 1/2\beta^2 \cos^2(\Theta/2) - \mathcal{O}(\beta^4)}{2\beta \sin(\Theta/2)}$$
$$\cdot [2\beta \sin(\Theta/2) + \beta^3 \sin(\Theta/2) \cos^2(\Theta/2) + 2/3\beta^3 \sin^3(\Theta/2) + \mathcal{O}(\beta^4)]$$
$$= [1 + 1/2\beta^2 \cos^2(\Theta/2) - \mathcal{O}(\beta^4)]$$
$$\cdot [1 + 1/2\beta^2 \cos^2(\Theta/2) + 1/3\beta^2 \sin^2(\Theta/2) + \mathcal{O}(\beta^4)]$$
$$= 1 + 1/2\beta^2 \cos^2(\Theta/2)1 + 1/2\beta^2 \cos^2(\Theta/2) + 1/3\beta^2 \sin^2(\Theta/2) + \mathcal{O}(\beta^4)$$
$$= 1 + 1/2\beta^2 - 1/2\beta^2 \sin^2(\Theta/2) + 1/3\beta^2 \sin^2(\Theta/2) + \mathcal{O}(\beta^4)$$
$$= 1 + \beta^2 - \beta^2 \sin^2(\Theta/2) + 1/3\beta^2 \sin^2(\Theta/2) + \mathcal{O}(\beta^4)$$
$$= 1 + \beta^2 - \frac{2}{3}\beta^2 \sin^2\frac{\Theta}{2} + \mathcal{O}(\beta^4), \tag{8.184}$$

where $\mathcal{O}$ means "order of".

The differential cross section for soft-photon emission in the nonrelativistic limit is

$$\boxed{\frac{d\bar{\sigma}}{d\Omega_f} = \left(\frac{d\sigma}{d\Omega_f}\right)_{\text{elastic}} \frac{2\alpha}{\pi} \ln\frac{\omega_{\max}}{\omega_{\min}} \left[\frac{4}{3}\beta^2 \sin^2\frac{\Theta}{2} + \mathcal{O}(\beta^4)\right] \quad \text{for} \quad \beta \ll 1} \; . \tag{8.185}$$

In the extreme-relativistic limit ($\beta \to 1$), the factor in front of the logarithm in equation 8.183 can be approximated by $2\sin^2(\Theta/2)$, and the numerator of the logarithm can be approximated by $2\sin(\Theta/2)$. The denominator of the logarithm requires some care. If we let $\beta = 1 - \delta$, where $\delta \to 0$ in the extreme-relativistic limit, we can Taylor expand the denominator of the logarithm as a power series in δ:

$$\sqrt{1 - \beta^2 \cos^2(\Theta/2)} = \left[1 - (1-\delta)^2 \cos^2(\Theta/2)\right]^{1/2}$$
$$= \left[1 - \cos^2(\Theta/2) + 2\delta \cos^2((\Theta)/2) - \mathcal{O}(\delta^2)\right]^{1/2}$$
$$= \left[\sin^2(\Theta/2) + 2\delta \cos^2(\Theta/2) - \mathcal{O}(\delta^2)\right]^{1/2}$$
$$= \sin(\Theta/2) \left[1 + 2\delta \cot^2(\Theta/2) - \mathcal{O}(\delta^2)\right]^{1/2}$$
$$= \sin(\Theta/2) \left[1 + \delta \cot^2(\Theta/2) - \mathcal{O}(\delta^2)\right] \qquad (8.186)$$

and

$$\sqrt{1 - \beta^2 \cos^2(\Theta/2)} - \beta \sin(\Theta/2) \approx \sin(\Theta/2) \left[1 + \delta \cot^2(\Theta/2) - (1-\delta)\right]$$
$$= \frac{\delta}{\sin(\Theta/2)} + \mathcal{O}(\delta^2). \qquad (8.187)$$

The integral (equation 8.183) in the extreme-relativistic limit becomes

$$I = \frac{1}{2 \sin^2(\Theta/2)} \ln\left(\frac{2 \sin^2(\Theta/2)}{\delta}\right) [1 + \mathcal{O}(\delta)]. \qquad (8.188)$$

We will express δ in terms of the momentum transfer

$$q^2 = (p_f - p_i)^2 = m^2 + m^2 - 2E_f E_i + 2\vec{p}_f \cdot \vec{p}_i,$$
$$\approx -2E_f E_i (1 - \cos\Theta)$$
$$\approx -4E^2 \sin^2 \frac{\Theta}{2} \qquad (8.189)$$

and

$$-\frac{q^2}{m^2} = \frac{4}{1 - \beta^2} \sin^2 \frac{\Theta}{2}$$
$$\approx \frac{2}{\delta} \sin^2 \frac{\Theta}{2}. \qquad (8.190)$$

The differential cross section for soft-photon emission in the extreme relativistic limit is

$$\boxed{\frac{d\bar{\sigma}}{d\Omega_f} = \left(\frac{d\sigma}{d\Omega_f}\right)_{\text{elastic}} \frac{2\alpha}{\pi} \ln \frac{\omega_{max}}{\omega_{min}} \left[\ln \frac{-q^2}{m^2} - 1 + \mathcal{O}\left(\frac{m^2}{q^2}\right)\right] \quad \text{for} \quad \beta \to 1.}$$
$$(8.191)$$

The divergence in the limit as $\omega \to 0$ is evident in all the bremsstrahlung formula. The infrared divergence has been cut off by using ω_{min} as the lower limit in the momentum integration. This value will be determined by the resolution of the measuring apparatus. In the limit $\omega_{\min} \to 0$, we have to include radiative corrections in the calculation of the electron elastic scattering in order to get a finite result (see section 8.13.2.6).

The above formula for the cross sections was derived under the assumption that the field of the nucleus is a pure Coulomb field. The question arises as to whether the screening of the Coulomb field due to the charge distribution of the outer electrons necessitates any important alterations. To address this question one would ask, in a classical treatment, whether the field is screened appreciably for those impact parameters r which give the main contribution to the effect. In quantum theory, the idea of impact parameter has no exact meaning because the electron is represented by a plane wave. We can consider $r \sim \hbar c / q$ as the most important impact parameter. Small momentum transfers correspond to large impact parameters. For high electron energies, $E_i \gg m$, the minimum value of q is given by

$$
\begin{aligned}
|\vec{q}_{\min}| &= |\vec{p}_i| - |\vec{p}_f| - |\vec{k}| \\
q_{\min} &= |\vec{p}_i| - |\vec{p}_f| - \omega \\
&= |\vec{p}_i| - |\vec{p}_f| - (E_f - E_i) \\
&= E_i - \frac{m^2}{2E_i} - \left(E_f - \frac{m^2}{2E_f} \right) - (E_i - E_f) \\
&= \frac{m^2 \omega}{2 E_i E_f}.
\end{aligned} \tag{8.192}
$$

Therefore we can obtain a large contribution to the cross section for distances of the order

$$
r_{\max} = \frac{1}{q_{\min}} = \frac{2 E_i E_f}{m^2 \omega}. \tag{8.193}
$$

We see that for high energies, the screening of the Coulomb field by the outer electrons will lead to a decrease in the cross section. For soft photons, screening will even occur for somewhat smaller energies.

The value $r_{\max}$ has to be compared with the extent of the atomic shell a. According to the Thomas-Fermi model, a is of the order $Z^{-1/3}$ times the Bohr radius of a hydrogen atom

$$
a = \frac{1}{m\alpha} Z^{-1/3}. \tag{8.194}
$$

We can deduce that atomic screening will significantly reduce the radiation intensity at distances exceeding $r_{\max} > a$, or

$$
E > \frac{m}{Z^{1/3}\alpha}, \tag{8.195}
$$

where we have taken $E_i \sim E_f \sim \omega$. If this condition is satisfied, the pure Coulomb field should be replaced by a screened Coulomb field, and the entire calculation should be repeated.

8.8   Photon-Electron Scattering

Quantum electrodynamic processes can be classified by the number and type of particles in the initial state. Our goal in this chapter is to consider the two-particle initial states that lead to scattering processes. Photon-electron scattering will be considered in this section, while the photon-photon and electron-electron systems will be considered in subsequent sections. Our use of the word "electron" is generic and includes both electrons and positrons, in general.

The photon-electron system can have two kinds of final states, those in which there is only one electron present (and one or more photons), and those in which there is also one or more electron-positron pairs present. Processes leading to the former kind of final states are called photon-electron scattering, whereas processes in which pairs are produced are referred to as pair production in photon-electron collisions.

Photon-electron scattering with only one photon in the final state is the lowest-order photon-electron process involving a real incident photon. The separation of the lowest-order from the higher-order contributions is an idealization which does not correspond to physical reality. In any measurement, the energy of the final state can only be determined to within the energy resolution of the detector. It is therefore impossible to determine with certainty whether the final state contains exactly one photon or whether it contains an additional number of very soft (low energy) photons. These multiple-photon final states are suppressed relative to the lowest-order single-photon final state by at least order α.

The lowest order photon-electron scattering process is second order in α and is called Compton scattering. The Compton scattering process differs from bremsstrahlung in that the incoming photon in Compton scattering is real, while the non-outgoing photon in the bremsstrahlung process is due to a the static Coulomb field or a particle current.

Because of charge conjugation invariance of the S-matrix, the cross section for the photon-electron process is equal to the cross section for the photon-positron process.

Figure 8.17 shows the two diagrams leading to Compton scattering. It is important to realize that only the sum of the two diagrams in figure 8.17 describes photon-electron scattering. The separation of the matrix element into terms corresponds to the individual diagrams, though extremely useful, has in general no physical meaning. Only the sum of both diagrams is observable.

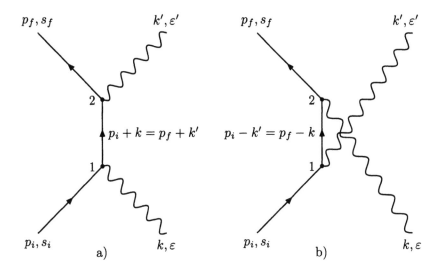

FIGURE 8.17: Feynman diagrams for Compton scattering.

In the first diagram (figure 8.17a), the incident photon (k, ε) is absorbed by the incident electron (p_i, s_i) and then the electron emits a photon (k', ε') into the final state. In the second diagram (figure 8.17b), the incident electron (p_i, s_i) emits a photon (k', ε') before it absorbs the incident photon (k, ε). Since each photon is associated with a different momentum four-vector and each electron path can be labeled as referring to the first, second or third electrons, we have two distinct Feynman diagrams in figure 8.17. The two diagrams are different since they differ in the sequence of the emitted and absorbed photons as one follows the arrows in the electron paths.

One can draw the second diagram (figure 8.18a) such that the intermediate electron is horizontal, or even reversed, and the two photons do not cross in the diagram, as shown in figures 8.18b and 8.18c. It is irrelevant whether or not point 2 is later in time than point 1. The relative position of the vertices in a diagram is not significant, since the diagram stands for a typical integrand, and x_1 and x_2 are dummy variables. The diagrams in figure 8.18 are not topologically different and will not be considered further, since the propagator includes all of them.

Let $A^\mu(x; k)$ represent an incident photon, which is absorbed by an electron at one vertex, and $A'_\mu(x'; k')$ represent a final photon emitted at the second vertex:

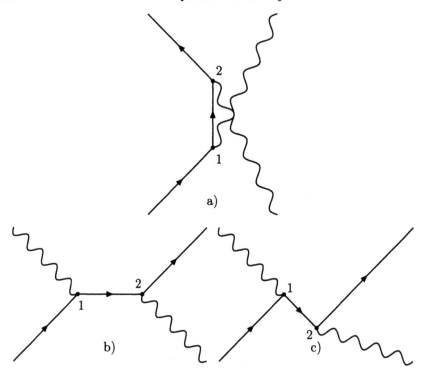

FIGURE 8.18: Three topologically equivalent diagrams of photon-electron scattering.

$$A^\mu(x;k) = \frac{\varepsilon^\mu(\vec{k},\lambda)}{\sqrt{2\omega V}}(e^{-ik\cdot x} + e^{ik\cdot x}) \qquad (8.196)$$

and

$$A'_\mu(x';k') = \frac{\varepsilon^*_\mu(\vec{k}',\lambda')}{\sqrt{2\omega' V}}(e^{-ik'\cdot x'} + e^{ik'\cdot x'}), \qquad (8.197)$$

where $\omega = k_0$ and $\omega' = k'_0$. In this calculation, we shall be more general and not immediately restrict the photon polarizations to be real.

The second-order Compton amplitude is

$$S_{fi} = \int d^4y\, d^4x\, \overline{\psi}_f(y)[ie\,\slashed{A}(y;k')iS_F(y-x)ie\,\slashed{A}(x;k)$$
$$+ ie\,\slashed{A}(y;k)iS_F(y-x)ie\,\slashed{A}(x;k')]\psi_i(x)$$
$$= \int d^4y\, d^4x\, \sqrt{\frac{m}{E_f V}}\,\overline{u}(p_f,s_f)e^{ip_f\cdot y}\left[\frac{ie\,\slashed{\varepsilon}'^*}{\sqrt{2\omega' V}}(e^{-ik'\cdot y} + e^{ik'\cdot y})\right.$$

$$\cdot \int \frac{d^4q}{(2\pi)^4} e^{-iq\cdot(y-x)} \frac{i}{\rlap{/}{q} - m} \frac{ie\,\rlap{/}{\varepsilon}}{\sqrt{2\omega V}} (e^{-ik\cdot x} + e^{ik\cdot x})$$

$$+ \frac{ie\,\rlap{/}{\varepsilon}}{\sqrt{2\omega V}} (e^{-ik\cdot y} + e^{ik\cdot y}) \int \frac{d^4q}{(2\pi)^4} e^{-iq\cdot(y-x)} \frac{i}{\rlap{/}{q} - m}$$

$$\cdot \frac{ie\,\rlap{/}{\varepsilon}'^*}{\sqrt{2\omega'V}} (e^{-ik'\cdot x} + e^{ik'\cdot x}) \Big] \sqrt{\frac{m}{E_i V}} u(p_i, s_i) e^{-ip_i\cdot x}. \qquad (8.198)$$

Each term in equation 8.198 represents one of the eight diagrams shown in figure 8.19. Not every term that occurs in the scattering amplitude is physically relevant to the process considered. The first two diagrams (figures 8.19a and 8.19b) are the processes we are interested in when studying Compton scattering. The second pair of diagrams (figures 8.19c and 8.19d) have the photon momenta k and k' interchanged. This interchange of momenta corresponds to the scattering of an incident photon with momentum k' to a final photon with momentum k. The process represented by these diagrams is not the physical process that we are interested in, and has energy-momentum conserving conditions that are incompatible with Compton scattering. We thus drop these terms from the scattering amplitude since they are not the process under study. The third and fourth pairs of diagrams have two photons in the final state (figures 8.19e and 8.19f) and two photons in the initial state (figures 8.19g and 8.19h), respectively. These processes are not kinematically allowed. Such terms in the scattering amplitude contain delta functions with an argument describing these kinematically forbidden processes. The delta functions cause these terms to vanish when the momenta are integrated over.

For the diagrams we are interested in, we retain from the incident photon wave function only the first term, $e^{-ik\cdot x}$, which corresponds to absorption at x of a photon of four-momentum k from the radiation field, and retain from the final photon wave function only the second term, $e^{ik'\cdot x'}$, which represents the emission at x' of a photon with four-momentum k'.

The scattering amplitude now becomes

$$S_{fi} = \int d^4y\, d^4x \sqrt{\frac{m}{E_f V}} \overline{u}(p_f, s_f) e^{ip_f\cdot y}$$

$$\cdot \Big[\frac{ie\,\rlap{/}{\varepsilon}'^*}{\sqrt{2\omega'V}} e^{ik'\cdot y} \int \frac{d^4q}{(2\pi)^4} e^{-iq\cdot(y-x)} \frac{i}{\rlap{/}{q} - m} \frac{ie\,\rlap{/}{\varepsilon}}{\sqrt{2\omega V}} e^{-ik\cdot x}$$

$$+ \frac{ie\,\rlap{/}{\varepsilon}}{\sqrt{2\omega V}} e^{-ik\cdot y} \int \frac{d^4q}{(2\pi)^4} e^{-iq\cdot(y-x)} \frac{i}{\rlap{/}{q} - m} \frac{ie\,\rlap{/}{\varepsilon}'^*}{\sqrt{2\omega'V}} e^{ik'\cdot x} \Big]$$

$$\cdot \sqrt{\frac{m}{E_i V}} u(p_i, s_i) e^{-ip_i\cdot x}$$

$$= \int d^4y\, d^4x \frac{d^4q}{(2\pi)^4} \sqrt{\frac{m^2}{E_f E_i V^2}} \frac{1}{\sqrt{2\omega 2\omega' V^2}}$$

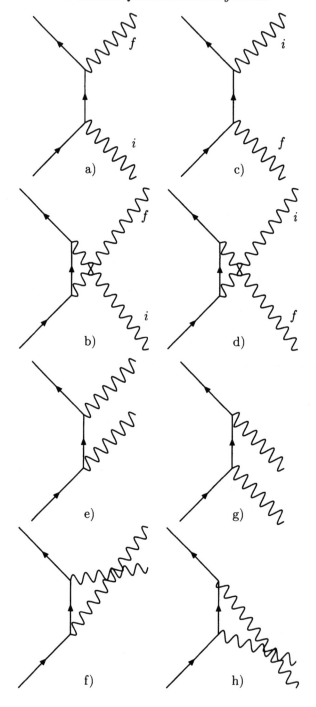

FIGURE 8.19:   Possible terms in the S-matrix for Compton scattering.

$$\cdot \, \bar{u}(p_f, s_f) \left[e^{iy \cdot (p_f + k' - q)} e^{ix \cdot (q - k - p_i)} (ie \, \rlap{/}{\epsilon}'^*) \frac{i}{\rlap{/}{q} - m} (ie \, \rlap{/}{\epsilon}) \right.$$

$$\left. + e^{iy \cdot (p_f - k - q)} e^{ix \cdot (q + k' - p_i)} (ie \, \rlap{/}{\epsilon}) \frac{i}{\rlap{/}{q} - m} (ie \, \rlap{/}{\epsilon}'^*) \right] u(p_i, s_i)$$

$$= \int (2\pi)^4 d^4 q \sqrt{\frac{m^2}{E_f E_i V^2}} \frac{1}{\sqrt{2\omega 2\omega' V^2}}$$

$$\cdot \, \bar{u}(p_f, s_f) \left[\delta^4(p_f + k' - q) \delta^4(q - k - p_i)(ie \, \rlap{/}{\epsilon}'^*) \frac{i}{\rlap{/}{q} - m} (ie \, \rlap{/}{\epsilon}) \right.$$

$$\left. + \delta^4(p_f - k - q) \delta^4(q + k' - p_i)(ie \, \rlap{/}{\epsilon}) \frac{i}{\rlap{/}{q} - m} (ie \, \rlap{/}{\epsilon}'^*) \right] u(p_i, s_i)$$

$$= \sqrt{\frac{m^2}{E_f E_i V^2}} \frac{1}{\sqrt{2\omega 2\omega' V^2}} (2\pi)^4 \delta^4(p_f + k' - p_i - k) \bar{u}(p_f, s_f)$$

$$\cdot \left[(ie \, \rlap{/}{\epsilon}'^*) \frac{i}{\rlap{/}{p}_i + \rlap{/}{k} - m} (ie \, \rlap{/}{\epsilon}) + (ie \, \rlap{/}{\epsilon}) \frac{i}{\rlap{/}{p}_i - \rlap{/}{k}' - m} (ie \, \rlap{/}{\epsilon}'^*) \right] u(p_i, s_i).$$

$$(8.199)$$

To allow us to use the Feynman diagram technique, we write

$$S_{fi} = i(2\pi)^4 \delta^4(p_f + k' - p_i - k) \sqrt{\frac{m^2}{E_f V E_i V}} \mathcal{M}_{fi}^{\mu\nu}(k, k') \frac{\epsilon_\mu^*(\vec{k}', \lambda') \epsilon_\nu(\vec{k}, \lambda)}{\sqrt{2\omega V 2\omega' V}},$$

$$(8.200)$$

where

$$\mathcal{M}_{fi}^{\mu\nu}(k, k')$$

$$= -i\bar{u}(p_f, s_f) \left[(ie\gamma^\mu) \frac{i}{\rlap{/}{p}_i + \rlap{/}{k} - m} (ie\gamma^\nu) + (ie\gamma^\nu) \frac{i}{\rlap{/}{p}_i - \rlap{/}{k}' - m} (ie\gamma^\mu) \right] u(p_i, s_i).$$

$$(8.201)$$

The invariant matrix element $i\mathcal{M}_{fi}^{\mu\nu}(k, k')$ in this case is a second-rank tensor depending on the kinematic variables of all four particles. Notice that S_{fi} is symmetric under interchange of k and ε with $-k'$ and ε', respectively. The two diagrams in figure 8.17 are thus related by this symmetry. This is known as crossing symmetry (see section 8.5), and it persists as an exact symmetry to all orders in α.

We form the differential cross section $d\sigma$ by squaring the scattering amplitude, then dividing by $(2\pi)^4 \delta^{(4)}(0) = VT$ to form a rate per volume. We then divide the rate by an incident flux $|\bar{v}_{\text{rel}}|/V$, where $\bar{v}_{\text{rel}}$ is the relative velocity of the photons with respect to the electrons. Then we divide by the number of target particles per unit volume $1/V$. Finally, summing over phase space of the final particles $(V^2/(2\pi)^6) d^3 p_f d^3 k'$, $d\sigma$ becomes

$$d\sigma = \frac{|S_{fi}|^2}{(2\pi)^4\delta^4(0)V} \frac{V}{|\vec{v}_{rel}|} \frac{Vd^3p_f}{(2\pi)^3} \frac{Vd^3k'}{(2\pi)^2}$$

$$= \frac{m}{2\omega E_i|\vec{v}_{rel}|}(2\pi)^4\delta^4(p_f + k' - p_i - k)|\varepsilon^{*\mu}\mathcal{M}_{\mu\nu}\varepsilon^{\nu}| \frac{m}{E_f}\frac{d^3p_f}{(2\pi)^3}\frac{1}{2\omega'}\frac{d^3k'}{(2\pi)^3}.$$

$$(8.202)$$

We use $d^3k' = \omega'^2\,d\omega'd\Omega$ to write the differential cross section per unit solid angle for scattering into a differential angular interval between θ and $\theta + d\theta$, and ϕ and $\phi + d\phi$. The angle θ is defined in figure 8.20. The integral over all recoil electron momenta can be evaluated with the aid of equation 8.102.

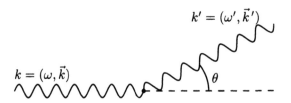

FIGURE 8.20: Definition of the scattering angle for Compton scattering in the rest frame of the electron.

The differential cross section now becomes

$$\frac{d\sigma}{d\Omega} = \frac{m^2}{2(2\pi)^2\omega E_i|\vec{v}_{rel}|}\int_0^{\infty} d\omega'\omega'$$

$$\cdot \int_{-\infty}^{+\infty} d^4p_f|\varepsilon^{*\mu}\mathcal{M}_{\mu\nu}\varepsilon^{\nu}|^2\delta^4(p_f + k' - p_i - k)\delta(p_f^2 - m^2)\theta(p_{f0})$$

$$= \frac{m^2}{2(2\pi)^2\omega E_i|\vec{v}_{rel}|}\int_0^{\infty} d\omega'\omega'$$

$$\cdot |\varepsilon^{*\mu}\mathcal{M}_{\mu\nu}\varepsilon^{\nu}|^2\delta[(p_i + k - k')^2 - m^2]\theta(E_i + \omega - \omega')$$

$$= \frac{m^2}{2(2\pi)^2\omega E_i|\vec{v}_{rel}|}\int_0^{E_i+\omega} d\omega'\omega'|\varepsilon^{*\mu}\mathcal{M}_{\mu\nu}\varepsilon^{\nu}|^2\delta[2p_i \cdot (k - k') - 2k \cdot k'],$$

$$(8.203)$$

where the kinematic variables in the square of the matrix element $|\varepsilon^{*\mu}\mathcal{M}_{\mu\nu}\varepsilon^{\mu}|^2$ must now obey the condition $p_i + k = p_f + k'$.

The cross section simplifies considerably if we calculate it in the rest frame of the initial or final electron. Since electrons in atoms move nonrelativistically, the laboratory frame for high-energy (x-ray and gamma-ray) photon-electron scattering experiments is usually, though not always, one in which the initial electron can be taken to be at rest. In this frame, $p_i = (m, 0)$ and the incident beam consists of photons with unit velocity such that $|\vec{v}_{rel}| = 1$. The differential cross section now becomes

$$
\begin{aligned}
\frac{d\sigma}{d\Omega} &= \frac{m}{2(2\pi)^2\omega} \int_0^{E_i+\omega} d\omega' \omega' |\varepsilon^{*\mu} \mathcal{M}_{\mu\nu} \varepsilon^\nu|^2 \delta[2m(\omega - \omega') - 2\omega\omega'(1 - \cos\theta)] \\
&= \frac{m}{2(2\pi)^2\omega} |\varepsilon^{*\mu} \mathcal{M}_{\mu\nu} \varepsilon^\nu|^2 \frac{\omega'}{|2m + 2\omega(1 - \cos\theta)|} \\
&= \frac{1}{4(2\pi)^2} \left(\frac{\omega'}{\omega}\right)^2 \varepsilon^{*\mu} |\mathcal{M}_{\mu\nu} \varepsilon^\nu|^2,
\end{aligned}
\tag{8.204}
$$

where θ is the angle between the initial- and final-state photons.

The last line in equation 8.204 was obtained by using the root of the delta function to relate ω to ω':

$$
\omega' = \frac{\omega}{1 + (\omega/m)(1 - \cos\theta)} = \frac{\omega}{1 + (2\omega/m)\sin^2(\theta/2)}.
\tag{8.205}
$$

This is known as the Compton condition. This kinematic relationship takes on a familiar form if one uses the wavelength, $\lambda = 2\pi/\omega$:

$$
\lambda' = \lambda + \frac{2\pi}{m}(1 - \cos\theta).
\tag{8.206}
$$

This is the familiar Compton formula. The wavelength of the scattered photon is increased by an amount of order $1/m$, and $\hbar/mc$ is called the Compton wavelength.

The differential cross section for electron-photon scattering with specific initial- and final-state polarizations is now

$$
\frac{d\sigma}{d\Omega} = \alpha^2 \left(\frac{\omega'}{\omega}\right)^2 \left| \bar{u}(p_f, s_f) \left[\slashed{\varepsilon}'^* \frac{1}{\slashed{p}_i + \slashed{k} - m} \slashed{\varepsilon} + \slashed{\varepsilon} \frac{1}{\slashed{p}_i - \slashed{k}' - m} \slashed{\varepsilon}'^* \right] u(p_i, s_i) \right|^2.
\tag{8.207}
$$

We can simplify the spinor matrix element considerably by choosing the special gauge in which both initial- and final-state photons are transversely polarized in the rest frame of the electron. We choose

$$
\varepsilon^\mu = (0, \vec{\varepsilon}) \quad \text{so that} \quad \vec{\varepsilon} \cdot \vec{k} = 0 \quad \text{and}
\tag{8.208}
$$

$$
\varepsilon'^{*\mu} = (0, \vec{\varepsilon}'^*) \quad \text{so that} \quad \vec{\varepsilon}'^* \cdot \vec{k}' = 0.
\tag{8.209}
$$

Since the electron is initially at rest, it follows that

$$\varepsilon \cdot p_i = \varepsilon'^* \cdot p_i = 0. \tag{8.210}$$

This amounts to choosing the "radiation gauge" in which the electromagnetic potential has no time component. However, the condition in equation 8.210 can be imposed in any given frame of reference. This can be shown by applying a gauge transformation to any arbitrary set of polarization vectors ε and ε'^*. The normalization and transversality conditions are not affected by the transformation. Thus without restricting the generality of our calculation, we will impose the condition in equation 8.210.

Because of our choice of gauge, $\not k$, $\not p_i$, and $\not p_i'$ anticommute with $\not\varepsilon$, $\not k'$, and $\not\varepsilon'^*$, respectively. The invariant matrix element thus becomes

$$
\varepsilon^{*\mu}\mathcal{M}_{\mu\nu}\varepsilon^{\nu} = -e^2\overline{u}(p_f, s_f)\left[\not\varepsilon'^*\frac{\not p_i + \not k + m}{2k\cdot p_i}\not\varepsilon + \not\varepsilon\frac{\not p_i - \not k' + m}{-2k'\cdot p_i}\not\varepsilon'^*\right]u(p_i, s_i)
$$

$$
= -e^2\overline{u}(p_f, s_f)\left[\not\varepsilon'^*\not\varepsilon\frac{-\not p_i - \not k + m}{2k\cdot p_i} + \not\varepsilon\not\varepsilon'^*\frac{-\not p_i + \not k' + m}{-2k'\cdot p_i}\right]u(p_i, s_i)
$$

$$
= e^2\overline{u}(p_f, s_f)\left[\frac{\not\varepsilon'^*\not\varepsilon\not k}{2k\cdot p_i} + \frac{\not\varepsilon\not\varepsilon'^*\not k'}{2k'\cdot p_i}\right]u(p_i, s_i), \tag{8.211}
$$

where the energy-projection operator $(-\not p + m)u(p, s) = 0$ has been used in the last step.

We now consider the case when the electrons are unpolarized but the initial- and final-state photons may be polarized with polarizations λ and λ'. We thus average over the initial electron spins and sum over the final electron spins only:

$$
\frac{d\bar\sigma}{d\Omega}(\lambda, \lambda') = \frac{1}{2}\sum_{\pm s_i, \pm s_f}\frac{d\sigma}{d\Omega}. \tag{8.212}
$$

Applying the usual trace techniques, we have

$$
\frac{d\bar\sigma}{d\Omega}(\lambda, \lambda') = \frac{\alpha^2}{2}\left(\frac{\omega'}{\omega}\right)^2 \cdot \text{Tr}\left[\frac{\not p_f + m}{2m}\left(\frac{\not\varepsilon'^*\not\varepsilon\not k}{2k\cdot p_i} + \frac{\not\varepsilon\not\varepsilon'^*\not k'}{2k'\cdot p_i}\right)\right.
$$
$$
\left.\cdot\frac{\not p_i + m}{2m}\left(\frac{\not k\not\varepsilon\not\varepsilon'^*}{2k\cdot p_i} + \frac{\not k'\not\varepsilon'^*\not\varepsilon}{2k'\cdot p_i}\right)\right], \tag{8.213}
$$

where we have used the rule given by equation 8.73 $\overline{\not a\not b\not c} = \not c\not b\not a$ (see problem 8.7) in the last factor.

Equation 8.213 contains traces with six and eight gamma matrices in them. In general, traces of products of six or eight gamma matrices would be given by a sum of 15 or 105 terms, respectively. To reduce the traces which contain the same vectors, we anticommute the gamma matrices until the identical vectors are alongside each other, then the identity $\not a\not a = a^2$ removes two gamma matrices. We also make use of $k^2 = 0$, $\varepsilon^2 = (\varepsilon'^*)^2 = -1$, $k \cdot p_f = k' \cdot p_i$, and $k \cdot \varepsilon'^* = p_f \cdot \varepsilon'^*$. Evaluating the traces one by one, we have

$$
\begin{aligned}
T_1 &= \mathrm{Tr}[(\slashed{p}_f + m)\, \slashed{\varepsilon}'^*\, \slashed{\varepsilon}\, \slashed{k}(\slashed{p}_i + m)\, \slashed{k}\, \slashed{\varepsilon}\, \slashed{\varepsilon}'^*] \\
&= \mathrm{Tr}[\slashed{p}_f\, \slashed{\varepsilon}'^*\, \slashed{\varepsilon}\, \slashed{k}\, \slashed{p}_i\, \slashed{k}\, \slashed{\varepsilon}\, \slashed{\varepsilon}'^*] + m^2\mathrm{Tr}[\slashed{\varepsilon}'^*\, \slashed{\varepsilon}\, \slashed{k}\, \slashed{k}\, \slashed{\varepsilon}\, \slashed{\varepsilon}'^*] \\
&= 2k \cdot p_i\,\mathrm{Tr}[\slashed{p}_f\, \slashed{\varepsilon}'^*\, \slashed{\varepsilon}\, \slashed{k}\, \slashed{\varepsilon}\, \slashed{\varepsilon}'^*] \\
&= 2k \cdot p_i\,\mathrm{Tr}[\slashed{p}_f\, \slashed{\varepsilon}'^*\, \slashed{k}\, \slashed{\varepsilon}'^*] \\
&= 2k \cdot p_i(\mathrm{Tr}[\slashed{p}_f\, \slashed{k}] + 2k \cdot \varepsilon'^*\mathrm{Tr}[\slashed{p}_f\, \slashed{\varepsilon}'^*]) \\
&= 8k \cdot p_i(p_f \cdot k + 2k \cdot \varepsilon'^* p_f \cdot \varepsilon'^*) \\
&= 8k \cdot p_i[k' \cdot p_i + 2(k \cdot \varepsilon'^*)^2].
\end{aligned}
\tag{8.214}
$$

Using the symmetry we had earlier $(\varepsilon, k \leftrightarrow \varepsilon'^*, -k')$, we write

$$
\begin{aligned}
T_2 &= \mathrm{Tr}[(\slashed{p}_f + m)\, \slashed{\varepsilon}\, \slashed{\varepsilon}'^*\, \slashed{k}'(\slashed{p}_i + m)\, \slashed{k}'\, \slashed{\varepsilon}'^*\, \slashed{\varepsilon}] \\
&= 8k' \cdot p_i[k \cdot p_i - 2(k' \cdot \varepsilon)^2].
\end{aligned}
\tag{8.215}
$$

We show that the two cross-terms are equal:

$$
\begin{aligned}
T_3 &= \mathrm{Tr}[\slashed{p}_f + m)\, \slashed{\varepsilon}'^*\, \slashed{\varepsilon}\, \slashed{k}(\slashed{p}_i + m)\, \slashed{k}'\, \slashed{\varepsilon}'^*\, \slashed{\varepsilon}] \\
&= \mathrm{Tr}[\slashed{\varepsilon}\, \slashed{\varepsilon}'^*\, \slashed{k}'(\slashed{p}_i + m)\, \slashed{k}\, \slashed{\varepsilon}\, \slashed{\varepsilon}'^*(\slashed{p}_f + m)] \\
&= \mathrm{Tr}[(\slashed{p}_f + m)\, \slashed{\varepsilon}\, \slashed{\varepsilon}'^*\, \slashed{k}'(\slashed{p}_i + m)\, \slashed{k}\, \slashed{\varepsilon}\, \slashed{\varepsilon}'^*].
\end{aligned}
\tag{8.216}
$$

Using energy-momentum conservation we have

$$
\begin{aligned}
T_3 &= \mathrm{Tr}[(\slashed{p}_f + m)\, \slashed{\varepsilon}'^*\, \slashed{\varepsilon}\, \slashed{k}(\slashed{p}_i + m)\, \slashed{k}'\, \slashed{\varepsilon}'^*\, \slashed{\varepsilon}] \\
&= \mathrm{Tr}[(\slashed{p}_i + \slashed{k} - \slashed{k}' + m)\, \slashed{\varepsilon}'^*\, \slashed{\varepsilon}\, \slashed{k}(\slashed{p}_i + m)\, \slashed{k}'\, \slashed{\varepsilon}'^*\, \slashed{\varepsilon}] \\
&= \mathrm{Tr}[(\slashed{p}_i + m)\, \slashed{\varepsilon}'^*\, \slashed{\varepsilon}\, \slashed{k}(\slashed{p}_i + m)\, \slashed{k}'\, \slashed{\varepsilon}'^*\, \slashed{\varepsilon}] + \mathrm{Tr}[(\slashed{k} - \slashed{k}')\, \slashed{\varepsilon}'^*\, \slashed{\varepsilon}\, \slashed{k}\, \slashed{p}_i\, \slashed{k}'\, \slashed{\varepsilon}'^*\, \slashed{\varepsilon}] \\
&= \mathrm{Tr}[(\slashed{p}_i + m)\, \slashed{k}(\slashed{p}_i + m)\, \slashed{k}'\, \slashed{\varepsilon}'^*\, \slashed{\varepsilon}\, \slashed{\varepsilon}'^*\, \slashed{\varepsilon}] - \mathrm{Tr}[\slashed{\varepsilon}'^*\, \slashed{k}\, \slashed{\varepsilon}\, \slashed{k}\, \slashed{p}_i\, \slashed{k}'\, \slashed{\varepsilon}'^*\, \slashed{\varepsilon}] \\
&\quad - 2k \cdot \varepsilon'^*\mathrm{Tr}[\slashed{k}\, \slashed{p}_i\, \slashed{k}'\, \slashed{\varepsilon}'^*] + \mathrm{Tr}[\slashed{\varepsilon}'^*\, \slashed{\varepsilon}\, \slashed{k}\, \slashed{p}_i\, \slashed{k}'\, \slashed{\varepsilon}'^*\, \slashed{k}'\, \slashed{\varepsilon}] + 2k' \cdot \varepsilon\,\mathrm{Tr}[\slashed{\varepsilon}\, \slashed{k}\, \slashed{p}_i\, \slashed{k}'] \\
&= -\mathrm{Tr}[(\slashed{p}_i - m)\, \slashed{k}(\slashed{p}_i + m)\, \slashed{k}'\, \slashed{\varepsilon}'^*\, \slashed{\varepsilon}\, \slashed{\varepsilon}'^*\, \slashed{\varepsilon}] + 2k \cdot p_i\,\mathrm{Tr}[\slashed{p}_i\, \slashed{k}'\, \slashed{\varepsilon}'^*\, \slashed{\varepsilon}\, \slashed{\varepsilon}'^*\, \slashed{\varepsilon}] \\
&\quad - 8k \cdot \varepsilon'^*[k \cdot \varepsilon'^* p_i \cdot k'] + 8k' \cdot \varepsilon[\varepsilon \cdot k' k \cdot p_i] \\
&= -2k \cdot p_i\,\mathrm{Tr}[\slashed{p}_i\, \slashed{k}'] + 4k \cdot p_i \varepsilon \cdot \varepsilon'^*\mathrm{Tr}[\slashed{p}_i\, \slashed{k}'\, \slashed{\varepsilon}'^*\, \slashed{\varepsilon}] \\
&\quad - 8(k \cdot \varepsilon'^*)^2 k' \cdot p_i + 8(k' \cdot \varepsilon)^2 k \cdot p_i \\
&= -8[k \cdot p_i p_i \cdot k'] + 16k \cdot p_i \varepsilon \cdot \varepsilon'^*[p_i \cdot k' \varepsilon'^* \cdot \varepsilon] \\
&\quad - 8(k \cdot \varepsilon'^*)^2 k' \cdot p_i + 8(k' \cdot \varepsilon)^2 k \cdot p_i \\
&= 8(k \cdot p_i)(k' \cdot p_i)[2(\varepsilon'^* \cdot \varepsilon)^2 - 1] - 8(k \cdot \varepsilon'^*)^2 k' \cdot p_i + 8(k' \cdot \varepsilon)^2 k \cdot p_i.
\end{aligned}
\tag{8.217}
$$

Therefore the differential cross section becomes

$$\frac{d\bar{\sigma}}{d\Omega}(\lambda, \lambda') = \frac{\alpha^2}{4m^2} \left(\frac{\omega'}{\omega}\right)^2 \left[\frac{k' \cdot p_i + 2(k \cdot \varepsilon'^*)^2}{k \cdot p_i} + \frac{k \cdot p_i - 2(k' \cdot \varepsilon)^2}{k' \cdot p_i}\right.$$
$$\left. + 2[2(\varepsilon'^* \cdot \varepsilon)^2 - 1] - 2\frac{(k \cdot \varepsilon'^*)^2}{k \cdot p_i} + 2\frac{(k' \cdot \varepsilon)^2}{k' \cdot p_i}\right]$$
$$= \frac{\alpha^2}{4m^2} \left(\frac{\omega'}{\omega}\right)^2 \left[\frac{k' \cdot p_i}{k \cdot p_i} + \frac{k \cdot p_i}{k' \cdot p_i} + 4(\varepsilon'^* \cdot \varepsilon)^2 - 2\right]. \qquad (8.218)$$

The calculation of the invariant matrix element has so far been covariant. In the rest frame of the initial electron, the differential cross section becomes

$$\boxed{\frac{d\bar{\sigma}}{d\Omega}(\lambda, \lambda') = \frac{\alpha^2}{4m^2} \left(\frac{\omega'}{\omega}\right)^2 \left[\frac{\omega'}{\omega} + \frac{\omega}{\omega'} + 4(\varepsilon'^* \cdot \varepsilon)^2 - 2\right]}, \qquad (8.219)$$

which is the Klein-Nishina formula for Compton scattering.

In the low-energy limit of $\omega \to 0$, equation 8.205 shows that $\omega'/\omega \to 1$, and the cross section reduces to the classical Thomson scattering cross section

$$\boxed{\left(\frac{d\bar{\sigma}}{d\Omega}(\lambda, \lambda')\right)_{\omega \to 0} = \frac{\alpha^2}{m^2}(\varepsilon \cdot \varepsilon'^*)^2}, \qquad (8.220)$$

where

$$r_0 \equiv \frac{\alpha}{m} = \frac{e^2}{4\pi mc^2} = 2.8 \times 10^{-13} \text{ cm} \qquad (8.221)$$

is the classical electron radius. The Thomson cross section for an electron

$$\sigma_{\text{Thom}} = \frac{8\pi r_0^2}{3} \sim 10^{-24} \text{ cm}^2 \qquad (8.222)$$

was originally determined using classical mechanics and electromagnetism by calculating the reradiation of light by a nonrelativistic point charge in a plane-wave electromagnetic field. For an electron, the charge radius and cross section are far larger than the classical charge radius and cross section for a target proton of $\sigma_p \sim 10^{-31}$ cm^2. Since the classical limit is recovered for low energies, the Thomson cross section must be the low-energy limit for radiation scattering off any charged object, depending only upon the ratio of the object's charge squared to it mass.

For forward scattering, $\theta \to 0$ and according to equation 8.205 $\omega \to \omega'$. The Thomson cross section is thus also valid for forward scattering at all energies.

Returning to the general expression for the cross section (equation 8.219), we can sum over final-state photon polarizations ε_{λ}^*, and average over initial-state polarizations ε_{λ} to obtain the unpolarized cross section, as follows:

$$\frac{d\bar{\sigma}}{d\Omega} = \frac{1}{2} \sum_{\lambda,\lambda'=1}^{2} \frac{d\bar{\sigma}}{d\Omega}(\lambda,\lambda')$$

$$= \frac{\alpha^2}{8m^2} \left(\frac{\omega'}{\omega}\right)^2 \sum_{\lambda,\lambda'=1}^{2} \left[\frac{\omega'}{\omega} + \frac{\omega}{\omega'} + 4(\varepsilon_\lambda \cdot \varepsilon^*_{\lambda'})^2 - 2\right]$$

$$= \frac{\alpha^2}{2m^2} \left(\frac{\omega'}{\omega}\right)^2 \left[\frac{\omega'}{\omega} + \frac{\omega}{\omega'} + \sum_{\lambda,\lambda'=1}^{2} (\varepsilon_\lambda \cdot \varepsilon^*_{\lambda'})^2 - 2\right]. \qquad (8.223)$$

We evaluate the remaining spin sum by choosing the incident photon to arrive along the z-direction, while the final photon departs into the solid angle $d\Omega$ described by polar angles θ and ϕ, such that

$$\hat{k} = (0,0,1), \qquad (8.224)$$
$$\hat{k}' = (\sin\theta\cos\phi, \sin\theta\sin\phi, \cos\theta). \qquad (8.225)$$

We may select the associated polarization vectors to be

$$\vec{\varepsilon}_{(1)} = (1,0,0), \qquad \varepsilon'_{(1)} = (\sin\phi, -\cos\phi, 0), \qquad (8.226)$$
$$\vec{\varepsilon}_{(2)} = (0,1,0), \qquad \varepsilon'_{(2)} = (\cos\theta\cos\phi, \cos\theta\sin\phi, -\sin\theta). \qquad (8.227)$$

Since we are now using linear polarization, the polarization vectors are real. It is easy to show that this choice of vectors satisfies all the required normalization and orthogonality relationships. We obtain

$$\sum_{\lambda,\lambda'=1}^{2} (\varepsilon_\lambda \cdot \varepsilon_{\lambda'})^2 = \sin^2\phi + \cos^2\theta\cos^2\phi + \cos^2\phi + \cos^2\theta\sin^2\phi = 1 + \cos^2\theta.$$

$$(8.228)$$

The cross section thus becomes

$$\boxed{\frac{d\bar{\sigma}}{d\Omega} = \frac{\alpha^2}{2m^2} \left(\frac{\omega'}{\omega}\right)^2 \left(\frac{\omega'}{\omega} + \frac{\omega}{\omega'} - \sin^2\theta\right)}. \qquad (8.229)$$

The low-energy or forward-scattering limit (classical limit) now becomes

$$\boxed{\left(\frac{d\bar{\sigma}}{d\Omega}\right)_{\text{class}} = \frac{r_0^2}{2}(1 + \cos^2\theta)}. \qquad (8.230)$$

To integrate the differential cross section, we simplify the notation by introducing $z = \cos\theta$, and use equation 8.205 to write

$$\bar{\sigma} = \frac{\pi\alpha^2}{m^2} \int_{-1}^{1} dz \left[\frac{1}{[1+(\omega/m)(1-z)]^3} \right.$$
$$\left. + \frac{1}{1+(\omega/m)(1-z)} - \frac{1-z^2}{[1+(\omega/m)(1-z)]^2} \right]. \qquad (8.231)$$

To perform the integration, we define $x = 1 - z$, then

$$\bar{\sigma} = \frac{\pi\alpha^2}{m^2} \int_{0}^{2} dx \left[\frac{1}{[1+(\omega/m)x]^3} + \frac{1}{1+(\omega/m)x} + \frac{x^2-2x}{[1+(\omega/m)x]^2} \right]. \qquad (8.232)$$

Using the integrals (b is an arbitrary constant)

$$\int \frac{dx}{1+bx} = \frac{1}{b}\ln(1+bx)\bigg|_0^2 = \frac{1}{b}\ln(1+2b), \qquad (8.233)$$

$$\int \frac{dx}{(1+bx)^3} = -\frac{1}{2b(1+bx)^2}\bigg|_0^2 = \frac{1}{2b}\left[1 - \frac{1}{(1+2b)^2}\right], \qquad (8.234)$$

$$\int \frac{x^2\,dx}{(1+bx)^2} = \frac{1}{b^3}\left[1+bx - 2\ln(1+bx) - \frac{1}{1+bx}\right]\bigg|_0^2$$
$$= \frac{1}{b^3}\left[2b - 2\ln(1+2b) - \frac{1}{1+2b} + 1\right], \qquad (8.235)$$

$$\int \frac{x\,dx}{(1+bx)^2} = \frac{1}{b^2}\left[\ln(1+bx) + \frac{1}{1+bx}\right]\bigg|_0^2$$
$$= \frac{1}{b^2}\left[\ln(1+2b) + \frac{1}{1+2b} - 1\right], \qquad (8.236)$$

we write

$$\bar{\sigma} = \frac{\pi\alpha^2}{m^2} \left(\frac{m}{\omega}\right)^3 \left\{ 1 - \frac{1}{1+2(\omega/m)} - 2\ln\left(1+2\frac{\omega}{m}\right) \right.$$
$$+ 2\left(\frac{\omega}{m}\right)\left[2 - \frac{1}{1+2(\omega/m)} - \ln\left(1+2\frac{\omega}{m}\right)\right]$$
$$\left. + \frac{1}{2}\left(\frac{\omega}{m}\right)^2\left[1 - \frac{1}{[1+2(\omega/m)]^2} + 2\ln\left(1+2\frac{\omega}{m}\right)\right] \right\}, \qquad (8.237)$$

which is valid for all initial photon energies ω.

For low energies, $\omega/m \to 0$ and

$$\bar{\sigma} \approx \frac{\pi \alpha^2}{m^2} \left(\frac{m}{\omega} \right)^3 \left\{ -2 \left(\frac{\omega}{m} \right) + \frac{8}{3} \left(\frac{\omega}{m} \right)^3 + \cdots \right.$$

$$\left. + 2 \left(\frac{\omega}{m} \right) \left[1 - 2 \left(\frac{\omega}{m} \right)^2 + \cdots \right] + \frac{1}{2} \left(\frac{\omega}{m} \right)^2 \left[8 \left(\frac{\omega}{m} \right) + \cdots \right] \right\}$$

$$\approx \frac{8\pi}{3} \frac{\alpha^2}{m^2} = \frac{8\pi}{3} r_0^2, \qquad (8.238)$$

which is the total classical Thomson cross section $\sigma_0 = 8\pi r_0^2 / 3$.

At high energies, $m/\omega \to 0$ and

$$\bar{\sigma} \approx \frac{\pi \alpha^2}{m^2} \left(\frac{m}{\omega} \right)^3 \left\{ 2 \left(\frac{\omega}{m} \right) \left[-\ln \frac{2\omega}{m} + \cdots \right] + \frac{1}{2} \left(\frac{\omega}{m} \right)^2 \left[1 + 2\ln \frac{2\omega}{m} + \cdots \right] \right\}$$

$$\approx \frac{\pi \alpha^2}{\omega m} \left[\ln \frac{2\omega}{m} + \frac{1}{2} + \mathcal{O} \left(\frac{m}{\omega} \ln \frac{\omega}{m} \right) \right]$$

$$= \pi r_0^2 \frac{m}{\omega} \left[\ln \frac{2\omega}{m} + \frac{1}{2} + \mathcal{O} \left(\frac{m}{\omega} \ln \frac{\omega}{m} \right) \right]$$

$$= \sigma_0 \frac{8}{3} \frac{m}{\omega} \left[\ln \frac{2\omega}{m} + \frac{1}{2} + \mathcal{O} \left(\frac{m}{\omega} \ln \frac{\omega}{m} \right) \right]. \qquad (8.239)$$

Thus for very high energies, the cross section decreases with increasing photon energy. This is the reason why the penetrating power of gamma-rays increases with increasing energy, as long as no other absorption processes, such as pair production, are important.

An experimental test of the total cross section in equation 8.237 is provided by measurements of the total absorption coefficient of x-rays or gamma-rays in various materials. For a comparison of the theory with experiment one has to correct for two effects. One, for x-rays the total absorption is not only due to scattering but also to the photoelectric effect, which gives a strong absorption, but, however, decreases rapidly with photon energy. Two, for gamma-rays, the absorption is largely due to pair production, which increases with photon energy. The theoretical predictions fit the experimental data excellently, and thus confirm the Klein-Nishina formula.

We have considered the electron as free. This would no longer be true for softer radiation, for which the binding of the electrons has to be taken into account. Our result is based on first-order perturbation theory. Corrections (radiative and damping corrections) exist. They are small for all energies.

8.9   Electron-Positron Annihilation into Two Photons

There are three essentially different kinds of electron-positron pair annihilations. The first kind is the annihilation of a free positron with a free electron in relative motion towards each other; we shall call this the annihilation of free electron-positron pairs and shall discuss it here. Since the probability for the occurrence of this process increases with decreasing relative velocity, as we shall see, it will compete with the process of capture into the bound positronium state, which in turn can decay into photons. This positronium annihilation most likely occurs from one of the lowest energy levels (S-state) and will always be governed by selection rules. Finally, the third kind of annihilation takes place in the presence of an external field, e.g. the Coulomb field of a nucleus. This pair annihilation in an external field is of importance when positrons annihilate with tightly bound electrons whose binding to the nucleus cannot be neglected. This latter process is the only one in which energy-momentum conservation does not exclude one-photon annihilation, i.e. pair annihilation with the emission of only one photon (see problem 8.19).

Consider the process of annihilation of an electron-positron pair into two gamma-rays, as shown in figure 8.21. This is the lowest order in e^2 in which this process can occur, since pair annihilation to a single photon cannot conserve energy and momentum simultaneously: $(p_- + p_+)^2 = 2m^2 + 2E_- E_+ (1 - \beta_- \beta_+ \cos\theta) > 0$ but $k^2 = 0$. The diagrams can be viewed as Compton scattering turned on their sides.

Using our previously determined Feynman rules, we can write down the relevant S-matrix element in momentum space from inspection of the diagram in figure 8.21:

$$S_{fi} = \frac{e^2}{V^2} \sqrt{\frac{m^2}{E_+ E_- 2\omega_1 2\omega_2}} (2\pi)^4 \delta^4(k_1 + k_2 - p_+ - p_-) \bar{v}(p_+, s_+)$$

$$\cdot \left[(-i\,\not{\varepsilon}_2) \frac{i}{\not{p}_- - \not{k}_1 - m} (-i\,\not{\varepsilon}_1) + (-i\,\not{\varepsilon}_1) \frac{i}{\not{p}_- - \not{k}_2 - m} (-i\,\not{\varepsilon}_2) \right]$$

$$\cdot u(p_-, s_-). \tag{8.240}$$

We notice that the S-matrix element is symmetric under interchange of the two photons, which is required by Bose statistics. Both diagrams must be included in order to ensure this required symmetry.

Compared to Compton scattering (table 8.1 the first two rows), the substitutions

$$\text{Compton} \leftrightarrow \text{Pair Annihilation}$$

$$\varepsilon, k \leftrightarrow \varepsilon_1, -k_1, \tag{8.241}$$

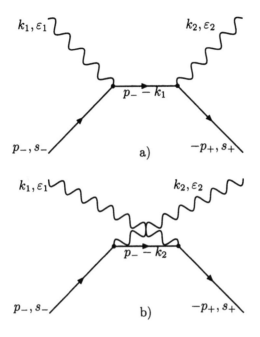

FIGURE 8.21:   Feynman diagrams for pair annihilation into gamma-rays.

$$\varepsilon', k' \leftrightarrow \varepsilon_2, +k_2, \tag{8.242}$$

$$p_i, s_i \leftrightarrow p_-, s_-, \tag{8.243}$$

$$p_f, s_f \leftrightarrow -p_+, s_+, \tag{8.244}$$

transform the two amplitudes into each other. This is an example of a general substitution rule which is valid to arbitrary order and which relates processes of the type $AB \to CD$ to processes of the type $A\bar{C} \to \bar{B}D$, where $\bar{B}$ denotes the antiparticle of B, and $\bar{C}$ the antiparticle of C.

By familiar steps, we proceed from the matrix element to a differential cross section

$$
\begin{aligned}
d\sigma &= \frac{|S_{fi}|^2}{VT} \frac{V}{|\vec{J}_{\text{inc}}|} \frac{Vd^3k_1}{(2\pi)^3} \frac{Vd^3k_2}{(2\pi)^3} \\
&= \frac{e^4}{(2\pi)^2} \int \frac{m^2}{E_+ E_- |\vec{v}_+ - \vec{v}_-|} |\mathcal{M}_{fi}|^2 \frac{d^3k_1}{2\omega_1} \frac{d^3k_2}{2\omega_2} \delta^4(k_1 + k_2 - p_+ - p_-).
\end{aligned}
\tag{8.245}
$$

The two common frames of reference for this process are the rest frame of one of the particle or the center-of-mass frame. The calculation in the center-of-mass frame can be found in problem 8.20. For an electron at rest,

$$d\sigma = \frac{e^4}{(2\pi)^2} \int \frac{m}{E_+\beta_+} |\mathcal{M}_{fi}|^2 \frac{d^3k_1}{2\omega_1} \frac{d^3k_2}{2\omega_2} \delta^4(k_1 + k_2 - p_+ - p_-), \qquad (8.246)$$

where $\beta_+ = p_+/E_+$ is the incident positron velocity. The invariant amplitude is

$$
\begin{aligned}
\mathcal{M}_{fi} &= \bar{v}(p_+, s_+) \left[(-i\,\displaystyle{\not}{\epsilon}_2) \frac{i}{\displaystyle{\not}{p}_- - \displaystyle{\not}{k}_1 - m} (-i\,\displaystyle{\not}{\epsilon}_1) \right. \\
&\qquad \left. + (-i\,\displaystyle{\not}{\epsilon}_1) \frac{i}{\displaystyle{\not}{p}_- - \displaystyle{\not}{k}_2 - m} (-i\,\displaystyle{\not}{\epsilon}_2) \right] u(p_-, s_-) \\
&= i\bar{v}(p_+, s_+) \left[\displaystyle{\not}{\epsilon}_2 \frac{\displaystyle{\not}{p}_- - \displaystyle{\not}{k}_1 + m}{2p_- \cdot k_1} \displaystyle{\not}{\epsilon}_1 + \displaystyle{\not}{\epsilon}_1 \frac{\displaystyle{\not}{p}_- - \displaystyle{\not}{k}_2 + m}{2p_- \cdot k_2} \displaystyle{\not}{\epsilon}_2 \right] u(p_-, s_-) \\
&= -i\bar{v}(p_+, s_+) \left[\frac{\displaystyle{\not}{\epsilon}_2 \displaystyle{\not}{k}_1 \displaystyle{\not}{\epsilon}_1}{2p_- \cdot k_1} + \frac{\displaystyle{\not}{\epsilon}_1 \displaystyle{\not}{k}_2 \displaystyle{\not}{\epsilon}_2}{2p_- \cdot k_2} \right] u(p_-, s_-), \qquad (8.247)
\end{aligned}
$$

where we have used $(\displaystyle{\not}{p}_- + m)\displaystyle{\not}{\epsilon}\,u(p_-, s_-) = \displaystyle{\not}{\epsilon}(-\displaystyle{\not}{p}_- + m)u(p_-, s_-) = 0$ and have chosen the special transverse gauge in which $p_- \cdot \varepsilon_1 = p_- \cdot \varepsilon_2 = 0$.

For an unpolarized positron incident on an unpolarized electron, we average over the initial spins s_- and s_+. Representing the quantity in square brackets in equation 8.247 as Γ, we write

$$
\begin{aligned}
\frac{1}{4} \sum_{s_-, s_+} |\mathcal{M}_{fi}|^2 &= \frac{1}{4} \sum_{s_-, s_+} \bar{v}(p_+, s_+)_\alpha \Gamma_{\alpha\beta} u(p_-, s_-)_\beta \bar{u}(p_-, s_-)_\delta \bar{\Gamma}_{\delta\rho} v(p_+, s_+)_\rho \\
&= \frac{1}{4} \sum_{s_+} \bar{v}(p_+, s_+)_\alpha \Gamma_{\alpha\beta} \left(\frac{\displaystyle{\not}{p}_- + m}{2m} \right)_{\beta\delta} \bar{\Gamma}_{\delta\rho} v(p_+, s_+)_\rho \\
&= -\frac{1}{4} \left(\frac{m - \displaystyle{\not}{p}_+}{2m} \right)_{\rho\alpha} \Gamma_{\alpha\beta} \left(\frac{\displaystyle{\not}{p}_- + m}{2m} \right)_{\beta\delta} \bar{\Gamma}_{\delta\rho}
\end{aligned}
$$

$$
= -\frac{1}{4} \mathrm{Tr} \left[\frac{m - \displaystyle{\not}{p}_+}{2m} \left(\frac{\displaystyle{\not}{\epsilon}_2 \displaystyle{\not}{k}_1 \displaystyle{\not}{\epsilon}_1}{2p_- \cdot k_1} + \frac{\displaystyle{\not}{\epsilon}_1 \displaystyle{\not}{k}_2 \displaystyle{\not}{\epsilon}_2}{2p_- \cdot k_2} \right) \frac{\displaystyle{\not}{p}_- + m}{2m} \left(\frac{\displaystyle{\not}{\epsilon}_1 \displaystyle{\not}{k}_1 \displaystyle{\not}{\epsilon}_2}{2p_- \cdot k_1} + \frac{\displaystyle{\not}{\epsilon}_2 \displaystyle{\not}{k}_2 \displaystyle{\not}{\epsilon}_1}{2p_- \cdot k_2} \right) \right].
$$
$$(8.248)$$

The unpolarized cross section becomes

$$
\begin{aligned}
d\bar{\sigma} = -\frac{e^4}{(2\pi)^2} \int \frac{m}{E_+\beta_+} \frac{1}{4} \\
\cdot \mathrm{Tr} \left[\frac{m - \displaystyle{\not}{p}_+}{2m} \left(\frac{\displaystyle{\not}{\epsilon}_2 \displaystyle{\not}{k}_1 \displaystyle{\not}{\epsilon}_1}{2p_- \cdot k_1} + \frac{\displaystyle{\not}{\epsilon}_1 \displaystyle{\not}{k}_2 \displaystyle{\not}{\epsilon}_2}{2p_- \cdot k_2} \right) \frac{\displaystyle{\not}{p}_- + m}{2m} \left(\frac{\displaystyle{\not}{\epsilon}_1 \displaystyle{\not}{k}_1 \displaystyle{\not}{\epsilon}_2}{2p_- \cdot k_1} + \frac{\displaystyle{\not}{\epsilon}_2 \displaystyle{\not}{k}_2 \displaystyle{\not}{\epsilon}_1}{2p_- \cdot k_2} \right) \right] \\
\cdot \frac{d^3k_1}{2\omega_1} \frac{d^3k_2}{2\omega_2} \delta^4(k_1 + k_2 - p_- - p_+).
\end{aligned}
$$
$$(8.249)$$

Notice the overall minus sign, which comes from our normalization of the positron spinors. The simplified form of the matrix element is due to the choice of transverse gauge $\varepsilon_1 \cdot p_- = \varepsilon_2 \cdot p_- = 0$ and is the same gauge used in the Compton scattering calculation. We thus obtain the trace directly from our previous result using the substitutions above (see problem 8.21):

$$
\begin{aligned}
d\bar{\sigma} &= \frac{e^4}{(2\pi)^2} \int \frac{m}{E_+ \beta_+} \frac{-1}{4} \frac{1}{2m^2} \left[\frac{\omega_2}{-\omega_1} + \frac{-\omega_1}{\omega_2} + 4(\varepsilon_1 \cdot \varepsilon_2)^2 - 2 \right] \frac{d^3 k_1}{2\omega_1} \frac{d^3 k_2}{2\omega_2} \\
&\qquad \cdot \delta^4(k_1 + k_2 - p_- - p_+) \\
&= \frac{\alpha^2}{2m} \int \frac{1}{p_+} \left[\frac{\omega_2}{\omega_1} + \frac{\omega_1}{\omega_2} + 2 - 4(\varepsilon_1 \cdot \varepsilon_2)^2 \right] \frac{d^3 k_1}{2\omega_1} \frac{d^3 k_2}{2\omega_2} \\
&\qquad \cdot \delta^4(k_1 + k_2 - p_- - p_+).
\end{aligned}
\tag{8.250}
$$

It remains to reduce the delta function for laboratory kinematics:

$$
\begin{aligned}
\int \frac{d^3 k_1}{2\omega_1} \frac{d^3 k_2}{2\omega_2} &\delta^4(k_1 + k_2 - p_+ - p_-) \\
&= \int \frac{d^3 k_1}{2\omega_1} \int_{-\infty}^{\infty} d^4 k_2 \delta^4(k_1 + k_2 - p_+ - p_-)\delta(k_2^2)\theta((k_2)_0) \\
&= \int \frac{d^3 k_1}{2\omega_1} \delta[(p_+ + p_- - k_1)^2]\theta(E_+ + E_- - \omega_1) \\
&= \frac{1}{2} \int_0^{\infty} \omega_1 d\omega_1 d\Omega_{k_1} \delta[(p_+ + p_-)^2 - 2k_1 \cdot (p_+ + p_-)]\theta(E_+ + E_- - \omega_1) \\
&= \frac{d\Omega_{k_1}}{2} \int_0^{E_+ + m} \omega_1 d\omega_1 \delta[2m^2 + 2mE_+ - 2\omega_1(m + E_+ - p_+ \cos\theta)].
\end{aligned}
\tag{8.251}
$$

The delta function requires

$$
\omega_1 = \frac{m(m + E_+)}{m + E_+ - p_+ \cos\theta}
\tag{8.252}
$$

and the property of the delta function given in equation 2.32 requires us to calculate

$$
f'(\omega_1) = -2(m + E_+ - p_+ \cos\theta).
\tag{8.253}
$$

We now obtain

$$
\int \frac{d^3 k_1}{2\omega_1} \frac{d^3 k_2}{2\omega_2} \delta^4(k_1 + k_2 - p_+ - p_-) = \frac{1}{4} \frac{m(m + E_+)}{[m + E_+ - p_+ \cos\theta]^2} d\Omega_{k_1}.
\tag{8.254}
$$

The delta functions allow us to write ($z \equiv \cos\theta$, $\beta = p_+/E_+$, $\gamma = E_+/m$)

$$\frac{\omega_1}{m} = \frac{(m + E_+)}{m + E_+ - p_+ \cos\theta} = \frac{1/\gamma + 1}{1/\gamma + 1 - \beta z} = \frac{1 + \gamma}{1 + \gamma(1 - \beta z)}, \quad (8.255)$$

$$\frac{\omega_2}{m} = 1 + \frac{E_+}{m} - \frac{\omega_1}{m} = \frac{(1 + \gamma)\gamma(1 - \beta z)}{1 + \gamma(1 - \beta z)}, \quad (8.256)$$

$$\frac{\omega_1 + \omega_2}{m} = \frac{m + E_+}{m} = 1 + \gamma, \quad (8.257)$$

$$\frac{\omega_2}{\omega_1} = \gamma(1 - \beta z). \quad (8.258)$$

Combining the above expressions gives

$$\frac{d\bar{\sigma}}{d\Omega_{k_1}} = \frac{\alpha^2}{8m^2\beta\gamma}\left(\frac{\omega_1}{m}\right)^2 \frac{m}{\omega_1 + \omega_2}\left[\frac{\omega_1}{\omega_2} + \frac{\omega_2}{\omega_1} + 2 - 4(\varepsilon_1 \cdot \varepsilon_2)^2\right]. \quad (8.259)$$

The angle between the two photons Θ can be found from

$$(k_1 + k_2)^2 = (p_- + p_+)^2,$$
$$2k_1 \cdot k_2 = 2m^2 + 2mE,$$
$$\omega_1\omega_2(1 - \cos\Theta) = m(m + E),$$
$$1 - \cos\Theta = \frac{m(m + E)}{\omega_1\omega_2} = m\frac{\omega_1 + \omega_2}{\omega_1\omega_2} = m\left(\frac{1}{\omega_1} + \frac{1}{\omega_2}\right). \quad (8.260)$$

This leads to

$$\frac{d\bar{\sigma}}{d\Omega_{k_1}} = \frac{\alpha^2}{8m^2\beta\gamma}\left(\frac{\omega_1}{\omega_2}\right)\frac{1}{1 - \cos\Theta}\left[\frac{\omega_1}{\omega_2} + \frac{\omega_2}{\omega_1} + 2 - 4(\varepsilon_1 \cdot \varepsilon_2)^2\right]. \quad (8.261)$$

The sum over the final polarization states in the rest frame gives (cf. equation 8.228)

$$\sum(\varepsilon_1 \cdot \varepsilon_2)^2 = 1 + \cos^2\Theta, \quad (8.262)$$

and a factor of 4 for those terms which are independent of ε_1 and ε_2. We now have

$$\frac{d\bar{\sigma}}{d\Omega_{k_1}} = \frac{\alpha^2}{2m^2\beta\gamma}\left(\frac{\omega_1}{\omega_2}\right)\frac{1}{1 - \cos\Theta}\left[\frac{\omega_1}{\omega_2} + \frac{\omega_2}{\omega_1} + 1 - \cos^2\Theta\right]$$

$$= \frac{\alpha^2}{2m^2\beta\gamma}\left(\frac{\omega_1}{\omega_2}\right)\frac{1}{1 - \cos\Theta}\left[(\omega_1 + \omega_2)\left(\frac{1}{\omega_2} + \frac{1}{\omega_1}\right) - 1 - \cos^2\Theta\right]$$

$$= \frac{\alpha^2}{2m^2\beta\gamma}\left(\frac{\omega_1}{\omega_2}\right)\frac{1}{1 - \cos\Theta}\left[(1 + \gamma)(1 - \cos\Theta) - (1 + \cos^2\Theta)\right]$$

$$= \frac{\alpha^2}{2m^2\beta\gamma}\frac{\omega_1}{\omega_2}\left[1 + \gamma - \frac{1 + \cos^2\Theta}{1 - \cos\Theta}\right]. \quad (8.263)$$

We now express the cross section in terms of the energies of the two photons. Simple kinematics gives

$$\cos \Theta = 1 - \frac{m}{\omega_1} - \frac{m}{\omega_2},$$

$$\cos^2 \Theta = 1 + \left(\frac{m}{\omega_1}\right)^2 + \left(\frac{m}{\omega_2}\right)^2 - 2\frac{m}{\omega_1} - 2\frac{m}{\omega_2} + 2\frac{m}{\omega_1}\frac{m}{\omega_2}, \tag{8.264}$$

$$\frac{1 + \cos^2 \Theta}{1 - \cos \Theta} = \frac{\omega_1 \omega_2}{m(\omega_1 + \omega_2)} \left[2 + \left(\frac{m}{\omega_1}\right)^2 + \left(\frac{m}{\omega_2}\right)^2 - 2\frac{m}{\omega_1} - 2\frac{m}{\omega_2} + 2\frac{m}{\omega_1}\frac{m}{\omega_2}\right]$$

$$= \frac{m}{\omega_1 + \omega_2} \left[2\frac{\omega_1}{m}\frac{\omega_2}{m} + \frac{\omega_2}{\omega_1} + \frac{\omega_1}{\omega_2} - 2\frac{\omega_1}{m} - 2\frac{\omega_2}{m} + 2\right]$$

$$= \frac{1}{1 + \gamma} \left[\frac{2\gamma(1 + \gamma)^2(1 - \beta z)}{[1 + \gamma(1 - \gamma z)]^2} + \gamma(1 - \beta z) + \frac{1}{\gamma(1 - \beta z)}\right]$$

$$\quad -2(1 + \gamma) + 2]$$

$$= \frac{1}{1 + \gamma} \left[\frac{2\gamma(1 + \gamma)^2(1 - \beta z)}{[1 + \gamma(1 - \gamma z)]^2} + \frac{[1 + \gamma(1 - \beta z)]^2}{\gamma(1 - \beta z)}\right]$$

$$\quad -2(1 + \gamma)]. \tag{8.265}$$

The differential cross section in terms of photon's energy becomes

$$\frac{d\bar{\sigma}}{d\Omega_{k_1}} = \frac{\alpha^2}{2m^2\beta\gamma^2(1 - \beta z)} \left[\gamma + 3 - \frac{[1 + \gamma(1 - \beta z)]^2}{\gamma(1 + \gamma)(1 - \beta z)} - \frac{2\gamma(1 + \gamma)(1 - \beta z)}{[1 + \gamma(1 - \beta z)]^2}\right]. \tag{8.266}$$

For the total cross section, we integrate over the solid angle. The final state contains two identical particles (photons). One of the photons emerges in $d\Omega_{k_1}$, and because of their indistinguishability, this can be either of the two photons. If we were to integrate $d\bar{\sigma}/d\Omega_{k_1}$ over the entire 4π solid angle, we would be counting each distinguishable state exactly twice. We therefore take one-half of this integral in forming a total cross section:

$$\bar{\sigma} = \frac{1}{2} \int \frac{d\bar{\sigma}}{d\Omega_{k_1}} d\Omega_{k_1}. \tag{8.267}$$

We now integrate the differential cross section using

$$\int_{-1}^{1} \frac{dz}{[1 + \gamma(1 - \beta z)]^2} = \frac{1}{1 + \gamma}, \tag{8.268}$$

$$\int_{-1}^{1} \frac{dz}{(1 - \beta z)^2} = 2\gamma^2, \tag{8.269}$$

$$\int_{-1}^{1} \frac{dz}{1 - \beta z} = -\frac{1}{\beta} \ln \frac{1 - \beta}{1 + \beta}. \tag{8.270}$$

Using these integrals, we have

$$
\begin{aligned}
\sigma &= \frac{\pi\alpha^2}{2m^2\beta\gamma^2}\left[-\frac{\gamma+3}{\beta}\ln\frac{1-\beta}{1+\beta} - \frac{2\gamma}{1+\gamma} + \frac{2}{\beta(1+\gamma)}\ln\frac{1-\beta}{1+\beta} - \frac{2\gamma}{1+\gamma} - 2\gamma\right] \\
&= \frac{\pi\alpha^2}{2m^2\beta^2\gamma^2(\gamma+1)}\left[(\gamma^2+4\gamma+1)\ln\frac{1+\beta}{1-\beta} - 2\beta\gamma(\gamma+3)\right] \\
&= \frac{\pi\alpha^2}{m^2\beta^2\gamma(\gamma+1)}\left[\left(\gamma+4+\frac{1}{\gamma}\right)\ln\sqrt{\frac{1+\beta}{1-\beta}} - \beta(\gamma+3)\right] \\
&= \frac{\pi\alpha^2}{m^2\beta^2\gamma(\gamma+1)}\left[\left(\gamma+4+\frac{1}{\gamma}\right)\ln(\gamma+\sqrt{\gamma^2-1}) - \beta(\gamma+3)\right]. \quad (8.271)
\end{aligned}
$$

For incident positrons of low energy, $\gamma\to 1$ and $\beta\to 0$ gives

$$
\begin{aligned}
\bar\sigma &\approx \frac{\pi\alpha^2}{2m^2\beta^2}\left[\left(6-\frac{1}{2}\beta^4+\cdots\right)\left(\beta-\frac{1}{4}\beta^4+\cdots\right) - 4\beta - \frac{1}{2}\beta^3\cdots\right] \\
&\approx \frac{\pi\alpha^2}{2m^2\beta^2}\left[2\beta-\frac{1}{2}\beta^3+\cdots\right] \\
&= \frac{\pi\alpha^2}{m^2\beta}[1+\mathcal{O}(\beta)] \quad \text{for} \quad \beta\ll 1. \quad (8.272)
\end{aligned}
$$

This expression is seen to approach infinity as β approaches zero. But the number of annihilations per unit time remains finite, since the current of the incoming positron $e\beta$ approaches zero in this limit.

In the extreme-relativistic limit, $\gamma\gg 1$ and $\beta\to 1$ gives

$$
\begin{aligned}
\bar\sigma &= \frac{\pi\alpha^2}{m^2\gamma^2}[(\gamma+4\cdots+)\ln(\gamma+\gamma(1-\cdots)) - \gamma] \\
&= \frac{\pi\alpha^2}{m^2\gamma}\left[\ln 2\gamma - 1 + \frac{4}{\gamma}\ln 2\gamma + \cdots\right] \\
&= \frac{\pi\alpha^2}{mE_+}\left[\ln\frac{2E_+}{m} - 1 + \mathcal{O}\left(\frac{m}{E_+}\ln\frac{E_+}{m}\right)\right] \quad \text{for} \quad \gamma\gg 1. \quad (8.273)
\end{aligned}
$$

The cross section has a maximum and tends to zero for large energies.

8.10 Electron-Positron Pair Production

An electron-positron pair can be created from a quantum of energy. For the process to occur we must have at least $2mc^2$ of energy. Any excess energy will appear as the kinetic energy of the electron and positron. The energy needed to create an electron-positron pair can be supplied through the absorption of a photon or by the impact of a particle with kinetic energy greater than $2mc^2$.

In the case of photon absorption, energy and momentum can only be conserved simultaneously if another particle is present. The additional particle can be a nucleus, an electron (or positron), and even another photon. The case of a photon interacting with the electromagnetic field of a nucleus is the most common in atomic physics. Electron-positron pairs are created by the passage of gamma-rays or fast particles through matter.

The process of pair production by a photon in the Coulomb field of a nucleus is shown in figure 8.22. It is very closely related to the process of bremsstrahlung (see problem 8.22). A comparison shows that the corresponding diagrams are related by the substitution law.

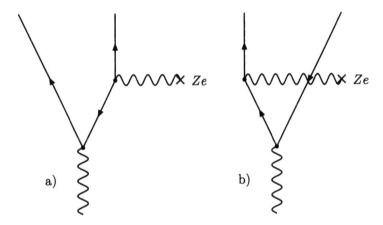

FIGURE 8.22: Pair production in a Coulomb field.

Other methods of pair production are known. Pairs can be produced in photon-electron collisions. The process is like Compton scattering but the emitted photon splits into an electron-positron pair. In addition, a single virtual bremsstrahlung photon can split into an electron-positron pair since its momentum is off mass shell. These are higher-order processes, which involve at least three vertices.

A theoretical process for producing pairs is $\gamma\gamma \to e^+e^-$. The process is exactly the inverse process of electron-positron annihilation to two photons. The cross section for this process is reasonably large (see problem 8.23). However, pair production by two *real* photons has not yet been observed experimentally, since it is difficult to prepare two colliding beams of high-energy photons with high intensity. Experimentally, there are two ways around this. One approach is to use high-intensity laser photons. Another approach is to use particle accelerators by observing the collision of virtual photons which produce electron-positron pairs – the virtual photons are the radiation produced by the colliding beam particles.

8.11   Electron-Electron Scattering

Electron-electron systems are specified by their initial states: two electrons, two positrons, or one electron and one positron. The last case is of special interest, since it permits bound states; these states, however, are not stable. The bound structure which consists of one electron and one positron is called positronium.

As in the photon-electron system, the scattering of two electrons into a final state of exactly two electrons and no photon is only an approximate description of the process. This is so because the scattering of a charged particle through a finite angle will always be accompanied by an undetermined number of very low-energy photons. We obtain correct results only in the approximation in which the possible emission of photons in the final state can be consistently neglected. This occurs only in lowest order and when the scattering angle is not too small. The radiative corrections to this approximation require the knowledge of the energy resolution of the experiment, which determines the minimum energy of possible photons in the final state. The affect of bremsstrahlung in electron-electron collisions is therefore strictly not separable from that of electron-electron scattering.

Electron-electron scattering is like electron-proton scattering but with an important addition. Since the scattering particles are the same type, there is no way to tell which of the two emerging electrons is the incident one and which is the target particle. This ambiguity leads to an additional exchange contribution to the scattering. The two diagrams for electron-electron scattering are shown in figure 8.23. The second diagram arises because of the identity of the electrons. This process is also known as Møller scattering.

The S-matrix element, with spin labels suppressed, is

$$S_{fi} = \frac{-m^2}{V^2 \sqrt{E_1 E_2 E_1' E_2'}} \left[i \frac{\overline{u}(p_1')(ie\gamma_\mu)u(p_1)\overline{u}(p_2')(ie\gamma^\mu)u(p_2)}{(p_1 - p_1')^2} \right.$$
$$\left. -i \frac{i\overline{u}(p_1')(ie\gamma_\mu)u(p_2)\overline{u}(p_2')(ie\gamma^\mu)u(p_1)}{(p_1 - p_2')^2} \right] (2\pi)^4 \delta^4(p_1' + p_2' - p_1 - p_2).$$

$$(8.274)$$

We have written down the amplitude in momentum space directly, using the Feynman rules, since we have encountered all pieces in the diagrams before. Compared to electron-proton scattering (equations 8.92 and 8.93) there is a

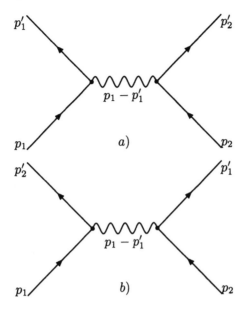

FIGURE 8.23: Feynman diagrams for electron-electron scattering: a) direct and b) exchange.

change in sign because the proton has the opposite charge to the electron. The second, or exchange, term can be neglected for scattering in the forward direction with small momentum transfer $(p_1 - p_2')^2$. In this limit, the scattering reduces to Coulomb scattering.

The relative minus sign between the direct and exchange terms is due to Fermi statistics, which requires the amplitude to be antisymmetric under interchange of the two identical electrons – either the exchange of the final-state electrons, or the initial-state electrons. By a similar argument, the scattering amplitude to or from a state containing two identical bosons must be symmetric under their interchange.

No additional normalization factors, such as $1/\sqrt{2}$ or 2, are introduced when the exchange term is added. The rules for constructing differential cross sections from the S-matrix are not altered by the presence of identical particles in the initial or final states. We must only take care that a factor of $1/2$ is included when integrating the differential cross section over the solid angle, if two identical particles appear in the final state. No special factors appear for identical particles in the initial state, since the incident flux is unchanged.

The differential cross section for the scattering of unpolarized electrons can be constructed:

$$do = \int \frac{|S_{fi}|^2}{VT} \frac{V}{J_{\text{inc}}} \frac{V d^3 p_1'}{(2\pi)^3} \frac{V d^3 p_2'}{(2\pi)^3}$$

$$= \int \frac{1}{(2\pi)^2} \frac{m^4}{|\vec{v}_1 - \vec{v}_2|} \frac{d^3 p_1' d^3 p_2'}{E_1 E_2 E_1' E_2'} |\mathcal{M}_{fi}|^2 \delta^4(p_1' + p_2' - p_1 - p_2).$$

$$(8.275)$$

We now evaluate the invariant matrix element

$$|\mathcal{M}_{fi}|^2 = e^4 \left[\frac{\overline{u}(p_1')\gamma_\mu u(p_1)\overline{u}(p_2')\gamma^\mu u(p_2)\overline{u}(p_2)\gamma^\nu u(p_2')\overline{u}(p_1)\gamma_\nu u(p_1')}{(p_1 - p_1')^4} \right.$$

$$- \frac{\overline{u}(p_1')\gamma_\mu u(p_1)\overline{u}(p_2')\gamma^\mu u(p_2)\overline{u}(p_1)\gamma^\nu u(p_2')\overline{u}(p_2)\gamma_\nu u(p_1')}{(p_1 - p_1')^2 (p_1 - p_2')^2}$$

$$- \frac{\overline{u}(p_1')\gamma_\mu u(p_2)\overline{u}(p_2')\gamma^\mu u(p_1)\overline{u}(p_2)\gamma^\nu u(p_2')\overline{u}(p_1)\gamma_\nu u(p_1')}{(p_1 - p_2')^2 (p_1 - p_1')^2}$$

$$+ \left. \frac{\overline{u}(p_1')\gamma_\mu u(p_2)\overline{u}(p_2')\gamma^\mu u(p_1)\overline{u}(p_1)\gamma^\nu u(p_2')\overline{u}(p_2)\gamma_\nu u(p_1')}{(p_1 - p_2')^4} \right].$$

$$(8.276)$$

We only need to calculate the first two terms since the last two terms can be obtained from the first two terms by the substitution $p_1' \leftrightarrow p_2'$. Averaging over initial spin states and summing over final spin states, we evaluate

$$\sum_{s_1, s_2, s_1', s_2'} \overline{u}(p_1')_\alpha (\gamma_\mu)_{\alpha\beta} u(p_1)_\beta \overline{u}(p_2')_\gamma (\gamma^\mu)_{\gamma\delta} u(p_2)_\delta$$

$$\cdot \overline{u}(p_2)_\epsilon (\gamma^\nu)_{\epsilon\phi} u(p_2')_\phi \overline{u}(p_1)_\rho (\gamma_\nu)_{\rho\theta} u(p_1')_\theta$$

$$= \left(\frac{\not{p}_1' + m}{2m} \right)_{\theta\alpha} (\gamma_\mu)_{\alpha\beta} \left(\frac{\not{p}_1 + m}{2m} \right)_{\beta\rho} (\gamma^\mu)_{\gamma\delta}$$

$$\cdot \left(\frac{\not{p}_2' + m}{2m} \right)_{\phi\gamma} (\gamma^\nu)_{\epsilon\phi} \left(\frac{\not{p}_2 + m}{2m} \right)_{\delta\epsilon} (\gamma_\nu)_{\rho\theta}$$

$$= \text{Tr} \left[\frac{\not{p}_1' + m}{2m} \gamma_\mu \frac{\not{p}_1 + m}{2m} \gamma_\nu \right] \text{Tr} \left[\frac{\not{p}_2 + m}{2m} \gamma^\nu \frac{\not{p}_2' + m}{2m} \gamma^\mu \right] \quad (8.277)$$

and

$$\sum_{s_1, s_2, s_1', s_2'} \overline{u}(p_1')_\alpha (\gamma_\mu)_{\alpha\beta} u(p_1)_\beta \overline{u}(p_2')_\gamma (\gamma^\mu)_{\gamma\delta} u(p_2)_\delta$$

$$\cdot \overline{u}(p_1)_\epsilon (\gamma^\nu)_{\epsilon\phi} u(p_2')_\phi \overline{u}(p_2)_\rho (\gamma_\nu)_{\rho\theta} u(p_1')_\theta$$

$$= \left(\frac{\not{p}_1' + m}{2m} \right)_{\theta\alpha} (\gamma_\mu)_{\alpha\beta} \left(\frac{\not{p}_1 + m}{2m} \right)_{\beta\epsilon} (\gamma^\nu)_{\epsilon\phi}$$

$$\cdot \left(\frac{\not{p}_2 + m}{2m} \right)_{\phi\gamma} (\gamma^\mu)_{\gamma\delta} \left(\frac{\not{p}_2 + m}{2m} \right)_{\delta\rho} (\gamma_\nu)_{\rho\theta}$$

$$= \text{Tr} \left[\frac{\not{p}_1 + m}{2m} \gamma_\mu \frac{\not{p}_1 + m}{2m} \gamma_\nu \frac{\not{p}_2 + m}{2m} \gamma^\mu \frac{\not{p}_2 + m}{2m} \gamma^\nu \right]. \quad (8.278)$$

Two of the traces have been evaluated previously. The third trace can be simplified for relativistic energies $E \gg m$, in which case we can neglect terms proportional to m^2.

$$\text{Tr} \left[\frac{\not{p}_1 + m}{2m} \gamma_\mu \frac{\not{p}_1 + m}{2m} \gamma_\nu \right]$$
$$= \frac{1}{m^2} [(p_1')_\mu (p_1)_\nu + (p_1)_\mu (p_1')_\nu - g_{\mu\nu}(p_1 \cdot p_1' - m^2)], \quad (8.279)$$

$$\text{Tr} \left[\frac{\not{p}_2 + m}{2m} \gamma^\mu \frac{\not{p}_2 + m}{2m} \gamma^\nu \right]$$
$$= \frac{1}{m^2} [(p_2')^\mu (p_2)^\nu + (p_2)^\mu (p_2')^\nu - g^{\mu\nu}(p_2 \cdot p_2' - m^2)], \quad (8.280)$$

$$\text{Tr} \left[\frac{\not{p}_1 + m}{2m} \gamma_\mu \frac{\not{p}_1 + m}{2m} \gamma_\nu \right] \text{Tr} \left[\frac{\not{p}_2 + m}{2m} \gamma^\mu \frac{\not{p}_2 + m}{2m} \gamma^\nu \right]$$
$$= \frac{2}{m^4} [p_1' \cdot p_2' p_1 \cdot p_2 + p_1' \cdot p_2 p_1 \cdot p_2'], \quad (8.281)$$

$$\text{Tr} \left[\frac{\not{p}_1 + m}{2m} \gamma_\mu \frac{\not{p}_1 + m}{2m} \gamma_\nu \frac{\not{p}_2 + m}{2m} \gamma^\mu \frac{\not{p}_2 + m}{2m} \gamma^\nu \right]$$
$$= \frac{1}{16m^4} \text{Tr}[\not{p}_1' \gamma_\mu \not{p}_1 \gamma_\nu \not{p}_2' \gamma^\mu \not{p}_2 \gamma^\nu] + \mathcal{O}(m^2)$$
$$= -\frac{1}{8m^4} \text{Tr}[\not{p}_1' \gamma_\mu \not{p}_1 \not{p}_2' \gamma^\mu \not{p}_2]$$
$$= -\frac{1}{2m^4} p_1 \cdot p_2 \text{Tr}[\not{p}_1' \not{p}_2']$$
$$= -\frac{2}{m^4} p_1 \cdot p_2 p_1' \cdot p_2'. \quad (8.282)$$

We see that traces corresponding to the two interference terms, which are the complex conjugates of each other, are identical since they are real.

In the center-of-mass frame, neglecting terms to order m^2, we define the kinematics according to figure 8.24. We obtain

$$E_1 = E_2 = E_1' = E_2' \equiv E, \quad (8.283)$$
$$|\vec{v}_1| = |\vec{v}_2| \equiv \beta, \quad (8.284)$$
$$|\vec{v}_1 - \vec{v}_2| = 2\beta, \quad (8.285)$$

where E is the center-of-mass energy of each electron and β their velocities. At relativistic energies, the relative velocity approaches twice the velocity of light. This is the relative velocity as observed in the center-of-mass system. The velocity of one electron as seen from the other electron is given by the relativistic velocity-addition formula and can never exceed the speed of light. There is thus no contradiction with special relativity.

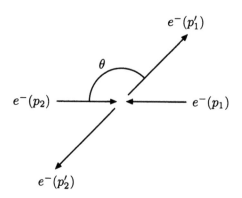

FIGURE 8.24: Definition of the kinematics and scattering angle for $e^-e^- \to e^-e^-$ in the center-of-mass frame.

The dot products of the four-momenta become

$$p_1 \cdot p_2 = p_1' \cdot p_2' \approx 2E^2, \tag{8.286}$$

$$p_1 \cdot p_2' = p_1' \cdot p_2 \approx 2E^2 \cos^2(\theta/2), \tag{8.287}$$

$$p_1 \cdot p_1' = p_2' \cdot p_2 \approx 2E^2 \sin^2(\theta/2), \tag{8.288}$$

$$(p_1' - p_1)^2 \approx -2p_1 \cdot p_1' = -4E^2 \sin^2(\theta/2), \tag{8.289}$$

$$(p_2' - p_1)^2 \approx -2p_1 \cdot p_2' = -4E^2 \cos^2(\theta/2). \tag{8.290}$$

The traces are

$$\mathrm{Tr}\left[\frac{\not{p}_1' + m}{2m}\gamma_\mu \frac{\not{p}_1 + m}{2m}\gamma_\nu\right] \mathrm{Tr}\left[\frac{\not{p}_2' + m}{2m}\gamma^\mu \frac{\not{p}_2 + m}{2m}\gamma^\nu\right]$$

$$= 8\left(\frac{E}{m}\right)^4\left[1 + \cos^4\left(\frac{\theta}{2}\right)\right] \tag{8.291}$$

and

$$\mathrm{Tr}\left[\frac{\not{p}_1' + m}{2m}\gamma_\mu \frac{\not{p}_1 + m}{2m}\gamma_\nu \frac{\not{p}_2' + m}{2m}\gamma^\mu \frac{\not{p}_2 + m}{2m}\gamma^\nu\right] = -8\left(\frac{E}{m}\right)^4. \tag{8.292}$$

The differential cross section and invariant matrix element thus become

$$d\bar{\sigma} = \frac{m^4}{8(2\pi)^2} \int \frac{d^3p_1' d^3p_2'}{\beta E^4} |\mathcal{M}_{fi}|^2 \delta^4(p_1' + p_2' - p_1 - p_2), \qquad (8.293)$$

where

$$|\mathcal{M}_{fi}|^2 = \frac{2(2\pi)^2\alpha^2}{m^4} \left[\frac{1+\cos^4(\theta/2)}{\sin^4(\theta/2)} + \frac{2}{\sin^2(\theta/2)\cos^2(\theta/2)} + \frac{1+\sin^4(\theta/2)}{\cos^4(\theta/2)} \right].$$
$$(8.294)$$

The first and third terms are the square of the matrix elements for the two diagrams, and the second term is the interference contribution.

Using equation 8.102 and integrating the delta function over p_2', we have

$$d\bar{\sigma} = \frac{m^4}{2(2\pi)^2 E^2} \int \frac{d^3p_1'}{2E'} \frac{d^3p_2'}{2E'} |\mathcal{M}_{fi}|^2 \delta^4(p_1' + p_2' - p_1 - p_2)$$

$$= \frac{m^4}{2(2\pi)^2 E^2} \int \frac{d^3p_1'}{2E'} d^4p_2' \delta[(p_2')^2 - m^2]\theta(p_{2_0}') |\mathcal{M}_{fi}|^2 \delta^4(p_1' + p_2' - p_1 - p_2)$$

$$= \frac{m^4}{2(2\pi)^2 E^2} \int \frac{d^3p_1'}{2E'} \delta((p_1 + p_2 - p_1')^2 - m^2)\theta(2E - E') |\mathcal{M}_{fi}|^2. \qquad (8.295)$$

Using $d^3p_1' = |\vec{p}_1'|^2 d|\vec{p}_1'| d\Omega = |\vec{p}_1'| E' dE' d\Omega$, we have

$$\frac{d\bar{\sigma}}{d\Omega} = \frac{m^4}{4(2\pi)^2 E^2} \int_0^{2E} |\vec{p}_1'| dE' \delta[(p_1 + p_2)^2 - 2(p_1 + p_2) \cdot p_1'] |\mathcal{M}_{fi}|^2$$

$$= \frac{m^4}{4(2\pi)^2 E^2} \int_0^{2E} |\vec{p}_1'| dE' \delta[(2E)^2 - 2(2E)E'] |\mathcal{M}_{fi}|^2$$

$$= \frac{m^4}{4(2\pi)^2 E^2} \frac{|\vec{p}_1'|}{|-4E|} |\mathcal{M}_{fi}|^2$$

$$= \frac{m^4}{16(2\pi)^2 E^2} |\mathcal{M}_{fi}|^2,$$

$$\boxed{\frac{d\bar{\sigma}}{d\Omega} = \frac{\alpha^2}{8E^2} \left[\frac{1+\cos^4(\theta/2)}{\sin^4(\theta/2)} + \frac{2}{\sin^2(\theta/2)\cos^2(\theta/2)} + \frac{1+\sin^4(\theta/2)}{\cos^4(\theta/2)} \right].}$$
$$(8.296)$$

We have obtained the high-energy limit of the differential cross section for Møller scattering in the center-of-mass frame.

The cross section can not be integrated over all angles, because the integrals diverge at $\theta = 0$ and π. This divergence is connected with the physical unrealizable requirement that the two electrons scatter without emission of photons. Very low energy photon emission cannot be neglected when the momentum transfers become very small ($\theta \to 0$). However, because the two electrons are indistinguishable, the case $\theta \to \pi$ will lead to the same difficulty.

Now that we have examined the contribution from the separate terms, we can use trigonometric expressions to write the cross section in a simpler form (see problem 8.24):

$$\frac{d\bar{\sigma}}{d\Omega} = \frac{\alpha^2}{4E^2} \frac{(3+\cos^2\theta)^2}{\sin^4\theta}. \tag{8.297}$$

8.12 Electron-Positron Scattering

The Feynman diagrams for the electron-positron scattering process are shown in figure 8.25. This process is also known as Bhabha scattering[6]. The direct scattering diagram (figure 8.25a) for Bhabha scattering is similar to the direct scattering diagram (figure 8.23a) for Møller scattering, with the replacement of an electron by a positron. Comparing the second diagram (figure 8.23b) for Møller scattering with the second diagram (figure 8.25b) for Bhabha scattering reveals a significant difference. The exchange effect in electron-electron scattering has been replaced by an annihilation effect in electron-positron scattering.

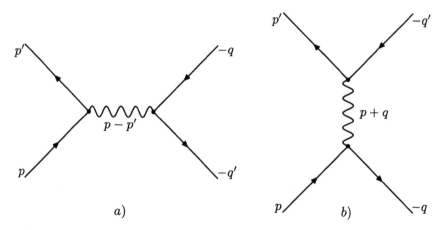

FIGURE 8.25: Feynman diagrams for electron-positron scattering: a) exchange and b) annihilation.

[6]H.J. Bhabha, "The Scattering of Positrons by Electrons with Exchange on Dirac's Theory of the Positron", Proc. Roy. Soc. **154** (1936) 195-206.

Apart from the exchange of spinors, the annihilation amplitude has one qualitative difference compared to the scattering processes we have studied up to now: the virtual photon is time-like, i.e. the square of its four-momentum is greater than zero. This is most easily seen in the center-of-mass frame, where for annihilation $p = (E, \vec{p})$ and $q = (E, -\vec{p})$ gives a four-momentum transfer of $p + q = (2E, 0)$, which is time-like. Whereas for scattering $p = (E, \vec{p})$ and $p' = (E, \vec{p}')$ gives a four-momentum transfer of $p - p' = (0, \vec{p} - \vec{p}')$, which is space-like.

We invoke the substitution rules to obtain the Bhabha cross section from the Møller formula. Using the substitutions in the last two rows of table 8.1

$$p_1 \leftrightarrow p, \tag{8.298}$$

$$p_1' \leftrightarrow p', \tag{8.299}$$

$$p_2 \leftrightarrow -q', \tag{8.300}$$

$$p_2' \leftrightarrow -q, \tag{8.301}$$

and changing of sign in the S-matrix, we find

$$S_{fi} = +\frac{m^2}{V^2} \frac{1}{\sqrt{E_p E_{p'} E_q E_{q'}}} \left[i \frac{\bar{u}(p')(ie\gamma_\mu)u(p)\bar{v}(q)(ie\gamma^\mu)v(q')}{(p - p')^2} \right. $$
$$\left. - i \frac{\bar{u}(p')(ie\gamma_\mu)v(q')\bar{v}(q)(ie\gamma^\mu)u(p)}{(p + q)^2} \right] (2\pi)^4 \delta^4(p' + q' - p - q). \tag{8.302}$$

The overall relative minus sign between Bhabha and Møller scattering comes from the sign difference between the S-matrix for electron and positron scattering. The first term represents direct electron-positron scattering. The second term represents annihilation. When integrating the differential cross section over the solid angle, there will be no factor of $1/2$ as in Møller scattering, because the particles are distinguishable – by their charge.

The relative minus sign between the two terms in Bhabha scattering comes from applying the substitution rules to the S-matrix for Møller scattering. The antisymmetry of Møller scattering under the interchange of the two final-state, or initial-state, electrons becomes in Bhabha scattering, an antisymmetry between an incoming positive-energy electron (p) and an incoming negative-energy electron ($-q'$) running backwards in time, or between outgoing electrons (p') and outgoing positrons ($-q$).

In order to obtain the cross section for electron-positron scattering in the center-of-mass system, we apply the substitution rules and carry out the traces as for Møller scattering. Only the invariant matrix element changes. The four-momentum transfer in the denominator of the terms in the invariant matrix element become, according to figure 8.26,

$$p \cdot q = p' \cdot q' = 4E^2, \tag{8.303}$$

$$p \cdot p' = q \cdot q' = 4E^2 \sin^2(\theta/2), \tag{8.304}$$

$$p \cdot q' = p' \cdot q = 4E^2 \cos^2(\theta/2), \tag{8.305}$$

$$(p - p')^2 \approx -2p \cdot p' = -4E^2 \sin^2(\theta/2), \tag{8.306}$$

$$(p + q)^2 \approx 2p \cdot q = 4E^2. \tag{8.307}$$

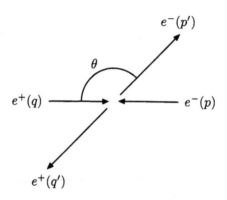

FIGURE 8.26: Definition of the kinematics and scattering angle for $e^+e^- \to e^+e^-$ in the center-of-mass frame.

The traces become

$$\text{Tr}\left[\frac{\not{p}' + m}{2m}\gamma_\mu \frac{\not{p} + m}{2m}\gamma_\nu\right] \text{Tr}\left[\frac{-\not{q} + m}{2m}\gamma^\mu \frac{-\not{q}' + m}{2m}\gamma^\nu\right]$$

$$= \frac{2}{m^4}[p' \cdot q \, p \cdot q' + p' \cdot q' p \cdot q]$$

$$= 8\left(\frac{E}{m}\right)^4 [1 + \cos^4(\theta/2)], \tag{8.308}$$

$$\text{Tr}\left[\frac{-\not{q}' + m}{2m}\gamma_\mu \frac{\not{p} + m}{2m}\gamma_\nu\right] \text{Tr}\left[\frac{\not{p}' + m}{2m}\gamma^\mu \frac{-\not{q}' + m}{2m}\gamma^\nu\right]$$

$$= \frac{2}{m^4}[q \cdot p' p \cdot q' + q \cdot q' p \cdot p']$$

$$= 8\left(\frac{E}{m}\right)^4 [\sin^4(\theta/2) + \cos^4(\theta/2)]$$

$$= 8 \left(\frac{E}{m}\right)^4 \left[\left(\frac{1-\cos\theta}{2}\right)^2 + \left(\frac{1+\cos\theta}{2}\right)^2\right]$$

$$= 4 \left(\frac{E}{m}\right)^4 [1 + \cos^2\theta], \qquad (8.309)$$

and

$$\mathrm{Tr}\left[\frac{\not{p}'+m}{2m}\gamma^\mu\frac{\not{p}+m}{2m}\gamma^\nu\frac{-\not{q}+m}{2m}\gamma_\mu\frac{-\not{q}'+m}{2m}\gamma_\nu\right]$$

$$= -\frac{2}{m^2}p\cdot q'p'\cdot q$$

$$= -8\left(\frac{E}{m}\right)^4\cos^4(\theta/2). \qquad (8.310)$$

Thus the invariant matrix element is

$$|\mathcal{M}|^2_{fi} = \frac{1}{2m^4}\left[\frac{1+\cos^4(\theta/2)}{\sin^4(\theta/2)} - \frac{2\cos^4(\theta/2)}{\sin^2(\theta/2)} + \frac{1+\cos^2\theta}{2}\right]. \qquad (8.311)$$

In the extreme-relativistic limit, the differential cross section becomes

$$\boxed{\frac{d\bar\sigma}{d\Omega} = \frac{\alpha^2}{8E^2}\left[\frac{1+\cos^4(\theta/2)}{\sin^4(\theta/2)} - \frac{2\cos^4(\theta/2)}{\sin^2(\theta/2)} + \frac{1+\cos^2\theta}{2}\right].} \qquad (8.312)$$

We have obtained the high-energy limit of the differential cross section for Bhabha scattering in the center-of-mass frame. We see that the first term in the Møller and Bhabha cross sections are identical. The last term is the exchange term in Møller scattering and it is the annihilation term in Bhabha scattering. The middle term in each formula is the respective interference term, since both exchange and annihilation effects are added in the scattering amplitude, rather than in the scattering probability.

Like Møller scattering, the differential cross section cannot be integrated completely since it diverges as the scattering angle approaches zero. The physical reason for this divergence is as before. There is no difficulty at $\theta = \pi$ for Bhabha scattering, in contradistinction to Møller scattering.

For muon pair creation, $e^+e^- \to \mu^+\mu^-$, only the annihilation diagram can contribute because the initial and final states of particle-antiparticle pairs are of different types. In the extreme-relativistic limit $(E > m_\mu)$ the differential cross section for muon pair creation thus becomes (given by the last term in equation 8.312)

$$\frac{d\bar\sigma}{d\Omega} = \frac{\alpha^2}{16E^2}(1 + \cos^2\theta). \qquad (8.313)$$

Integrating over the solid angle gives the total cross section:

$$\bar{\sigma} = \frac{\pi\alpha^2}{3E^2}. \tag{8.314}$$

This form has been used to analyze the reaction wherein an electron and positron annhilate to a quark-antiquark pair. The fact that the experimental cross section behaves as $(1 + \cos^2\theta)$ supports the fact that quarks are spin-1/2 objects.

Finally, we note that quantum electrodynamics no longer gives a correct description of annihilation processes of the type $e^+e^- \rightarrow \ell^+\ell^-$ (where ℓ refers to any lepton) if the available energy comes close to the mass of the neutral intermediate vector boson Z (91 GeV). This particle can be produced on mass shell as a resonance (width 2.5 GeV) and then it completely dominates over the contribution from the virtual photon.

The theory of electroweak interactions gives a unified description of both contributions. If the theoretically well understood contributions from the weak interaction are introduced, the predictions of quantum electrodynamics compare well with experimental data. From this, we conclude that the electron is a point-like elementary particle. An extended composite object would be described by a momentum-dependent form factor $F(q^2)$. Experiments tell us that for the electron $F(q^2) = F(0)$ up to momentum transfers of several hundreds GeV. This implies that the electron is a point-like object down to a distance of at least $r_e < \hbar c/q = 10^{-16}$ cm. The same conclusion also holds for the heavy leptons μ and τ.

8.13   Beyond Tree Diagrams

So far, we have considered the calculation of scattering amplitudes to lowest order in α. The lowest-order calculations give reasonable results because of the small value of α. We should be able to improve the accuracy of our results by including calculations to higher-order in α. When we do, we find the paradoxical result that the higher-order contributions often diverge at high and low energies. It seems that there is a flaw with our perturbation-expansion approach or there is a serious inconsistency in the theory, as we have developed it.

Feynman diagrams represent terms in a perturbation expansion of physical amplitudes. Terms of a given order all involve the same power of e. In practice it often turns out that the relevant parameter is actually the square of the coupling constant $\alpha = e^2/4\pi$. Equivalently, an expansion in terms of the number of vertices appearing in the diagram can be made, since one power of e is associated with each vertex (see appendix C). By increasing the number of vertices, and thus propagators, that we have used in the lowest-order Feynman diagrams, we can draw more complicated diagrams with the same initial and final states. If in electron-electron scattering, for example, one includes higher-order Feynman diagrams, one obtains correction terms in powers of the fine-structure constant α. Some of these diagrams are of a completely new form, whereas others can be shown to be modifications of the electron propagator, the photon propagator, or the electron-electron-photon vertex.

One example of a diagram with a completely new form is shown for electron-electron scattering in figure 8.27. All higher-order processes beyond lowest-order involve "loops". By applying four-momentum conservation at each vertex one finds there is a free loop-momentum k. The correct procedure for handling this momentum is to integrate over all possible values including $k = 0$ and $k = \infty$. Unfortunately, most of the integrals that arise in the calculation of such loop diagrams are divergent. Depending upon the number of momentum factors at each vertex in the loop and the geometry of the closed loop, the closed-loop integral can diverge at $k = \infty$ (ultraviolet divergence). If one of the internal loop particles has zero mass, then the loop integral can also diverge at $k = 0$ (infrared divergence), but only when one of the other particles at a vertex in the loop is external – on its mass shell. It thus becomes important to be aware of which diagrams give finite contributions and which infinite contributions. It is useful to distinguish between "tree" diagrams – ones with no loops – and "loop" diagrams – ones with one or more loops.

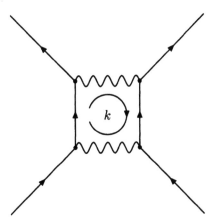

FIGURE 8.27: Two-photon exchange contribution to $2 \to 2$ scattering of order α^2 in the amplitude.

For theories in which the physical coupling constant is small, such as quantum electrodynamics, the tree diagram calculations are usually a good approximation to compare with experiment. Moreover, at the loop level, all the subtleties of handling infinities enter, and so for many purposes it is practical to just perform calculations at the tree level.

In this last section, we investigate the Feynman loop diagrams where lots of interesting physics shows up. The intention is to give the reader a general qualitative impression of the way in which higher-order corrections are calculated and the approach to handling the infinities. Unlike the previous sections of the book, we will not present detailed quantitative calculations.

8.13.1 Types of Divergences and Renormalization

In quantum electrodynamics many loop diagrams diverge not only as the loop momentum goes to infinity (ultraviolet divergence) but also as it goes to zero (infrared divergence). Ultraviolet divergences are generic. Infrared divergences can only arise when there are massless particles in the theory, such as photons, otherwise the particle mass in the propagator will always prevent any infinity at low k.

In a complete and satisfactory theory, we should be able to calculate all radiative processes. This is not possible without adding to the theory we have developed so far. The precise program for manipulating and "taming" these infinities is known as "renormalization" of the theory. It is so called because all the infinities are miraculously swept up into formal expressions for quantities like the physical mass and charge of the particle. If for a particular process one considers all diagrams up to some finite order n, then these are equivalent to the lowest-order diagrams plus some finite radiative corrections, mass and charge renormalization having been applied to eliminate all divergences.

The divergent diagrams are dealt with by various methods. In order to characterize these methods, it is convenient to classify the divergent diagrams into four types as follows:

- Vacuum fluctuations can simply be ignored.

- Certain closed-loop divergences associated with the photon self-energy can be handled with one of two methods. One can invoke the invariance of the theory under gauge transformations. We have already formulated the free fields and the coupling of the fields in a gauge-invariant manner. Consistency of the theory requires that all results also be gauge invariant. A second method is to introduce gauge invariance only for the coupled and renormalized fields. The free photons are treated as neutral vector meson fields of undetermined mass. With this approach the divergencies can be treated in exactly the same way as those of the next type.

- Divergences due to vacuum polarizations, electron self-energy, and vertex modifications can be handled by renormalization. In the first step, we separate the infinite from the finite observable parts of the matrix elements. The second step consists in showing that the infinite parts can all be combined with the two phenomenological constants m and e to which finite values are imposed a posteriori as observed experimental values. These divergences are then eliminated by a redefinition of the mass and charge. This procedure is called mass and charge renormalization.

- Infrared divergences can be tamed by careful consideration of the contribution to the physical cross section of amplitudes involving real emission of very low energy photons, along with infrared divergent virtual photon processes.

Therefore all divergences are removed from the theory and it then becomes possible to calculate radiative processes to any desired accuracy. Some of the aspects mentioned above are now elaborated on.

8.13.2 Radiative Corrections

There are three characteristic subgraphs which can occur at various places in a diagram. It suffices to consider the effects of these subgraphs on the propagator, vertices, and wavefunction once and for all, and then to substitute the results into any Feynman diagram.

It is useful to distinguish between reducible and irreducible diagrams. Diagrams that are really repeated diagrams of lower order, are called reducible. Diagrams that can not be split in two by cutting just one line are called irreducible.

8.13.2.1 Vacuum Fluctuations

A lowest-order diagram representing vacuum fluctuations is shown in figure 8.28. A vacuum diagram is one with no external lines of any kind, and is a disconnected diagram. These vacuum fluctuations take place all the time, independent of whether there are real particles present or not. Lowest-order scattering can take place, and independently in the vacuum a virtual pair can be created and annihilated some time later. The vacuum fluctuations can thus combine with lowest-order processes to give higher-order contributions.

FIGURE 8.28: Lowest-order vacuum diagram.

The vacuum diagrams when considered by themselves contribute only to the diagonal matrix element of the vacuum state. These diagrams do not give rise to any new transitions.

The S-matrix element for a process combined with the vacuum diagrams separates into a product of matrix elements for the connected part, which contains the external lines, and the disconnected vacuum loops. The vacuum loops simply have the effect of multiplying any element of the S-matrix by the same phase factor. Although the phase factor is infinite, this factor cannot lead to any observable effects, since physical effects are always expressed by the absolute value of the matrix elements. This circumstance makes it possible to ignore vacuum fluctuations altogether.

8.13.2.2 Furry's Theorem and the Tadpole Diagram

In the previous subsection, we considered a fermion loop with no external photon lines; in this section we consider a fermion loop with an arbitrary number of external photon lines. Furry's theorem states that a diagram, or part of a diagram, from which only photon lines emerge does not contribute to the matrix element if the number of these photon lines is odd. Figure 8.29 shows a plausible explanation for this result. In a closed loop there can be an electron as well as a positron circling around. These particles interact with the electromagnetic field with their opposite sign for the electric charge. Thus their contributions cancel each other for an odd number of vertices – odd power of $\pm e$. The existence of two contributions having equal absolute value has the additional consequence that for an even number of vertices, the contribution to the amplitude made by a loop diagram is doubled.

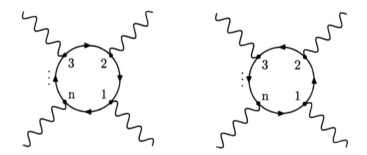

FIGURE 8.29: Two diagrams with opposite directions of the internal fermion loop.

In the case of two vertices, Furry's Theorem is incorrect since both diagrams in figure 8.29 represent exactly the same Feynman diagram, and are not topologically different. For the case of more than two vertices Furry's Theorem does apply since the loops turning left and turning right lead to topologically distinct diagrams which differ in the ordering of the vertices.

Figure 8.30 shows the case of a loop with a single vertex. Again there can be no cancellation in Furry's theorem. This so called "tadpole" diagram does not vanish automatically. The photon line can not refer to a free photon since it cannot simply disappear – violating energy-momentum conservation. The tadpole diagram will contribute in higher orders of perturbation theory where the loop is coupled via a virtual photon to an electron line in some more complicated diagram. This leads to a contribution to the self-energy of the electron. It turns out that the tadpole contribution has no physical observable consequence since its size is independent of the momentum of the electron, in contrast to the self-energy correction. The tadpole contribution can be fully absorbed into the renormalization constant which at the end drops out of any calculation. The same effect can be achieved more economically by simply leaving out any tadpole contributions from the onset.

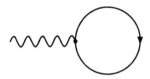

FIGURE 8.30: Tadpole diagram.

8.13.2.3   Fermion Self-Energy

An electron can interact with its own radiation field by emitting and re-absorbing photons. Figure 8.31 shows the lowest-order radiative correction to the spin-1/2 fermion line. This may be an external line, i.e. an incident or emitted fermion, or it may be an internal line, i.e. a fermion propagator. The modification describes a radiative correction that is second order in the coupling constant.

FIGURE 8.31: Fermion self-energy diagram.

These radiative corrections modify the properties of the electron propagator. The loop integral can be expanded into three terms: a term that appears linearly divergent but turns out to be logarithmically divergent, a logarithmically divergent term, and a finite piece. All the divergencies are at high values of the loop momentum (ultraviolet divergent). One of the logarithmically divergent terms causes the mass in the propagator to shift, the other logarithmically divergent term modifies the amplitude of the propagator, and the finite term gives an observable radiative correction of second order. In the case of an external fermion line, the loop integral only modifies the amplitude of the wave function.

The mass of the electron is changed from its free-space value by the presence of interactions with the vacuum; there are no other external sources present. The effective mass of the electron includes all vacuum interactions to all orders. Since the electron's own radiation field modifies the particle's energy, the process is called self-energy.

Since the electron propagator (internal line) is always located between two vertices, one can absorb the factor modifying the propagator amplitude into the electric charge. Similarly, since external lines always connect a vertex, the wave function correction can also be absorbed into the electric charge. Thus in the calculation of matrix elements, the wave function modification and propagator amplitude modification can be combined in a consistent way to represent just an overall modification to the electric charge.

Electron self-energy thus implies that the electron's mass and charge need to be renormalized. The radiative photons modify the propagation properties of the "bare" electron, giving rise to a physical electron. A physical electron is thus a bare electron surrounded by its photon cloud. An electron which originally had the bare mass m, now moves with the physical mass $m_R = m + \delta m$, if one takes into account the interaction with the self-generated electromagnetic field. Since it is completely impossible to switch off this interaction, the quantities m and δm separately do not have any physical significance, just as the bare charge has no physical significance in the case of charge renormalization. All observables, i.e. S-matrix elements, contain the renormalized mass $m_R \approx 0.511$ MeV. Thus we need not worry too much about the fact that δm is a divergent expression.

8.13.2.4 Vacuum Polarization

The lowest-order contribution to the photon self-energy is represented by the diagram in figure 8.32. The photon creates a virtual fermion-antifermion pair, which subsequently recombines to yield a photon once more. The virtual pairs act as dipoles of length $\sim 1/m$ which screen the bare electron charge as shown in figure 8.33. If an external electromagnetic field is present, these virtual pairs are polarized, much in the same way in which a dielectric is polarized by an applied electric field. For this reason, it is called a vacuum polarization effect. The modification describes a radiative correction that is second order in the coupling constant. The correction can be applied to an external photon line or a photon propagator.

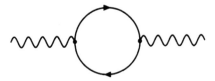

FIGURE 8.32: Vacuum Polarization.

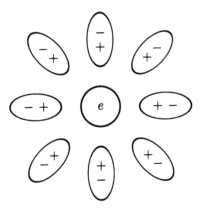

FIGURE 8.33: Virtual e^+e^- pairs acting as dipoles to screen the bare charge of the electron.

The two phenomena of vacuum polarization and self-energy are analogous and essentially describe the interaction of each particle (i.e. the electromagnetic field and the e^+e^- field) with the vacuum fluctuations of the other field.

The photon self-energy would give the same results as the fermion self-energy if the photon had mass, i.e. renormalization of the mass and electric charge, and a modification to the photon propagator to second order. The loop integral causes the propagator to change by a q^2-dependent term and modifies the amplitude of the propagator. The mass of the bare photon must vanish in order that the theory be gauge invariant. This causes us to essentially throw away the mass renormalization on grounds of gauge invariance. There exists alternative, perhaps more comfortable, approaches to handling this problem, for example, Pauli-Villars regularization.

In the case of an external photon line, the propagator corrections vanish because the particle is on-mass shell, but the amplitude of the photon wave function is modified. Again, modifications to external lines can be absorbed into the electric charge at the vertex connecting the external lines to the remainder of the diagram. It is thus sufficient to just drop the contribution of the vacuum polarization in the calculation of a diagram with external photon lines.

The loop in the propagator affects the attraction between the charges which it connects. A particle with the bare charge e for a distant observer seems to carry the renormalized charge e_R given by

$$e_R \equiv e\left(1 - \frac{\alpha}{3\pi} \ln \frac{\Lambda^2}{m_R^2}\right)^{1/2}, \qquad (8.315)$$

where Λ^2 is a cut-off to replace ∞ as the upper limit of integration. The bare charge e is the charge appearing in the lowest-order Feynman diagrams. The physical charge e_R is the charge measured in any long-range Coulomb experiment, and is the electric charge listed in tables of constants, that is, $e_R^2/4\pi = 1/137$. We say that the infinity associated with the cut-off $\Lambda \to \infty$ has been absorbed into e_R.

8.13.2.5 Vertex Correction

Just as a loop in the propagator effects the attraction between the electric charges which it connects, we can anticipate that a loop around a vertex will modify the structure of the electron current. Figure 8.34 shows the lowest-order radiative correction to the vertex. The modification describes a radiative correction that is second order in the coupling constant, and has the same structure as $\bar{u}\gamma^\mu A_\mu u$. It is called a vertex correction.

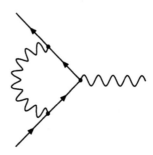

FIGURE 8.34: Vertex correction diagram.

The amplitude for the diagram, without the external lines is logarithmically divergent. The loop integral can be expanded into two terms, one which is logarithmically divergent and another which is finite. The divergent term modifies the electric charge. The finite term gives second-order radiative corrections.

This charge renormalization must be combined with the charge renormalization originating from the electron and photon self-energy effects. The divergent integral from the vertex correction exactly cancels the divergent integral from the self-energy. Thus the charge renormalization stems entirely from the vacuum polarization, not from the fermion self-energy or vertex corrections.

The result is truly general, not only to second order, and it also holds for spin-0 particles. This has the important consequence that the charge renormalization depends exclusively on the self-energy effects of the photon, i.e. on the vacuum polarization modifications inserted into photon lines. Hence the charge renormalization is the same for all elementary particles. It follows that the experimentally observed equality of the charges of the elementary particles implies the equality of their bare charges.

A computation of the finite piece of the diagram gives the generalized vertex function

$$\gamma_\mu F_1(q^2) + \frac{i}{2m_R}\sigma_{\mu\nu}q^\nu F_2(q),\qquad(8.316)$$

which we had postulated earlier (see problems 8.12 and 8.15) by general considerations of the interaction of a photon with a spin-1/2 particle. The functions $F_1(q^2)$ and $F_2(q^2)$ are called form factors. The electron acquires an apparent internal structure due to its interaction with the virtual radiation field and thus differs in its behavior from a *pure* Dirac particle.

For most cases of interest $-q^2 \to 0$,

$$F_2(0) = \frac{\alpha}{2\pi}\qquad(8.317)$$

and

$$F_1(q^2) \approx 1 + \frac{\alpha}{3\pi}\frac{q^2}{m_R^2}\left(\ln\frac{m_R}{\mu} - \frac{3}{8}\right).\qquad(8.318)$$

Neither ultraviolet nor infrared divergencies affect $F_2(0)$ to lowest order. This result was first obtained by J. Schwinger[7].

$F_1(0)$ shows a new feature arising since the loop also diverges for small (infrared) loop momenta. This problem is sidestepped by giving the photon a small fictitious mass μ. The prescription for rendering the integrals finite by introducing fictitious massive particles is known as Pauli-Villars regularization. The fictitious photon has no physical significance, and this method is only one of many for defining the divergent integrals. We must hope that the new parameter μ will not appear in our final result for the observable cross section. In this case, the divergence disappears when the contributions from the bremsstrahlung emission of soft photons is included in the cross section.

8.13.2.6 Infrared Catastrophe

The result for the form factor $F_1(q^2)$ of the electron in equation 8.318 is incomplete. The expression is infrared divergent as the fictitious photon mass

---

[7]J. Schwinger, "On Quantum-Electrodynamics and the Magnetic Moment of the Electron", Phys. Rev. **73** (1948) 416.

μ tends to zero. This divergence cancels when bremsstrahlung emission of soft photons to the same order in perturbation theory as the radiative corrections are included. The result is independent of the photon mass μ, which had been introduced only as means of computation.

To describe a scattering process to higher order, it is not sufficient to sum up only the diagrams for the radiative corrections from the fermion self-energy, vacuum polarization, and vertex correction. In addition to the purely elastic scattering there is also the possibility of bremsstrahlung radiation. Due to the finite energy resolution δE of the measurement apparatus, the emission of a real photon with $\omega < \delta E$ cannot be detected. Purely elastic scattering and scattering in which soft photons are emitted are indistinguishable, and consequently their cross sections must be added. The scattering amplitudes of the diagrams corresponding to elastic scattering are summed coherently; in addition, the amplitudes for the corresponding bremsstrahlung emission process are summed coherently. Both contributions must then be added incoherently because the quantum mechanical final states are different. The sum must then be integrated over all photon energies ω from the lower bound μ up to the value δE.

Including infrared real photons, the form factor $F_1(q^2)$ in equation 8.318 to lowest order in α becomes

$$F_1(q^2) = 1 + \frac{\alpha}{3\pi}\frac{q^2}{m_R^2}\left(\ln\frac{m_R}{2\delta E} + \frac{5}{6} - \frac{3}{8} - \frac{1}{5}\right). \qquad (8.319)$$

This is the infrared-corrected form factor. It is independent of the photon mass μ. The energy δE is determined by the experimental apparatus and the result depends on this energy resolution.

We have seen that in a careful analysis, the problem of the infrared catastrophe turns out to be fictitious. It arises if one does not account for the fact that the electron is always surrounded by a cloud of soft photons, which can be virtual as well as real (soft bremsstrahlung). Inconsistent results are obtained if one tries to separate the electron from its radiation cloud, for instance, by insisting on a final state without photons. It can be shown, that in every scattering process of charged particles an arbitrary number of soft photons are emitted. Thus pure elastic scattering does not exist in a theory of massless particles. In higher orders of perturbation theory, the infrared catastrophe can always be removed by combing internal and external radiative corrections.

8.13.3   Examples of Measurable Radiative Corrections

In spite of the difficulties with infinities, there is not one single experimental fact known today concerning radiative processes, which can not be quantitatively explained by the theory. The following are three examples of which the effects of radiative corrections are measurable. The effects are not only measurable, but for two of the cases the measurements agree with theory to unprecedented accuracy. With such good agreement it is very easy to sweep aside all the pathologies mentioned above and think of quantum electrodynamics as a very practical theory for performing calculations. The last example leads us to question the concepts of a fundamental constant: the electric charge.

8.13.3.1 Lamb Shift

According to the Dirac theory of the hydrogen atom, the $2s_{1/2}$ and $2p_{1/2}$ levels are degenerate. The effect of radiative corrections is to cause a shift in the energy levels, by different amounts for different states, leading to a splitting of the $2s_{1/2}$-$2p_{1/2}$ levels, as shown in figure 8.35. It is this phenomenon which is known as the Lamb shift of the hydrogen atom.

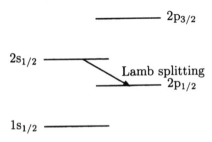

FIGURE 8.35: Lamb shift splitting.

High precision measurements, by Lamb and collaborators, of the Lamb shift in hydrogen gave the value[8]

$$\frac{\Delta E}{2\pi\hbar} \equiv \frac{\Delta E(2s_{1/2}) - \Delta E(2p_{1/2})}{2\pi\hbar} = (1057.77 \pm 0.10) \text{ MHz.} \qquad (8.320)$$

The calculation of the second-order radiative corrections due to the electron self-energy gives a value of 1011.45 MHz for $\Delta E/2\pi\hbar$, showing that the Lamb shift is mainly due to electron self-energy effects. The vacuum polarization and the vertex correction (anomalous magnetic moment) diagrams also contribute: -27.13 MHz and 67.82 MHz, respectively (see Jauch & Rohrlich [15]). Combining these results one obtains to second order in perturbation theory

$$\left(\frac{\Delta E}{2\pi\hbar}\right)_{\text{2nd order}} = 1052.14 \pm 0.08 \text{ MHz.} \qquad (8.321)$$

[8]S. Triebwasser, E.S. Dayhoff, W.E. Lamb, "Fine Structure of the Hydrogen Atom. V", Phys. Rev. **89** (1953) 98-106.

The agreement between the experimental and theoretical values is sufficiently good to give one considerable confidence in the renormalization theory outlined above. The remaining discrepancy can be removed if one takes into account fourth-order radiative corrections, small corrections due to the finite size and mass of the proton, etc. In this way, one obtains for the improved theoretical value of the Lamb shift[9]

$$\left(\frac{\Delta E}{2\pi\hbar}\right)_{\text{4th order}} = (1057.911 \pm 0.012)\ \text{MHz}. \tag{8.322}$$

8.13.3.2 Anomalous Magnetic Moment of the Electron

According to the Dirac theory, the electron possesses a magnetic moment. The electron's magnetic momentum is augmented by an anomalous contribution due to second-order, and higher, radiative corrections.

The magnetic moment of an electron is

$$\vec{\mu} = -\frac{e_R}{2m_R}\left(1 + \frac{\alpha}{2\pi}\right)2\frac{\vec{\sigma}}{2} = -g\frac{e_R}{2m_R}\vec{s}, \tag{8.323}$$

where $\vec{\sigma}$ is a Pauli matrix and $\vec{s} = \vec{\sigma}/2$. The contribution proportional to the fine-structure constant is identified with the anomalous magnetic moment of the electron. The modification to the g-factor is[10]

$$\frac{g-2}{2} = \frac{\alpha}{2\pi} = 0.00116. \tag{8.324}$$

When one includes corrections of order α^2, α^3, etc. which arise from higher-order diagrams, one finds the value, to eighth order,[11]

$$\frac{g-2}{2} = 0.5\left(\frac{\alpha}{\pi}\right) - 0.328478966\left(\frac{\alpha}{\pi}\right)^2 + (1176.11 \pm 42)\times 10^{-3}\left(\frac{\alpha}{\pi}\right)^3 + \cdots$$
$$= (1\ 159\ 652\ 140 \pm 28)\times 10^{-12}. \tag{8.325}$$

The result is in excellent agreement with the experimental value[12]

$$\left(\frac{g-2}{2}\right)_{\text{exp}} = (1\ 159\ 652\ 186.9 \pm 4.1)\times 10^{-12}. \tag{8.326}$$

---

[9]A. Peterman, "A New Value for the Lamb Shift", Phys. Lett. **38B** (1992) 330-332.

[10]J. Schwinger, "On Quantum-Electrodynamics and the Magnetic Moment of the Electron", Phys. Rev. **73** (1947) 416.

[11]T. Kinoshita & W.B. Lindquist, "Eight-order magnetic moment of the electron. V. Diagrams containing no vacuum-polarization loop", Phys. Rev. D **42** (1990) 636-655.

[12]P.J. Mohr & B.N. Taylor, "CODATA recommended values of the fundamental physical constants: 1998", Rev. Mod. Phys. **72** (2) (2000) 351-495.

We thus see that the agreement between experiment and theory is good to one part in 10^{12}.

The increase in the magnetic moment of the electron can be understood qualitatively as follows. The electron is continually emitting and reabsorbing photons and is thus surrounded by a cloud of photons. Thus, a certain amount of the electron's energy, and therefore mass, resides with these photons. Hence, the charge-to-mass ratio of the electron is effectively increased and this reveals itself in a measurement of the magnetic moment.

8.13.3.3 Running Coupling Constant

At large $q^2 = -Q^2$, we must include the q-dependent term in the vacuum polarization that gave equation 8.315. Including loops to all orders and defining $\alpha(Q^2) = e_R^2(Q^2)/4\pi$, we have

$$\alpha(Q^2) \equiv \frac{\alpha_0}{1 - \frac{\alpha_0}{3\pi}\ln\frac{Q^2}{\Lambda^2}} \tag{8.327}$$

for large Q^2.

To eliminate the explicit dependence of $\alpha(Q^2)$ on the cutoff Λ, we choose a renormalization momentum μ. The renormalization procedure is then to subtract $\alpha(\mu^2)$ from $\alpha(Q^2)$ to get

$$\alpha(Q^2) = \frac{\alpha(\mu^2)}{1 - \frac{\alpha(\mu^2)}{3\pi}\ln\left(\frac{Q^2}{\mu^2}\right)}. \tag{8.328}$$

The electric charge the experimenter measures depends on the Q^2 of the experiment and is referred to as the running coupling constant. The running coupling constant $\alpha(Q^2)$ describes how the effective charge depends on the separation of the two charged particles. By inserting numerical values, we find that for all practically attainable Q^2, the variation of α with Q^2 is extremely small; α increases from $1/137$ very slowly as Q^2 increases.

At high Q^2 not only can electron-positron pairs be produced in the vacuum but also other particles, such as, the heavy leptons and quarks. Including both the leptonic and hadronic (quark) contributions we have

$$\alpha(Q^2 = (50 \text{ GeV})^2) \approx \frac{1}{137} \cdot \frac{1}{0.94} \approx \frac{1}{129}. \tag{8.329}$$

We can understand the slowly varying fine structure constant as follows. In the vicinity of a test charge in the vacuum, charged pairs can be created. Pairs of particles of mass m can exist for a time of the order $\Delta t \sim \hbar/mc^2$. They can spread apart a distance of order $c\Delta t$ in this time, i.e. a distance of approximately $\hbar/mc$, which is the Compton wavelength. The polarized virtual pairs provide a vacuum screening effect around the original charged particle. The bare vacuum corresponds to no virtual pairs, while the physical vacuum contains virtual pairs. We cannot get outside the physical vacuum, so that we are really always dealing with effective charges that depend on r. The familiar e is the effective charge as $r \to \infty$ or $Q^2 \to 0$; or, in practice, the charge relevant for distances much larger than the particles' Compton wavelength. When q^2 moves to large space-like values, such that Q^2 is much greater than m^2, i.e. to distances well within the cloud, the fine structure constant has a Q^2 dependence.

This ends our phenomenological description of radiative corrections. Rather than applying the wavefunction approach to calculate radiative corrections, the reader is probably best served by moving on at this point to studying quantum field theory, armed with the practical calculational abilities provided by this book.

8.14 Problems

1. [12] In the center-of-mass frame for the process AB → CD, show that the phase-space element is

$$dQ = \frac{1}{4\pi^2} \frac{p_f}{4\sqrt{s}} d\Omega,$$

the flux is

$$F = 4p_i\sqrt{s},$$

and hence that the differential cross section is

$$\left.\frac{d\sigma}{d\Omega}\right|_{CM} = \frac{1}{64\pi^2 s} \frac{p_f}{p_i} |\mathcal{M}|^2,$$

where $d\Omega$ is the element of solid angle about $\vec{p}_C$, $s = (E_A + E_B)^2$, $|\vec{p}_A| = |\vec{p}_B| = p_i$, and $|\vec{p}_C| = |\vec{p}_D| = p_f$.

2. [4] Show that the plane-wave solutions, normalized as

$$\psi_i(x) = \sqrt{\frac{m}{E_i V}} u(p_i, s_i) e^{-ip_i \cdot x} \quad \text{and} \quad \overline{\psi}_f(x) = \sqrt{\frac{m}{E_f V}} \overline{u}(p_f, s_f) e^{ip_f \cdot x},$$

have the desired Lorentz transformation properties. In particular, include the effect of a Lorentz transformation on the box volume V to show that $\overline{\psi}(x)\psi(x)$ is a scalar and that $\psi^\dagger(x)\psi(x)$ is the time component of a four-vector, as desired.

3. Show that equation 8.37 reduces to the Rutherford scattering formula in the nonrelativistic limit.

4. Obtain an expression for the spin matrix element in equation 8.37 by using free-particle spinors.

5. Calculate the Coulomb scattering of a spin-0 particle of charge $-e$ from a static external field of charge $+Ze$.

6. Calculate the differential scattering cross section $\frac{d\sigma}{d\Omega}(\lambda_f, \lambda_i)$ for Coulomb scattering of electrons with longitudinal polarization.

7. Prove the following:

   (a) $\text{Tr}[\gamma_5 \slashed{A} \slashed{B}] = 0$,

   (b) $\text{Tr}[\gamma_5 \slashed{A} \slashed{B} \slashed{C} \slashed{D}] = 4i\epsilon_{\alpha\beta\gamma\delta}a^\alpha b^\beta c^\gamma d^\delta$,

   (c) $\gamma_\mu \slashed{A} \slashed{B} \slashed{C} \gamma^\mu = -2 \slashed{C} \slashed{B} \slashed{A}$,

   (d) $\gamma_\mu \slashed{A} \slashed{B} \slashed{C} \slashed{D} \gamma^\mu = 2[\slashed{D} \slashed{A} \slashed{B} \slashed{C} + \slashed{C} \slashed{B} \slashed{A} \slashed{D}]$,

   (e) $\overline{\slashed{A} \slashed{B} \slashed{C} \cdots \slashed{P}} = \slashed{P} \cdots \slashed{C} \slashed{B} \slashed{A}$.

8. Compute

   (a) $\gamma_\mu \gamma^\mu$, $\gamma_\mu \gamma_\alpha \gamma^\mu$, and $\gamma_\mu \gamma_\alpha \gamma_\beta \gamma^\mu$.

   (b) Simplify $\slashed{p} \slashed{p}$, $\slashed{p} \gamma^\mu \slashed{p}$, and $\slashed{p} \gamma^\mu \gamma^\nu \slashed{p}$.

9. Describe the substitution rules required to calculate the following:

   (a) Calculate electron-positron annihilation and electron-positron production, given electron-photon scattering.

   (b) Calculate electron-photon scattering and electron-positron annihilation, given electron-positron production.

   (c) Calculate electron-photon scattering and electron-positron production, given electron-positron annihilation.

10. Show that the amplitude given by equation 8.93 is proportional to

$$j^\mu(e)\frac{g_{\mu\nu}}{q^2}j^\nu(P).$$

Using the methods at the end of section 7.4, show that the above expression can be written as the exchange of a transversely polarized photon and an instantaneous coulomb interaction between charge densities.

11. Show equation 8.108.

12. [7] The realistic description of the scattering of an electron from a spin-1/2 hadron has to take into account the internal structure and anomalous magnetic moment of the hadron. To that end, one replaces the transition current in momentum space originating from the Dirac equation, with the more general bilinear expression

$$\bar{u}(P')\gamma_\mu u(P) \rightarrow \bar{u}(P')\Gamma_\mu(P', P)u(P).$$

(a) Show that the most general expression for a transition current that fulfills the conditions of Lorentz covariance, hermiticity, and gauge invariance can be written as

$$\bar{u}(P')\Gamma_\mu(P', P)u(P) = \bar{u}(P')\left[\gamma_\mu F_1(q^2) + \frac{i}{2M}F_2(q^2)q^\nu \sigma_{\mu\nu}\right]u(P).$$

Here $q = P' - P$ is the four-momentum transfer, and $F_1(q^2)$ and $F_2(q^2)$ are unspecified real functions called form factors.

(b) What is the physical meaning of $F_1(0)$ and $F_2(0)$? This can be deduced by studying the energy of the interaction with static electromagnetic fields in the nonrelativistic limit.

(c) Calculate the unpolarized cross section for electron scattering from a hadron with the above vertex function in the extreme relativistic limit.

13. Calculate the electron-proton scattering cross section in the center-of-mass reference frame. Calculate it in a frame in which the proton has energy 820 GeV and the electron has energy 30 GeV. How does the cross section change for positron-proton scattering?

14. Show that the cross section for elastic scattering of unpolarized electrons from spinless point-like particles at rest is

$$\frac{d\bar{\sigma}}{d\Omega} = \frac{\alpha^2}{4E^2 \sin^4 \frac{\theta}{2}} \frac{E'}{E} \cos^2 \frac{\theta}{2},$$

where we have neglected the mass of the electron.

15. [14] We derived the laboratory frame cross section for scattering of a high energy electron from an "ideal structureless" proton. However, the proton matrix element of the electromagnetic current was assumed to have a naïve form. In reality one should use the current matrix element

$$\langle P_f | J_\mu^{\text{em}} | P_i \rangle = \bar{u}_f(P_f)\left[\gamma_\mu F_1(q^2) - i\frac{\kappa}{2M}\sigma_{\mu\nu}q^\nu F_2(q^2)\right]u_i(P_i),$$

where $q = P_i - P_f$ is the four-momentum transfer, $\kappa = 1.79$ is the anomalous magnetic moment of the proton, and $F_1(q^2)$ and $F_2(q^2)$ are form factors which account for the finite hadronic size.

(a) Calculate the high-energy laboratory cross section for electron-proton scattering using the full current matrix element and show that

$$\frac{d\sigma}{d\Omega} = \frac{\alpha^2 \cos^2\theta/2}{4E_i^2 \sin^4\theta/2} \frac{1}{\left[1 + \frac{2E_i \sin^2\theta/2}{M}\right]} \left\{ |F_1(q^2)|^2 \right.$$
$$\left. + \frac{q^2}{4M^2}\left[2|F_1(q^2) + \kappa F_2(q^2)|^2 \tan^2\frac{\theta}{2} + \kappa^2|F_2(q^2)|^2\right]\right\},$$

where

$$q^2 = \frac{4E_i^2 \sin^2\theta/2}{1 + \frac{2E_i \sin^2\theta/2}{M}}.$$

(b) Verify that this expression reduces to that given earlier in the limit that $\kappa \to 0$ and $F_1(q^2) \to 1$.

16. [4] Construct the amplitude for bremsstrahlung in electron-proton scattering and show that the static limit reduces to

$$S_{fi} = \frac{-Ze^3}{V^{3/2}} 2\pi\delta(E_f + k - E_i)\frac{1}{\sqrt{2\omega}}\sqrt{\frac{m^2}{E_f E_i}}\frac{1}{|\vec{q}|^2}\bar{u}(p_f, s_f)$$
$$\left[(-i\,\not{\epsilon})\frac{i}{\not{p}_f + \not{k} - m}(-i\gamma_0) + (-i\gamma_0)\frac{i}{\not{p}_i - \not{k} - m}(-i\,\not{\epsilon})\right]u(p_i, s_i)$$
$$(8.330)$$

for bremsstrahlung in a Coulomb field. Show that there is the same correspondence in factors between these two cases as was found in

$$S_{fi} = iZe^2\frac{1}{V}\sqrt{\frac{m^2}{E_f E_i}}\frac{\bar{u}(p_f, s_f)\gamma^0 u(p_i, s_i)}{|\vec{q}|^2}2\pi\delta(E_f - E_i)$$

and

$$S_{fi} = \frac{-ie^2}{V^2}(2\pi)^4\delta^4(P_f - P_i + p_f - p_i)\sqrt{\frac{m^2}{E_f E_i}}\sqrt{\frac{M^2}{E_f' E_i'}}$$
$$\cdot [\bar{u}(p_f, s_f)\gamma_\mu u(p_i, s_i)]\frac{1}{(p_f - p_i)^2 + i\epsilon}[\bar{u}(P_f, S_f)\gamma^\mu u(P_i, S_i)]$$

for elastic scattering.

17. Derive the Bethe-Heitler cross section for bremsstrahlung of photons of arbitrary energy.

18. Using the matrix element in equation 8.150 show that $k_\mu \mathcal{M}_{fi}^\mu(k) = 0$.

19. Calculate the cross section for electron-positron pair annihilation to a single photon in the presence of the Coulomb field of a nucleus.

20. Derive the differential and total unpolarized cross section for pair annihilation $e^+e^- \to \gamma\gamma$ in the center-of-mass reference frame.

21. Show that equation 8.250 follows from previous results by using substitutions.

22. Calculate the cross section for electron-positron pair creation by an incoming photon in the field of a heavy nucleus with charge $+Ze$. *Hint:* the calculation can be considerably simplified by exploiting crossing symmetry which relates pair creation and bremsstrahlung.

    Show that the amplitude for this process is related to the bremsstrahlung amplitude in equation 8.330 by substitution rules. Write down the differential cross section as a five-fold differential in terms of the positron energy, and the solid angles of the electron and positron. Also write down the average squared Lorentz invariant matrix element in terms of a single trace. Do not evaluate the trace.

23. Derive the total unpolarized cross section for creation of an electron-positron pair by two colliding photons, $\gamma\gamma \to e^+e^-$. Express the result in terms of the velocity of the produced particles in the center-of-mass reference frame. *Hint:* use the result for the pair annihilation cross section.

24. Show that equation 8.297 follows from equation 8.296.

25. [4] Construct the differential cross section for electron-electron scattering in the lowest order Born approximation in terms of laboratory energies and scattering angles.

26. Repeat the Møller scattering cross section calculation in the rest frame of one of the electrons.

27. Evaluate the unpolarized cross section for Møller scattering in the non-relativistic limit.

28. [12] Using $e^+e^- \to e^+e^-$ in the s-channel process, verify that

$$s = 4(k^2 + m^2),$$
$$t = -2k^2(1 - \cos\theta),$$
$$u = -2k^2(1 + \cos\theta),$$

where θ is the center-of-mass scattering angle and $k = |\vec{k}_i| = |\vec{k}_j|$, where $\vec{k}_i$ and $\vec{k}_j$ are the momenta of the incident and scattered electrons in the center-of-mass frame. Show that the process is physically allowed provided $s \geq 4m^2$, $t \leq 0$, and $u \leq 0$. Note that $t = 0$ ($u = 0$) corresponds to forward (backwards) scattering.

29. [7]

(a) Show that the kinematics of any binary scattering process $ab \to cd$ can be expressed in terms of the three Lorentz-invariant Mandelstam variables.

$$s = (p_a + p_b)^2 = (p_c + p_d)^2,$$
$$t = (p_c - p_a)^2 = (p_d - p_b)^2,$$
$$u = (p_c - p_b)^2 = (p_d - p_a)^2.$$

Prove the identity

$$s + t + u = m_a^2 + m_b^2 + m_c^2 + m_d^2.$$

(b) Derive the differential cross section for electron-electron scattering and electron-positron scattering in terms of the Mandelstam variables. Do not neglect the electron mass in this calculation.

(c) Write down the explicit results for the Møller and Bhabha cross sections in the center-of-mass system and in the laboratory system.

30. [4] Calculate the cross section for the absorption of light by a bound electron in an atom with low atomic number Z, such that $Z\alpha = Z/137 \ll 1$ and $E_{\text{binding}} \ll mc^2$. Assume also that the frequency of the light is such that $\hbar\omega \gg E_{\text{binding}}$. Making these simplifying assumptions calculate the differential and total cross sections for the two limiting cases:

(a) $E_{\text{binding}} \ll \hbar\omega \ll mc^2$ nonrelativistic ,

(b) $\hbar\omega \gg mc^2$ extremely relativistic.

31. [27] The formula for electron scattering from a Dirac proton can be modified to represent ionization energy loss of charged particles in matter, dE/dx. Energetic protons traversing a medium lose energy $Q = E_p - E_p' = E_e' - m$ in collisions with atomic electrons,

$$p e^- (\vec{p}_e \approx 0) \to p e^-. \tag{8.331}$$

Show that a change of variables converts $d\sigma/d\Omega$ to

$$\frac{d\sigma}{dQ} = \frac{2\pi\alpha^2}{m\beta^2} \frac{1}{Q^2} \left(1 - \beta^2 \frac{Q}{Q_{\max}} + \frac{1}{2}\frac{Q^2}{E_{\mathrm{p}}^2}\right), \qquad (8.332)$$

where $Q_{\max} \approx 2m\gamma^2\beta^2$ for incident protons of velocity $\beta \ll 1$, and where the last term is absent for a beam of spinless particles. The mean free path is computed via the exponential dampening law $\exp(-n\sigma x)$, where n is the number density formula. Show that the ideal ionization loss is

$$\frac{dE}{dx} = n\int_{Q_{\min}}^{Q_{\max}} Q d\sigma \approx \frac{2\pi\alpha^2}{m\beta^2} n \left(\ln \frac{Q_{\max}}{Q_{\min}} - \beta^2\right), \qquad (8.333)$$

for $Q^2 \ll E_{\mathrm{p}}^2$. Realistic modifications of the above formula lead to the Bethe-Bloch formula.

32. To show your understanding of QED, develop spinless QED. Calculate

   (a) Compton scattering of bosons and

   (b) electro-production of pion pairs.

Part IV

Appendices

Appendix A

Lorentz-Invariant Flux Factor

The Lorentz-invariant flux factor is defined as

$$F = \frac{4\sqrt{(p_1 \cdot p_2)^2 - m_1^2 m_2^2}}{N_1 N_2}, \tag{A.1}$$

where $N_i = 1$ for photons, and $N_i = 2m_i$ for spin-1/2 fermions.

Table A.1 summarizes the flux factors for different cases in the center-of-mass frame and the laboratory frame. An outline of the calculations and meaning of the symbols follows.

TABLE A.1: Flux factors.

Reference Frame	No Photons	One Photon	Two Photons
center-of-mass	$2\gamma^2\beta$		$8\omega_1\omega_2$
laboratory	$\gamma\beta$	2ω	

For two photons $m_1 = m_2 = 0$, and

$$F_{\gamma_1\gamma_2} = 4p_1 \cdot p_2 = 4\omega_1\omega_2(1 - \cos\theta). \tag{A.2}$$

For collinear beams $\theta = \pi$, and

$$F_{\gamma_1\gamma_2} = 4p_1 \cdot p_2 = 8\omega_1\omega_2. \tag{A.3}$$

For one photon, $p_1 = \omega(1, \hat{k})$, and one fermion, $p_2 = E(1, \vec{\beta})$,

$$F_{\gamma e} = 4\frac{p_1 \cdot p_2}{2m} = 2\gamma\omega(1 - \beta\cos\theta), \tag{A.4}$$

where $\gamma = E/m$. For collinear beams

$$F_{\gamma e} = 2\gamma\omega(1 + \beta). \tag{A.5}$$

In the laboratory (LAB) frame $\beta = 0$, $\gamma = 1$, and

$$F_{\gamma e}^{\text{LAB}} = 2\omega. \tag{A.6}$$

For no photons

$$F_{ee} = 4\frac{\sqrt{E_1^2 E_2^2 (1 - \beta_1 \beta_2 \cos\theta)^2 - m_1^2 m_2^2}}{2m_1 2m_2} = \sqrt{\gamma_1^2 \gamma_2^2 (1 - \beta_1 \beta_2 \cos\theta)^2 - 1}.$$

$$(A.7)$$

For collinear beams

$$F_{ee} = \sqrt{\gamma_1^2 \gamma_2^2 (1 + \beta_1 \beta_2)^2 - 1}.$$

$$(A.8)$$

In the laboratory frame $\beta_2 = 0$, $\gamma_2 = 1$, and

$$F_{ee}^{LAB} = \sqrt{\gamma^2 - 1} = \gamma\beta,$$

$$(A.9)$$

where $\gamma = \gamma_1$ and $\beta = \beta_1$. For identical particles in the center-of-mass (CMS) frame $\gamma = \gamma_1 = \gamma_2$, $\beta = \beta_1 = \beta_2$, and

$$F_{ee}^{CMS} = \sqrt{\gamma^4 (1 + \beta^2)^2 - 1} = 2\gamma^2 \beta.$$

$$(A.10)$$

Appendix B

Lorentz-Invariant Phase Space

The element of n-body phase space is defined as

$$\mathrm{dLips}(s; p_i, \ldots, p_n) = (2\pi)^4 \delta^4 \left(\sqrt{s} - \sum_i^n p_i \right) \prod_i^n \frac{N_i}{(2\pi)^3} \frac{d^3 p_i}{2E_i}, \qquad (\mathrm{B.1})$$

where s is the square of the available four-momentum, $N_i = 1$ for photons, and $N_i = 2m_i$ for spin-1/2 fermions. The phase-space element is Lorentz invariant because each $d^3 p_i / E_i$ is Lorentz invariant. The most practical use of this formula, and the one encountered previously in equation 8.107, is for the $1 + 2 \to 3 + 4$ process:

$$\mathrm{dLips}(s; p_3, p_4) = (2\pi)^4 \delta^4 (p_1 + p_2 - p_3 - p_4) \frac{N_3}{(2\pi)^3} \frac{d^3 p_3}{2E_3}, \frac{N_4}{(2\pi)^3} \frac{d^3 p_4}{2E_4}, \qquad (\mathrm{B.2})$$

where $s = (p_1 + p_2)^2$.

We will now evaluate this phase-space element. First we perform the integration over particle 4 invariantly using equation 8.102:

$$\begin{aligned}
\mathrm{dLips}(s; p_3, p_4) &= \frac{N_3 N_4}{8\pi^2} \delta^4 (p_1 + p_2 - p_3 - p_4) \frac{d^3 p_3}{E_3} d^4 p_4 \delta(p_4^2 - m_4^2) \theta(p_4^0) \\
&= \frac{N_3 N_4}{8\pi^2} \frac{d^3 p_3}{E_3} \delta[(p_1 + p_2 - p_3)^2 - m_4^2] \theta(E_1 + E_2 - E_3).
\end{aligned} \qquad (\mathrm{B.3})$$

Using $d^3 p_3 = |\vec{p}_3|^2 d|\vec{p}_3| d\Omega = |\vec{p}_3| E_3 dE_3 d\Omega$, we have

$$\mathrm{dLips}(s; p_3, p_4) = \frac{N_3 N_4}{8\pi^2} |\vec{p}_3| dE_3 d\Omega \delta[(p_1 + p_2 - p_3)^2 - m_4^2] \theta(E_1 + E_2 - E_3). \qquad (\mathrm{B.4})$$

Using the property of the delta function given by equation 2.32, the delta function can be re-expressed:

$$\delta[(p_1 + p_2 - p_3^2)^2 - m_4^2] = \frac{\delta(E_3 - \tilde{E}_3)}{2 \left| W - \frac{\tilde{E}_3}{\sqrt{\tilde{E}_3^2 - m_3^2}} |\vec{p}_1 + \vec{p}_2| \cos\theta \right|}, \qquad (\mathrm{B.5})$$

where $W = E_1 + E_2$, and $\tilde{E}_3$ is the solution to

$$s + m_3^2 - m_4^2 - 2W\tilde{E}_3 + 2|\vec{p}_1 + \vec{p}_2|\cos\theta\sqrt{\tilde{E}_3^2 - m_3^2} = 0. \qquad \text{(B.6)}$$

Integrating over E_3, the Lorentz-invariant phase-space element in a very general, but impractical, form is

$$\text{dLips}(s; p_3, p_4) = \frac{N_3 N_4}{16\pi^2} \frac{\sqrt{\tilde{E}_3^2 - m_3^2}}{\left| W - \frac{\tilde{E}_3}{\sqrt{\tilde{E}_3^2 - m_3^2}}|\vec{p}_1 + \vec{p}_2|\cos\theta \right|} d\Omega. \qquad \text{(B.7)}$$

The result is general and includes no approximations. There are two common reference frames used in practical calculations: the center-of-mass frame, and the laboratory frame in which one of the initial particles is at rest. In the center-of-mass reference frame (CMS), $|\vec{p}_1 + \vec{p}_2| = 0$ and

$$\boxed{\text{dLips}(s; p_3, p_4) = \frac{N_3 N_4}{16\pi^2} \frac{|\vec{p}_3|}{\sqrt{s}} d\Omega \quad \text{center-of-mass frame}}, \qquad \text{(B.8)}$$

where

$$|\vec{p}_3| = \sqrt{\tilde{E}_3^2 - m_3^2} \quad \text{and} \quad \tilde{E}_3 = \frac{s + m_3^2 - m_4^2}{2\sqrt{s}}. \qquad \text{(B.9)}$$

In quantum electrodynamics, the result is often not this general. For two photons in the initial state

$$\text{dLips}(s; p_3, p_4) = \frac{N_3 N_4}{32\pi^2} \sqrt{1 - \frac{4m^2}{s}} d\Omega \quad \text{CMS: } \gamma + \gamma \to 3 + 4, \qquad \text{(B.10)}$$

where $m \equiv m_3 = m_4$. For two photons in the final state

$$\text{dLips}(s; p_3, p_4) = \frac{N_3 N_4}{32\pi^2} d\Omega \quad \text{CMS: } 1 + 2 \to \gamma + \gamma. \qquad \text{(B.11)}$$

For processes with no photons, the expression does not simplify unless at least $E_3 \approx |\vec{p}_3|$:

$$\text{dLips}(s; p_3, p_4) = \frac{N_3 N_4}{32\pi^2} \frac{s + m_3^2 - m_4^2}{s} d\Omega \quad \text{CMS: } E_3 \approx |\vec{p}_3|. \qquad \text{(B.12)}$$

Obvious additional simplifications follow, if some of the masses can be ignored. Equation B.8 can be easily adapted to two-body decays of a particle at rest:

$$\boxed{\text{dLips}(m_1; p_2, p_3) = \frac{N_2 N_3}{16\pi^2} \frac{|\vec{p}|}{m_1} d\Omega \quad \text{decay } 1 \to 2 + 3}. \qquad \text{(B.13)}$$

The other reference frame of interest is the laboratory rest frame. In the rest frame of particle 1, $\vec{p}_1 = 0$ and

$$\boxed{\text{dLips}(s; p_3, p_4) = \frac{N_3 N_4}{16\pi^2} \frac{\sqrt{\tilde{E}_3^2 - m_3^2}}{W - \frac{\tilde{E}_3}{\sqrt{\tilde{E}_3^2 - m_3^2}} |\vec{p}_2| \cos\theta} d\Omega \quad \text{laboratory frame}}.$$

(B.14)

This is still not a very practical form. If particle 3 is a photon, or very energetic such that $E_3 \approx |\vec{p}_3|$, we have

$$\text{dLips}(s; p_3, p_4) = \frac{N_3 N_4}{8\pi^2} \frac{\tilde{E}_3^2}{s + m_3^2 - m_4^2} d\Omega,$$

(B.15)

where

$$\tilde{E}_3 = \frac{s + m_3^2 - m_4^2}{2W - 2|\vec{p}_2| \cos\theta}.$$

(B.16)

Again, obvious additional simplifications follow, if some of the masses can be ignored. For example, in a case of elastic scattering in which the mass of the beam particle can be ignored, we have

$$\text{dLips}(s; p_3, p_4) = \frac{N_3 N_4}{16\pi^2 m_1} \frac{|\vec{p}_3|}{|\vec{p}_2|} d\Omega,$$

(B.17)

where

$$|\vec{p}_3| = \frac{|\vec{p}_2|}{1 + (2|\vec{p}_2|/m_1) \sin^2(\theta/2)}.$$

(B.18)

Appendix C

Feynman Rules for Tree Diagrams

In chapter 6 we developed propagator methods which allowed us to calculate transition amplitudes using the S-matrix. In chapter 8 we calculated the lowest-order S-matrix elements for several quantum electrodynamic processes. In each case, we also drew diagrams in space-time, or graphs, that allowed us to visualize the process. Based on observations, we identified elements of each diagram with factors in the S-matrix elements. We gained enough experience to extract a set of rules – the Feynman rules – which in principle allow the calculation of any quantum electrodynamic process no mater how complicated. The set of diagrams in which each vertex is connected to every other vertex by only one internal line are called tree diagrams. In this appendix, we summarize the Feynman rules for tree diagrams.

The Feynman rules can be stated in configuration or momentum space. In configuration space there is a direct correspondence to the S-matrix elements as we have developed them. However, for plane waves, these expressions can always be integrated over the coordinates to leave an expression which depends only on the momenta of the particles. Thus it is simpler to perform these integrations ahead of time and state the Feynman rules in momentum space. The momentum-space diagrams are usually preferred because of the simpler structure of the integrand, and because the initial- and final-state particle are usually known in terms of their momenta rather than their positions.

We have, and will, restrict our attention to final states f that are different from the initial state i such that no subset of particles in the state f has precisely the same four-momentum as some corresponding subset of particles in the state i, other than the whole state itself. This means that we are considering only, so called, connected diagrams. A state can be specified by giving the momentum and the internal degrees of freedom, e.g. spin for electrons and polarization for photons. Choosing plane waves for the initial and final states is usually sufficient for most applications involving lowest-order – tree level – diagrams. In this case, the S-matrix element can be written as

$$S_{fi} = i(2\pi)^4 \delta^4(p_i - p_f) \prod_{j=0}^{N} \sqrt{\frac{m_j}{E_j V}} \prod_{k=0}^{M} \frac{1}{\sqrt{2\omega_k V}} \mathcal{M}_{fi}, \qquad (C.1)$$

where p_i and p_f are the total momenta of the initial and final states. The square-root factors are due to the normalization of the incoming and outgoing plane waves. If the process of interest contains N spin-1/2 particles and M photons, there is a factor of $\sqrt{m/(EV)}$ for each fermion – electron or positron – and a factor of $\sqrt{1/(2\omega V)}$ for each photon. The delta function arises from the integration of the exponential factors of the plane waves over the configuration coordinates, using the exponential representation of the delta function equation 2.31. Further integrations over the momentum variables in the propagators lead to the single over-all four-momentum conserving delta function in equation C.1. If the process involves Coulomb scattering off an external fixed potential, the delta function conserves energy only, and there is only a single power of 2π in front of the delta function. Separating the factors of 2π from the factor $i\mathcal{M}_{fi}$ is a convention. The remaining factor $i\mathcal{M}_{fi}$ is called the Lorentz-invariant matrix element, or amplitude. It contains all the essential physics, while the other factors simply represent kinematics.

The most practical use of the S-matrix element is in the calculation of processes involving one or two initial-state particles, and two final-state particles. These two processes are represented by the following decay rate and differential cross section.

$1 \rightarrow 2 + 3$ decay rate:

$$d\Gamma = \frac{1}{2m_1}|\mathcal{M}|^2 d\mathrm{Lips}(m_1^2; p_2, p_3). \qquad (C.2)$$

$1 + 2 \rightarrow 3 + 4$ cross section:

$$d\sigma = \frac{1}{F}|\mathcal{M}|^2 d\mathrm{Lips}(s; p_3, p_4), \qquad (C.3)$$

where $s = (p_1 + p_2)^2$. Appendix A describes the flux factor F, and the Lorentz invariant phase space factor $d\mathrm{Lips}$ is described in appendix B. We see that the expression for the cross section is divided into two parts: the square of the invariant amplitude $|i\mathcal{M}|^2$, which is a Lorentz scalar containing the physics, and two kinematical factors, the incident particle flux and the final-state phase space, which are also Lorentz invariant.

To compare the cross section with experiment, one has to integrate the differential cross section $d\sigma$ over the phase-space intervals which are not distinguished in the measurements. In addition, one has to average over the initial spins and polarizations, and sum over the final spins and polarizations, if these polarizations are not measured.

If there are identical particles in the final state, we must take special precautions. Since configurations differing only by a permutation of the particles describe the same quantum-mechanical state, the phase-space must be reduced by a degeneracy factor. If there are n identical particles of type k in the final state, we must include a degeneracy factor $1/n_k!$ in the decay rate or cross section. The denominator is $n_k!$ because there are $n_k!$ possibilities of arranging (countering) these particles, but only one such arrangement is measured experimentally. For example, in the case of electron-electron scattering this means the cross section is multiplied by a factor of $1/2!$. Another example of practical importance is the process of multiple photon bremsstrahlung, where the denominator factor $n_k!$ can become very large.

Now the remaining task is just to write down the invariant amplitude $i\mathcal{M}_{fi}$ for the process of interest. The following explains how to draw the Feynman diagrams for the process of interest, how to read these diagrams, and apply the Feynman rules to obtain the invariant amplitude $i\mathcal{M}_{fi}$.

C.1 Drawing and Reading Feynman Diagrams

In the nth order of perturbation theory, one has to draw all possible topologically distinct and connected Feynman diagrams with n vertices that have the prescribed number of particles in the initial and final states: external lines. When drawing Feynman diagrams, only the topological structure is important. Since the theory was formulated in a relativistically covariant way, all possible time orderings are automatically taken into account. As long as the order of the vertices along the fermion lines is kept, the diagrams can be arbitrarily deformed without changing their meaning.

To convert the diagrams to an expression for the invariant amplitude $i\mathcal{M}_{fi}$, we assign multiplicative factors, using the Feynman rules given in the next section, to the various elements of each diagram. The complete perturbation series for $i\mathcal{M}_{fi}$ in powers of the coupling constant e is obtained by adding up the contributions from each Feynman diagram to the desired order. The amplitude associated with a particular Feynman diagram is determined as follows.

There are basically three elements of a Feynman diagram at the tree level: external lines, vertices, and internal lines. When the diagrams are read, factors are written from left to right in the order of movement along a continuous line against the direction of the arrows. Arrows always point in the direction a particle is moving, and opposite to the direction an antiparticle is moving. Arrows are omitted for external photon lines since a photon is its own antiparticle. The choice of direction for a virtual photon line is also immaterial: a change in its direction simply reverses the sign of the four-momentum k, which does not matter, since the propagator $D_F^{\mu\nu}(k)$ is an even function of k. For each vertex, the four-momenta of the three lines that meet at a point satisfy energy and momentum conservation. The four-momentum of the virtual particle is determined by the conservation of four-momentum at the vertex: the total momentum of the incoming lines equals the total momentum of the outgoing lines at any vertex. The factors associated with each element in a Feynman diagram are now stated.

C.2   Rules for Tree Diagrams in Momentum Space

External Particles

With each external line, we associate one of the following factors.

Spin 0: For each incoming or outgoing spin-0 boson assign a factor 1.

Spin 1/2: For each incoming or outgoing spin-1/2 fermion include a bispinor,

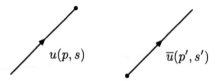

and for each incoming or outgoing spin-1/2 antifermion include a bispinor.

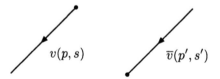

Photons: For each incoming or outgoing photon include a polarization vector.

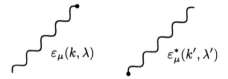

External Field: For each external field include the electromagnetic potential.

Propagators

With each internal line connecting two vertices, we associate one of the following propagators.

Spin-0:

$$i\Delta_F(p) = \frac{i}{p^2 - m + i\epsilon}$$

Spin-1/2:

$$iS_F(p) = \frac{i}{\not{p} - m + i\epsilon} = \frac{i(\not{p} + m)}{p^2 - m^2 + i\epsilon}$$

Photon:

$$iD_F^{\mu\nu}(k) = \frac{i}{k^2 + i\epsilon}\left[-g^{\mu\nu} + (1 - \zeta)\frac{k^\mu k^\nu}{k^2}\right]$$

for a general ζ-gauge. Calculations are usually performed in the Lorentz or Feynman gauge with $\zeta = 1$. In this gauge, the photon propagator is

$$iD_F^{\mu\nu}(k) = \frac{-ig^{\mu\nu}}{k^2 + \epsilon}.$$

Vertices

If two or more external lines meet at a point, we associate with it one of the following vertex factors.

Spin-0:
One-photon vertex

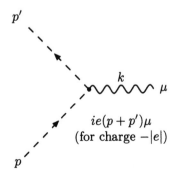

$$ie(p + p')\mu$$
(for charge $-|e|$)

Two-photon vertex

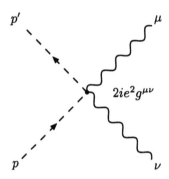

$2ie^2 g^{\mu\nu}$

Spin-1/2:

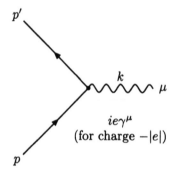

$ie\gamma^{\mu}$
(for charge $-|e|$)

The index μ (and ν) has to be multiplied with that of the photon line and summed over.

Additional Rules

The amplitudes corresponding to all diagrams have to be added coherently. The following relative phase factors between the amplitudes must be taken into account when forming the coherent sum.

A factor of -1 must be included for each incoming positron (outgoing electron with negative energy). This is the sign factor ϵ_f in the S-matrix element equation 6.124, which is positive for electrons and negative for positrons.

A relative factor of -1 must be included between two diagrams which differ only by the exchange of two external fermions lines. The rule also holds for the exchange of an incoming (outgoing) particle line with an outgoing (incoming) antiparticle line, since the latter is an incoming (outgoing) particle line with negative energy. This minus sign originates from the antisymmetry of the wave function required by Fermi-Dirac statistics.

Do not include any extra factor for boson external lines – according to Bose-Einstein statistics.

C.3   Steps in a Calculation

Based on our experience, we can now enumerate a set of simple steps to calculate processes at the tree level in quantum electrodynamics.

1. Draw the diagrams for the desired process.

2. Use the Feynman rules to write down the amplitude $\mathcal{M}_{fi}$.

3. Square the amplitude, and average or sum over spins and polarizations, using the completeness relationships.

4. Evaluate traces using the trace theorems, collect terms, and simplify the answer as much as possible.

5. Specialize to a particular frame of reference and draw a picture of the kinematic variables in that frame. Express all four-momentum vectors in terms of a suitable set of variables such as E and θ.

6. Plug the resulting expression for $|\overline{\mathcal{M}}_{fi}|$ into the cross section formula to obtain a differential over phase-space variables. Integrate over variables that are not measured to obtain a differential cross section in the desired form.

Appendix D

Trace Calculation Using FORM

The following are some example calculations using the symbolic-manipulation program FORM[1]. Equations 8.45, 8.114, 8.130, and 8.134 are solved symbolically. In addition, the last example calculates all the traces required for the photon-electron scattering cross section.

```
***************************************************************
FORM by J.Vermaseren,version 3.1(Oct 13 2002)
Run at: Sat Jun 11 15:01:48 2005
      *
      * Equation 8.45.
      *
      OFF STATS;
      SYMBOLS m, Ei, Ef;
      VECTORS pi, pf;
      *
      LOCAL TR = g_(1,0) * (g_(1,pi) + m)/(2*m)
                 * g_(1,0) * (g_(1,pf) + m)/(2*m);
      *
      TRACE4,1;
      *
      ID pi(0) = Ei;
      ID pf(0) = Ef;
      *
      BRACKET m;
      *
      PRINT;

   TR =
       + m^-2 * ( 2*Ei*Ef - pi.pf )

       + 1;

***************************************************************
FORM by J.Vermaseren,version 3.1(Oct 13 2002)
```

---

[1]J.A.M. Vermaseren, "New features of FORM", math-ph/0010025.

Run at: Sat Jun 11 14:50:12 2005
```
    *
    * Equation 8.114.
    *
    OFF STATS;
    INDICES mu, nu;
    SYMBOLS m, M;
    VECTORS pi, pf, Pi, Pf;
    *
    * Equation 8.110.
    LOCAL L = ((g_(1,pi) + m)/(2*m) * g_(1,mu)
                * (g_(1,pf) + m)/(2*m) * g_(1,nu))/2;
    * Equation 8.111.
    LOCAL H = ((g_(2,Pi) + M)/(2*M) * g_(2,mu)
                * (g_(2,Pf) + M)/(2*M) * g_(2,nu))/2;
    * Equation 8.114.
    LOCAL Mfi2 = L*H;
    *
    TRACE4,1;
    TRACE4,2;
    *
    BRACKET m, M;
    *
    PRINT;

L =
    + m^-2 * ( 1/2*pi(mu)*pf(nu) + 1/2*pi(nu)*pf(mu)
                - 1/2*d_(mu,nu)*pi.pf )

    + 1/2*d_(mu,nu);

H =
    + M^-2 * ( 1/2*Pi(mu)*Pf(nu) + 1/2*Pi(nu)*Pf(mu)
                - 1/2*d_(mu,nu)*Pi.Pf )

    + 1/2*d_(mu,nu);

Mfi2 =
    + m^-2*M^-2 * ( 1/2*pi.Pi*pf.Pf + 1/2*pi.Pf*pf.Pi )

    + m^-2 * ( - 1/2*pi.pf )

    + M^-2 * ( - 1/2*Pi.Pf )

    + 1;
```

```
******************************************************************
FORM by J.Vermaseren,version 3.1(Oct 13 2002)
Run at: Sat Jun 11 15:26:21 2005
   *
   * Equation 8.130.
   *
   OFF STATS;
   INDICES mu, nu;
   SYMBOLS m, M, E, Ep;
   VECTORS pi, pf, Pi, Pf;
   *
   LOCAL L = ((g_(1,pi) + m)/(2*m) * g_(1,mu)
               * (g_(1,pf) + m)/(2*m) * g_(1,nu))/2;
   LOCAL H = ((g_(2,Pi) + M)/(2*M) * g_(2,mu)
               * (g_(2,Pf) + M)/(2*M) * g_(2,nu))/2;
   LOCAL Mfi2 = L*H;
   *
   TRACE4,1;
   TRACE4,2;
   *
   ID Pf = Pi + pi - pf;
   ID Pi.Pi = M^2;
   ID pi.pi = m^2;
   ID pf.pf = m^2;
   ID Pi.pi = M*E;
   *
   ID Pi.pf = M*E;
   ID E/M = 0;
   *
   BRACKET m;
   *
   PRINT;

L =
    + m^-2 * ( 1/2*pi(mu)*pf(nu) + 1/2*pi(nu)*pf(mu)
    - 1/2*d_(mu,nu)*pi.pf )

    + 1/2*d_(mu,nu);

H =
    + 1/2*pi(mu)*Pi(nu)*M^-2 + 1/2*pi(nu)*Pi(mu)*M^-2
    - 1/2*pf(mu)*Pi(nu)*M^-2 - 1/2*pf(nu)*Pi(mu)*M^-2
    + Pi(mu)*Pi(nu)*M^-2;

Mfi2 =
```

```
+ m^-2 * ( E^2 - 1/2*pi.pf )

+ 1/2;
```

****
```
FORM by J.Vermaseren,version 3.1(Oct 13 2002)
Run at: Sat Jun 11 16:31:33 2005
    *
    * Equation 8.134.
    *
    OFF STATS;
    INDICES mu, nu;
    SYMBOLS m, M, E, Ep, q;
    VECTORS pi, pf, Pi, Pf;
    *
    Local L = ((g_(1,pi) + m)/(2*m) * g_(1,mu)
              * (g_(1,pf) + m)/(2*m) * g_(1,nu))/2
    ;
    Local H = ((g_(2,Pi) + M)/(2*M) * g_(2,mu)
              * (g_(2,Pf) + M)/(2*M) * g_(2,nu))/2
    ;
    Local Mfi2 = L*H;
    *
    TRACE4,1;
    TRACE4,2;
    *
    ID Pf = Pi + pi - pf;
    ID Pi.Pi = M^2;
    ID pi.pi = m^2;
    ID pf.pf = m^2;
    ID Pi.pi = M*E;
    ID Pi.pf = M*Ep;
    ID pi.pf = m^2 - q^2/2;
    *
    BRACKET m, E, Ep;
    *
    PRINT;

L =
    + m^-2 * ( 1/2*pi(mu)*pf(nu) + 1/2*pi(nu)*pf(mu)
    + 1/4*d_(mu,nu)*q^2 );

H =
    + E * (  - 1/2*d_(mu,nu)*M^-1 )
```

```
      + Ep * ( 1/2*d_(mu,nu)*M^-1 )

      + 1/2*pi(mu)*Pi(nu)*M^-2 + 1/2*pi(nu)*Pi(mu)*M^-2
      - 1/2*pf(mu)*Pi(nu)*M^-2 - 1/2*pf(nu)*Pi(mu)*M^-2
      + Pi(mu)*Pi(nu)*M^-2;

  Mfi2 =
      + m^-2*E*Ep * ( 1 )

      + m^-2*E * ( - 1/4*M^-1*q^2 )

      + m^-2*Ep * ( 1/4*M^-1*q^2 )

      + m^-2 * ( 1/4*q^2 )

      + E * ( - 1/2*M^-1 )

      + Ep * ( 1/2*M^-1 );

***************************************************************
FORM by J.Vermaseren,version 3.1(Oct 13 2002)
Run at: Sat Jun 11 16:33:44 2005
    *
    * Photon-Electron Scattering Traces.
    *
    OFF STATS;
    SYMBOLS m;
    VECTORS pi, pf, k, kp, e, ep;
    *
    LOCAL T11 = g_(1,pf,ep,e,k,pi,k,e,ep);
    LOCAL T12 = g_(1,pf,ep,e,k,   k,e,ep);
    LOCAL T13 = g_(1,   ep,e,k,pi,k,e,ep);
    LOCAL T14 = g_(1,   ep,e,k,   k,e,ep);
    LOCAL T1 = T11 + m*T12 + m*T13 + m^2*T14;
    *
    LOCAL T21 = g_(1,pf,e,ep,kp,pi,kp,ep,e);
    LOCAL T22 = g_(1,pf,e,ep,kp,   kp,ep,e);
    LOCAL T23 = g_(1,   e,ep,kp,pi,kp,ep,e);
    LOCAL T24 = g_(1,   e,ep,kp,   kp,ep,e);
    LOCAL T2 = T21 + m*T22 + m*T23 + m^2*T24;
    *
    LOCAL T31 = g_(1,pf,ep,e,k,pi,kp,ep,e);
    LOCAL T32 = g_(1,pf,ep,e,k,   kp,ep,e);
    LOCAL T33 = g_(1,   ep,e,k,pi,kp,ep,e);
    LOCAL T34 = g_(1,   ep,e,k,   kp,ep,e);
```

```
LOCAL T3 = T31 + m*T32 + m*T33 + m^2*T34;
*
LOCAL T41 = g_(1,pf,e,ep,kp,pi,k,e,ep);
LOCAL T42 = g_(1,pf,e,ep,kp,   k,e,ep);
LOCAL T43 = g_(1,   e,ep,kp,pi,k,e,ep);
LOCAL T44 = g_(1,   e,ep,kp,   k,e,ep);
LOCAL T4 = T41 + m*T42 + m*T43 + m^2*T44;
*
TRACE4,1;
*
ID pf    = pi + k - kp;
ID k.k   = 0;
ID kp.kp = 0;
ID e.e   = -1;
ID ep.ep = -1;
ID k.e   = 0;
ID kp.ep = 0;
ID pi.e  = 0;
ID pi.ep = 0;
ID k.kp  = k.pi - k.pf;
ID k.pf  = kp.pi;
ID pi.pi = m^2;
*
PRINT T1, T2, T3, T4;

T1 =
   8*pi.k*pi.kp + 16*pi.k*k.ep^2;

T2 =
   8*pi.k*pi.kp - 16*pi.kp*kp.e^2;

T3 =
   16*pi.k*pi.kp*e.ep^2 - 8*pi.k*pi.kp
   + 8*pi.k*kp.e^2 - 8*pi.kp*k.ep^2;

T4 =
   16*pi.k*pi.kp*e.ep^2 - 8*pi.k*pi.kp
   + 8*pi.k*kp.e^2 - 8*pi.kp*k.ep^2;
```

References

[1] I.J.R. Aitchison, *Relativistic Quantum Mechanics*, Macmillan, 1972.

[2] I.J.R. Aitchison and A.J.G. Hey, *Gauge Theories in Particle Physics*, Institute of Physics Publishing (2002) 0750308648.

[3] V.B. Berestetskii, E.M Lifshitz, L.P. Pitaevskii, *Quantum Electrodynamics*, 2nd Ed., Butterworth-Heinemann (1982) 0750633719.

[4] J.D. Bjorken and S.D. Drell, *Relativistic Quantum Mechanics*, 1st Ed., McGraw-Hill, (1998) 0072320028.

[5] A.Z. Capri, *Relativistic Quantum Mechanics and Introduction to Quantum Field Theory*, World Scientific (2002) 9812381376.

[6] H. Feshbach and F. Villars, *Elementary Relativistic Wave Mechanics of Spin 0 and Spin 1/2 Particles*, Rev. Mod. Phys., 30 (1958) pp. 24-45.

[7] W. Greiner, *Relativistic Quantum Mechanics*, 3rd Ed., Springer (2000) 3540674578.

[8] W. Greiner and J. Reinhardt, *Quantum Electrodynamics*, 3rd Ed., Springer (2003) 3540440291.

[9] D.J. Griffiths, *Introduction to Quantum Mechanics*, 2nd Ed., Person Prentice Hall, (2005) 0131118927.

[10] F. Gross, *Relativistic Quantum Mechanics and Field Theory*, Wiley-Interscience (1999) 0471353868.

[11] M. Hamermesh, *Group Theory and Its Applications to Physical Problems*, Dover (1990) 0486661814.

[12] F. Halzen and A.D. Martin, *Quarks & Leptons*, John Wiley & Sons (1984) 0471887412.

[13] W. Heitler, *The Quantum Theory of Radiation*, 3rd Ed., Dover (1984) 0486645584.

[14] B.R. Holstein, *Topics in Advanced Quantum Mechanics*, Addison-Wesley (1994) 0201410346.

[15] J.M. Jauch and F. Rohrlich, *The Theory of Photons and Electrons*, Addison-Wesley, 1955.

329

[16] L.D. Landau and E.M. Lifshitz, *Quantum Mechanics: Non-Relativistic Theory*, 3rd Ed., Butterworth-Heinemann, (1981) 0750635398.

[17] F. Mandl, *Quantum Mechanics*, 2nd Ed., Butterworths Scientific Publishers, 1957.

[18] F. Mandl, *Introduction to Quantum Field Theory*, Interscience Publishers, 1959.

[19] N.F. Mott and I.N. Sneddon, *Wave Mechanics and its Applications*, Oxford, 1948.

[20] N.F. Mott and M.S.W. Massey, *The Theory of Atomic Collisions*, Oxford, 1949.

[21] A. Messiah, *Quantum Mechanics*, John Wiley, 1958.

[22] A.I. Miller, *Early Quantum Electrodynamics*, Cambridge University Press (1994) 0521431697.

[23] M.E. Peskin and D.V. Schroeder, *An Introduction to Quantum Field Theory*, Perseus Books (1995) 0201503972.

[24] P. Roman, *Theory of Elementary Particles*, North-Holland, 1960.

[25] P. Roman, *Advanced Quantum Theory*, Addison-Wesley, 1965.

[26] L.I. Schiff, *Quantum Mechanics*, McGraw-Hill, 1955.

[27] M.D. Scadron, *Advanced Quantum Theory and its Applications Through Feynman Diagrams*, Springer (1979) 0387090452.

[28] F. Schwabl, *Advanced Quantum Mechanics*, 2nd Ed., Springer (2004) 3540401520.

[29] S.S. Schweber, *An Introduction to Relativistic Quantum Field Theory*, Harper & Row, 1962.

[30] S.S. Schweber, *QED and the Men Who Make It*, Princeton University Press (1994) 0691033277.

[31] J. Schwinger, *Selected Papers on Quantum Electrodynamics*, Dover, 1958.

[32] M. Veltman, *Diagrammatica: The Path to Feynman Diagrams*, Cambridge University Press (1994) 0521456924.

[33] S. Weinberg, *The Quantum Theory of Fields 00 Volume I Foundations*, Cambridge University Press (1995) 0521550017.

Index